Geologic Modeling and Mapping

COMPUTER APPLICATIONS IN THE EARTH SCIENCES

A series edited by Daniel F. Merriam

1969 — Computer Applications in the Earth Sciences

1970 — Geostatistics

1972 — Mathematical Models of Sedimentary Processes

1981 — Computer Applications in the Earth Sciences: An Update of the 70s

1988 — Current Trends in Geomathematics

1992 — Use of Microcomputers in Geology

1993 — Computerized Basin Analysis: The Prognosis of Energy and Mineral Resources

1996 — Geologic Modeling and Mapping

Geologic Modeling and Mapping

Edited by

Andrea Förster

GeoForschungsZentrum Potsdam
Potsdam, Germany

and

Daniel F. Merriam

University of Kansas
Lawrence, Kansas

Plenum Press • New York and London

Library of Congress Cataloging-in-Publication Data

On file

Proceedings of the 25th Anniversary Meeting of the International Association for Mathematical Geology, held October 10 – 14, 1993, in Prague, Czech Republic

ISBN-13: 978-1-4613-8017-7 e-ISBN-13: 978-1-4613-0363-3
DOI: 10.1007/978-1-4613-0363-3

Softcover reprint of the hardcover 1st edition 1996

A Division of Plenum Publishing Corporation
233 Spring Street, New York, N. Y. 10013

10 9 8 7 6 5 4 3 2 1

PREFACE

This volume is a compendium of papers on the subject, as noted in the book title, of modeling and mapping. They were presented at the 25th Anniversary meeting of the International Association for Mathematical Geology (IAMG) at Praha (Prague), Czech Republic in October of 1993. The Association, founded at the International Geological Congress (IGC) in Prague in 1968, returned to its origins for its Silver Anniversary celebration. All in all 146 papers by 276 authors were offered for the 165 attendees at the 3-day meeting convened in the Hotel Krystal. It was a time for remembrance and for future prognostication.

The selected papers in *Geologic Modeling and Mapping* comprise a broad range of powerful techniques used nowadays in the earth sciences. Modeling stands for reconstruction of geological features, such as subsurface structure, in space and time, as well as for simulation of geological processes both providing scenarios of geologic events and how these events might have occurred. Mapping stands for spatial analysis of data, a topic that always has been an extremely important part of the earth sciences. Because both modeling and mapping are used widely in conjunction, the book title should reflect the close relation of the subjects rather than a division.

Here, we bring together a collection of papers that hopefully contribute to the growing amount of knowledge on these techniques. All of the authors are experts in their field so the reader will have authoritative papers on the various aspects of the subject. Some time has elapsed since the Prague meeting, time used by the contributors of the book to incorporate additional and new ideas into their offerings. We want to thank all the authors for patience and help in preparing the book. We also thank the many reviewers who through their efforts improved the content and presentation of the papers. LeaAnn Davidson of the Kansas Geological Survey helped to prepare the final layout and her assistance is greatly appreciated. Cora Cowan, also of the Geological Survey, assisted with preparation of the index.

We hope the readers will enjoy the content and catch some of the flavor of the Silver Anniversary Meeting of the IAMG in Praha - those who attended enjoyed and learned much.

Andrea Förster

CONTENTS

INTRODUCTION

This book on *Geologic Modeling and Mapping* contains 15 papers by 38 authors. Although the book is divided into two parts - modeling and mapping - many of the papers could be in either part or both. This suggests that modeling and mapping go hand-in-hand in today's research and indeed they do. Both also are heavily dependent on graphics to convey the author's intent to the reader and these papers reflect that dependence.

The first paper on SEDSIM, a sedimentary simulation programming system, was used by **Wendebourg and Harbaugh** to describe petrophysical properties of subsurface flow simulations in 3D. This approach has practical applications in describing aquifers and oil and gas reservoirs. They conclude from their studies that '... interpretations of simulation results require careful consideration of assumptions and boundary conditions ... and an understanding of the ... geological framework ... represented by the simulation.'

Modeling of multicomponent, large-scale diagenetic systems is the subject of the paper by **Ondrak**. This approach, based on several assumptions and simplifications, helps us to understand the complex story of dissolution and precipitation of minerals in porous media. The model allows simulation of the time and place of porosity and mineral distribution in a sandstone; Ondrak illustrates the modeling technique with several theoretical examples representing different geological settings.

Watney, French, and Guy direct their attention to modeling of petroleum reservoirs with an example from the Pennsylvanian of the Midcontinent (USA). They note that this type of modeling is limited by the available data and knowledge of the reservoir. This approach to modeling can be qualitative, process oriented, or quantitative, and they conclude that a combination of these approaches probably provides the best model.

Bayer and others modeled the thermal conditions and heat flow for an ancient orogenetic front. They concentrated on the interaction of different factors and how they change the geothermal field and the uncertainties involved in extrapolation of near-surface conditions to depth. They concluded that modeling of complex situations under uncertainty allows a better understanding of various alternative scenarios.

Power law equations, related to transport properties in a two-component system, as applicable to the frequencies of the material and statistical measurements of anisotropy is the basic contention of **Kahle and Bayer**. Their two-component system provides a simple idealized model for complex anisotropic geological models.

Three-dimensional modeling of geological surfaces and bodies is the subject of the paper by **Alms, Klesper, and Siehl.** They illustrate their modeling by examples from the Cenozoic of the Lower Rhine Basin in Germany. The concepts and software systems used in the modeling for handling 2D and 3D data and the graphic output is outlined in their presentation. The graphics give an excellent 3D perspective to the results.

A three-dimensional method for the geometric reconstruction of geologic data is described and discussed by **Schaeben, Auerbach, and Schütz.** They use parallel cross sections or contour lines to accomplish the reconstruction, and then display the results graphically in 3D. Their method is automatic and requires no interaction by the user.

All of these papers stress the problems involved with modeling and the limits of available data, understanding the processes, and the wide range of interpretations that can be given to the results. All of the authors agree that this approach is useful and helpful in understanding geological conditions as known today and that with continued refinement and resolution, the simulation can be better and perhaps even more exact. Undoubtedly, we are the threshold of many exciting and valuable possibilities in the field of modeling.

The second part of the book is concerned with mapping or spatial analysis, and many of these papers stress practical (or economic) applications. As will be seen in these papers, as with modeling, graphics is an all important aspect.

Leading off in this part of the book is a paper on MSS data collected by LANDSAT, which were used to determine the temporal variations for an area in Nevada. **Yuan, Robinson, and Duggin** describe their statistical analysis of the remotely sensed data by discriminant analysis to recognize different lithologies and seasonal variations.

The paper by **Anderson and Brown** explores ways to reconstruct the location in the subsurface of the now gone Devonian salt units removed by dissolution. Thickness maps and seismic profiles were utilized in the reconstruction of the former extent of the salt. They used their model to interpret the processes and timing of the dissolution.

Shogenova uses a computer in a structural analysis of an Ordovician sedimentary basin in Estonia. Trend analysis was applied to a series of structural and stratigraphic maps to help in determining regional trends and isolating local features. From this analysis, she then was able to interpret better the basin's history.

Pairwise comparison of spatial (map) data is accomplished effectively by computing a similarity coefficient and constructing a resultant map according to **Merriam, Herzfeld, and Förster.** This gives insight into the predictability of one map to another, a quantitative descriptor for classification, and additional data for interpretation. They describe this approach and illustrate it by an example using geological, geophysical, and geothermal data for an area in southeastern Kansas.

Grunsky, Cheng, and Agterberg describe and apply spatial factor analysis to determine multivariate relationships in geochemical data, illustrated by examples from British Columbia in Canada. They determined that the regional geochemical trends are associated with the underlying lithological variations. Because of this association, the technique can be used as an aid in the exploration for mineral deposits in areas that have been adequately sampled.

Additional information on their continued study on exploration for mineral deposits is provided in the contribution by **Harff and others**. The additional information is to determine by discriminant analysis which geological observations are the most informative. This technique is illustrated by an example of the distribution of oil and gas fields in a petroleum area of western Kansas. They note that the method can be used for '... planning of exploration activities because it provides the geologist with information about favorable targets ...'

A Markov chain model is used by **Luo** for analyzing a set of spatial data described by discrete states. The model consists of five steps: definition of data; analysis; simulation; assessment of simulation results; and graphical representation of results. Methods of Markov analysis and case studies demonstrate this approach.

Plansky and others use an artificial intelligence (AI) and expert systems (ES) approach as a framework for constructing system models. AI/ES, and this includes neural networks, has been used more and more recently in the study of earth systems. They give a practical example from the mining industry as an application of their approach.

These papers give the reader a good broad background and detailed examples on the subject of *Geologic Modeling and Mapping*. Both academic and practical approaches are discussed by the authors. The references cited in the papers will give a sweep of available published material on the subjects for indepth reading. All in all the papers give a good idea as to the subjects covered during IAMG's Silver Anniversary meeting in Praha but the subjects are updated as of mid-1995.

To all users of this book - good reading, creative thinking, and productive results!

Daniel F. Merriam

Geologic Modeling and Mapping

SEDIMENTARY PROCESS SIMULATION: A NEW APPROACH FOR DESCRIBING PETROPHYSICAL PROPERTIES IN THREE DIMENSIONS FOR SUBSURFACE FLOW SIMULATIONS

Johannes Wendebourg
Stanford University, Stanford, California
and
Institut Français du Pétrole, Rueil-Malmaison, France

John W. Harbaugh
Stanford University, Stanford, California

ABSTRACT

Subsurface fluid flow is critically dependent on the 3D distribution of petrophysical properties in rocks. In sequences of sedimentary rocks these properties are strongly influenced by lithology and facies distribution that stem from the geologic processes that generated them. Three types of simulators are contrasted that represent variations of petrophysical properties: stochastic simulators, stratigraphic-form simulators, and sedimentary process simulators. The first two generally require closely spaced well information or seismic data and can be "conditioned" to accord with the data, but neither can represent the influence of depositional processes directly. By contrast, process simulators do not require closely spaced data, but they generally cannot be forced to accord closely with the data. The sedimentary process

Geologic Modeling and Mapping
Edited by A. Förster and D. F. Merriam, Plenum Press, New York, 1996

simulator described here, known as SEDSIM, provides 3D geometric forms and spatial distributions of grain sizes of alluvial, fluvial, and deltaic deposits that are controlled by information supplied as initial and boundary conditions, including the initial topographic surface, and fluvial and sediment discharge volumes through time. The resulting sediment distributions can be compared directly with maps and sections based on well data and seismic surveys, but they can also be transformed into estimates of porosity and permeability, thereby placing them in form for direct use with subsurface flow simulators. Several recent applications of SEDSIM in generating descriptions of groundwater aquifers and hydrocarbon reservoirs and their use in subsurface flow simulations are presented. Comparisons between complex actual and simulated 3D sequences suggest that statistical descriptions of simulated sequences could be used as input to stochastic simulators. This would combine the advantages of stochastic simulators that can condition simulations to field data, with the advantages of process simulators that treat geometric forms and flow properties of sequences interdependently and represent the development of sedimentary facies through space and time.

INTRODUCTION

Three-dimensional representation of petrophysical properties is needed for studies that involve subsurface fluid flow at various scales and stages. The main applications include the exploration for and production from oil and gas reservoirs, and in contaminant remediation of groundwater aquifers.

In hydrocarbon exploration, we need to know the properties of carrier beds if they exist and the migration paths through which hydrocarbons may have moved before accumulating in traps. Given enough information about the geologic structure and petrophysical properties of beds in the area, we could simulate the 3D motions of migrating hydrocarbons to forecast where they may be trapped.

In exploitation drilling, petrophysical properties of rocks interpreted from well logs and cores, and from production records guide the location and engineering features of wells. Opportunities to extract hydrocarbons from sedimentary sequences depend strongly on the 3D distributions of porosity and permeability within them, but rarely are these properties interpreted in appropriate 3D detail. Well logs and production records provide petrophysical information at and near well bores, but extrapolation in suitable detail beyond well bores is exceedingly difficult if conventional subsurface geological procedures are employed.

In aquifer remediation, small-scale sedimentary heterogeneities may determine the rates and transport direction of contaminants in the subsurface, and it is therefore important to understand their spatial distribution in choosing effective clean-up procedures. Contaminants not only pollute the groundwater, they also interact with the sediments where dispersion and adsorption processes are as important as advective transport. Without a detailed spatial description of the petrophysical properties that affect flow within the aquifer, these processes cannot be modeled correctly.

Simulators are needed to generate these distributions, for they are difficult or impossible to obtain in 3D detail otherwise. Three main types of simulators are available: (1) stochastic simulators, (2) geometric-form simulators, and (3) sedimentary process simulators. Once a simulator is selected, the challenge is to adapt it to obtain responses in suitable accord with information provided by real data such as existing wells.

Stochastic Simulators

Stochastic simulators use geostatistical procedures to generate constrained random variations of reservoir properties in two or three dimensions (Isaaks and Srivastava, 1989; Deutsch and Journel, 1992). They are useful in dealing with groundwater flow problems as well as in managing production from oil and gas reservoirs, and can be classified into discrete and continuous subtypes (Haldorsen and Damsleth, 1993).

Discrete or object-based simulators are useful for representing shapes and petrophysical properties of relatively large-scale sedimentary bodies such as channels, deltaic lobes, and shale barriers, and they usually employ rules for the dimensions, shapes, and orientations of these sedimentary bodies (Geehan and others, 1986). By contrast, continuous or sequence-based simulators are useful for representing small-scale variations in petrophysical properties that vary continuously in three-dimensions. They may be used to represent these properties within larger sedimentary bodies previously generated with discrete simulators (Rudkiewicz and others, 1990). Furthermore, both are capable of according closely with data provided by wells and seismic data (Doyen and others, 1991).

Both discrete and continuous stochastic simulators may assume statistical stationarity, which implies that the gross statistical properties of aquifer or reservoir beds do not change spatially. Stochastic simulators are also sensitive to the manner in which spatial continuity is represented, both in the choice of an appropriate mathematical model and in the choice of correct model parameters. For example, Murray (1992) represented permeability variations of selected oil reservoirs in the Powder River Basin of Wyoming with stochastic simulations whose results were conditional upon information from wells, but determined that none of the alternative and equally probable responses were reasonable because of the lack of continuity between beds with high permeabilities, suggesting that the geostatistical model was inadequate to represent variations in geological conditions. Murray then used an additional stochastic technique termed "simulated annealing" that introduced connectivity in the model results, thereby better approximating actual geological variations.

Geometric-form Simulators

Geometric-form simulators are purely deterministic, contrasting with stochastic simulators that depend on successions of random numbers. Geometric-form simulators are generally used to produce 2D vertical sections that display the geometrical forms of sedimentary sequences in accord with concepts of sequence stratigraphy that have been derived largely from interpretations of seismic sections (Strobel and others, 1990; Lawrence and others, 1990). Geometric-form simulators are well suited to represent the geometrical details of sedimentary units (parasequences) that usually are stacked one upon another and form as functions of the space available for deposition (accommodation space), which is a function of the interplay between eustatic variations in sealevel, tectonic subsidence, and sediment supply. Geometric-form simulators thus incorporate geological concepts directly, in contrast to stochastic simulators that generally do not represent geological concepts. The use of geometric-form simulators has stimulated the development of "inverse" methods for determining variations in eustatic sealevel and tectonic subsidence rates from the geometrical forms and facies relationships based on information from outcrops and wells (Lessenger, 1993).

Geometric-form simulators are not directly usable for use with flow simulators. First, they are generally 2D, which limits their use with 2D versions of flow simulators. Second, geometric-form simulators do not provide information directly about variations in porosity and permeability needed for flow simulators. These and other petrophysical properties must be assigned subsequently, usually being based on comparisons between actual facies and those generated by geometric-form simulators.

There is an additional limitation in that geometric-form simulators usually are operated to obtain responses that accord with continuous seismic sections, but the degree of resolution provided by seismic sections may be inadequate to permit accurate facies interpretations to be made, so that subsequent assignment of petrophysical properties may be very uncertain. This point has been elucidated by Shuster and Aigner (1994) who have generated synthetic seismic sections from geometric-form models whose input and boundary conditions were completely known. These synthetic sections were then interpreted by geologists in terms of their apparent geologic history and facies distribution, yielding a qualitative measure of the uncertainty involved in choosing model parameters, as well as in interpreting facies from seismic sections.

Sedimentary Process Simulators

Sedimentary process simulators link sedimentary processes with sedimentary features and materials and operate in accord with the physical laws and empirical relationships that govern transport, deposition, and erosion of clastic sediment. These processes are represented with equations that are solved numerically, and the numerical sequences obtained as solutions represent the responses of the processes through time and 3D space, producing sequences of beds whose properties can be represented by spatial variations in grain sizes. These variations in turn can be transformed into 3D distributions of porosities and permeabilities for use with 3D flow simulators. Sedimentary process simulators thus provide direct linkages between depositional environments and processes that create sedimentary sequences, and their subsequent ability to transmit pore fluids.

Some early sedimentary process simulators employed the diffusion equation to represent sediment transport, and involved the assumption that the rate of sediment transport is related to gradients of depositional slopes, yielding smoothly varying representations of sedimentary sequences in two or three dimensions (Harbaugh and Bonham-Carter, 1970). The diffusion equation for diffusion-based simulators is a macroscopic equation that can be derived from the integration of open-water flow equations together with empirical relationships for sedimentation and erosion, and that is applicable mainly to intermediate to large-scale descriptions of sedimentary sequences. The previously mentioned early simple simulators have been improved recently by using several lithologies (Rivenaes, 1992).

Process simulators based on the hydrodynamics of sediment transport treat the nonlinearities of sediment transport directly and therefore yield much more complex and realistic representations of sedimentary sequences than the other types of simulators (Tetzlaff and Harbaugh, 1989; Martinez and Harbaugh, 1993). However, the use of process simulators requires a fundamentally different approach in dealing with sedimentary sequences. Instead of employing reconstructions such as isopachous and lithofacies maps based on well logs and seismic sections, process simulators' responses are controlled by environmental data that are supplied as input, including topography at the beginning of a simulation, fluid and sediment discharge rates by streams during the experiment, and wave directions and intensities. Little or no environmental information can be extracted directly from well or seismic data, and therefore most of it must be estimated from modern depositional processes (Lee and others, 1991; Martinez and Harbaugh, 1993) or indirectly inferred from petrographic information (Wendebourg, 1991). Obtaining information for controlling process simulators is a major challenge and will require improved measurement of process rates in modern depositional environments.

The largest problem in using sedimentary process simulators is that they may not accord exactly with specific sedimentary sequences at known locations. For example, a large deltaic complex may be known from logs of scattered wells and intersecting seismic lines. We can use a process simulator to create a deltaic complex of similar form and level of detail, but the details are unlikely to match. The reason, of course, is that the process simulator responds in terms of the processes represented.

Thus, users of sedimentary process simulators are faced with the fact that the simulators cannot be conditioned to small-scale features observed in wells or seismic data. We can affect the processes that are represented in the model adjusting its boundary conditions so that the fluid and sediment discharge rates of inflowing rivers are changed, or sea-level rises or falls, or the heights and attack angles of incoming waves are changed, but we cannot command the processes to respond so that certain depositional features are reproduced in detail. Some of these process parameters are more or less stable whereas others need to be adjusted because of their sensitivity in controlling grain-size distributions and geometric forms of beds produced in experiments. By making judicious adjustments, users can focus on critical ranges for initial water depths, wave heights and directions, or fluvial discharge rates for water and sediment, and test their sensitivities with regard to simulated stratigraphic sequences.

Comparison of the Three Types of Simulators

In spite of their fundamental differences, the three types of simulators intergrade insofar as their performance is concerned. For example, although process simulators are totally deterministic in their formulation, they usually seem to behave randomly because of complex interactions between processes and materials, even though such apparent random behavior is not explicitly generated. By contrast, stochastic simulators yield random responses by design because of their dependence on sources of random numbers. Paradoxically, numerical procedures for generating sequences of random numbers (which are more properly termed pseudorandom numbers) are actually deterministic, further blurring the contrast between process simulators and stochastic simulators.

Stochastic simulators have an important advantage in that they can represent uncertainty in making 3D projections of facies and petrophysical properties that are expressed numerically. On the other hand, they have a disadvantage in that they lack geological "insight" by contrast with process simulators. Importantly, both stochastic simulators and process simulators share the quality of irreversibility and are therefore known as "forward" models. Their irreversibility stems either from their explicit random formulation or their representation of actual processes, both of which are inherently irreversible.

Geometric-form simulators may produce geometrical forms of sedimentary features similar to those of process simulators, but geometric-form simulators deal only with the gross geometrical forms of sedimentary bodies, and are not well suited for predicting petrophysical properties because they are usually based on larger-scale seismic data used for 2D representation.

SEDSIM AS A PROCESS SIMULATOR

The simulator described here for generating 3D sedimentary sequences is named SEDSIM, an abbreviation for SEDimentary Sequence SIMulation. SEDSIM represents most of the major processes that create and modify sedimentary systems in which clastic sediment predominates (Tetzlaff and Harbaugh, 1989; Martinez and Harbaugh,1993; Wendebourg and others, 1993). They include sediment erosion, transport, and deposition as embodied in fluvial and wave processes, eustatic sealevel change, isostatic compensation, tectonic uplift and subsidence, compaction, and submarine slope failure (Table 1). In eroding, transporting, and depositing clastic sediment in fluvial environments, SEDSIM employs the Navier-Stokes equations for open-channel flow that are solved using a Lagrangian discretization method (Tetzlaff and Harbaugh, 1989). Flow is coupled with empirical sediment-transport equations that control erosion, transport, and deposition of up to four grain sizes.

The processes represented in SEDSIM can be grouped in two main categories, namely autocyclic processes and allocyclic processes. Autocyclic processes are mutually interdependent upon each other and are "internal" to SEDSIM. Autocyclic responses are inherent in virtually all coupled systems, whether mechanical, biological, or geological. Basic autocyclic processes in SEDSIM include flow in open channels, erosion, transport, and deposition of clastic sediment, wave activity, and isostatic subsidence. More complex autocyclic processes represented by SEDSIM are derived from combinations of these basic processes, such as channel avulsion, delta-lobe switching, and slumping.

Allocyclic processes, by contrast, are imposed externally, and while they affect the autocyclic processes, they are not mutually affected in turn. Allocyclic processes represented in SEDSIM include eustatic sealevel change, tectonic subsidence, and climatic variations represented as changes in external boundary conditions, such as change in wind directions and intensity, and change in volume and geographic locations of the fluid and sediment discharge volumes of inflowing streams (Fig. 1).

Table 1. Overview of SEDSIM's modules listing their purposes, types of input, and spans of time involved when each module is used.

SEDSIM MODULE NAME	PURPOSE	INPUT	FREQUENCY OF ITERATIONS
FLOW	Flow of surface water and transport of sediment	Start and end time, flow timestep, empirical transport coefficients	Seconds to minutes
SOURCE	Discharge of water and sediment	Source location, volume of water and sediment and grain size distribution through time	Days to years
WAVE	Redistribute sediment by wave action	Main wind direction, transport rate, depth of mobile bed	Minutes - hours
SEALEVEL	Change eustatic sealevel	Relative sealevel curve	10^2-10^3 years
TECTONIC	Vertical faulting	Rate of elevation change of basement through time	10^2-10^3 years
SLOPE	Redistribute sediment by slumping	Maximum stable slope per grain size	Days to years
ISOSTA	Subside basement due to sediment load	Flexural rigidity of lithosphere	10^3-10^6 years
COMPACT	Reduce sediment thickness due to compaction	Porosity-load function for each grain size	10^3-10^6 years

In SEDSIM, sedimentary sequences and other operations are defined with respect to a gridwork of 3D cells that are indexed in the two horizontal directions by rows and columns. Each cell is of fixed size laterally during an experiment, with each cell's horizontal dimensions being the same, thus forming a square. The vertical height of an individual cell is variable and usually much less than that of the horizontal dimensions. The number of rows and columns, and the number of layers vertically may vary greatly, depending on the objectives of the experiment and the level of detail desired. In typical studies, gridworks range from 20 columns and rows and a few layers to more than 100 rows and columns and more than 50 layers thus they may involve hundreds of thousands of grid cells. Depositional sequences are stored in files as 3D

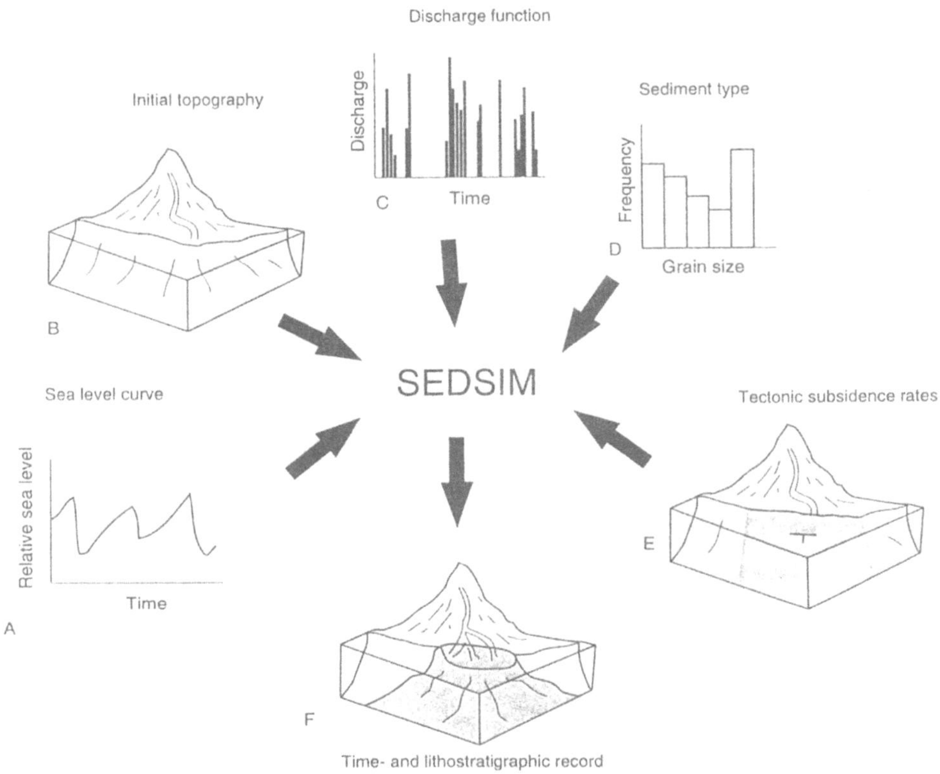

Figure 1. Schematic diagram showing types of inputs for SEDSIM simulations, clockwise from lower left: A, eustatic sealevel fluctuations over span of time represented in experiment; B, initial topography of area of simulation; C, discharge rates for water and clastic sediment over time span of experiment; D, frequency distribution of grain sizes of clastic sediment supplied during experiment (they may represent geological characteristics of watershed); E, orientation of faults and tectonic subsidence rates; F, SEDSIM's output represented by lithostratigraphic record over time span represented in experiment.

numerical records that provide both lithostratigraphic and timestratigraphic descriptions that may be displayed in 3D and in color on modern graphics workstations (Ramshorn and others, 1992).

SEDSIM has been used over a range of scales, with horizontal resolutions ranging from several tens of meters to several hundreds of meters. Vertical resolutions are usually much finer, ranging from a few centimeters to tens of meters. The selection of horizontal and vertical scales and the number of cells employed depends on the objectives and detail desired, including the amount of geologic time to be represented by each sedimentary layer that is generated. The aggregate geologic time represented in individual experiments may range widely, from several years to several hundreds of thousands of years. Comparisons of the geologic time represented in simulated sedimentary sequences with that of actual sedimentary sequences is difficult. The time represented in actual sequences may be substantially greater because SEDSIM represents only those intervals required for individual depositional events, such as the formation of channel deposits during floods, thereby ignoring intervals between floods when little erosion or deposition occurs.

SEDSIM has been applied to modern fluvial deposits (Lee and others, 1991), to beach environments (Martinez and Harbaugh, 1993), to Quaternary alluvial and deltaic complexes in California (Koltermann and Gorelick, 1992; Wendebourg, 1994), Wisconsin (Webb and Anderson, 1991), and Norway (Tuttle and others, 1995), to Cretaceous and Tertiary Gulf Coast sediments (Tetzlaff and Harbaugh, 1989; Wendebourg, 1991; Martinez and Harbaugh, 1993), and to Jurassic clastic hydrocarbon reservoirs in the North Sea (Griffiths, 1994a, 1994b).

INITIAL AND BOUNDARY CONDITIONS REQUIRED FOR SEDSIM'S OPERATION

In an experiment, SEDSIM must be provided with initial conditions at the outset of the experiment, as well as with boundary conditions that may vary over the span of geological time represented by the experiment. The most important boundary conditions include the form of the topographic surface at the outset of the experiment and variations in the volumes of fluid and sediment volumes supplied by inflowing streams during the span of the experiment.

Evidence for the importance of initial topography is provided by ultra-high resolution seismic surveys involving young deltaic environments. Chiocci (1994) describes Holocene deposits on the Calabrian shelf of the Eastern Tyrrhenian Sea, Italy, where postglacial sealevel rise and local tectonic uplift have influenced the deposition of deltaic and nearshore sediments whose geometric forms and facies distribution have been strongly dependent on the underlying topography.

Here, the topography consisted of a narrow, steep shelf on which depocenters marked the locations of river mouths, which moved landward during transgression. In addition, scattered bioherms served as dams that locally trapped suspended sediment.

An example of how a submerged topography has influenced geometric form and grain-size distribution is provided by the Cretaceous Woodbine Formation of East Texas (Wendebourg, 1994). Topographic features, such as the Edwards reef trend and subtle undulations of the submerged shelf, influenced the deposition of isolated lens-shaped bodies of sediment that today serve as oil and gas reservoirs (Fig. 2). In SEDSIM's simulation of these sediments, the Edwards reef trend served as an important sedimentation boundary, influencing variations in bed thickness and in the areal distribution of grain sizes (Fig. 3).

In applying SEDSIM to an actual area, initial topography for the stratigraphic sequence being simulated can be determined from well data (Koltermann and Gorelick, 1992) or 3D seismic data (Griffiths, 1994b). Although the present structural form of a sedimentary sequence may be defined in detail by seismic reflectors, changes that stem from postdepositional compaction and structural deformation must be subtracted if a reflecting horizon is to represent its topographic form during deposition (Wendebourg, 1994).

An ancient topographic surface may be readily defined if it involves the boundary between bedrock and overlying poorly consolidated sediment. An example is provided by coarse-grained sediments deposited as an ice-contact delta in the Holocene at Gardermoen in southern Norway (Tuttle and others, 1995). About 9500 BC during an interval spanning about 100 years, the retreating Scandinavian glacier dumped about 4.4 km^3 sediments ranging in grain size from cobbles to silt, thereby drowning an area of about 50 km^2 whose original topography involved differences in elevation of about 150 m (Fig. 4).

To yield meaningful results, SEDSIM must be provided with appropriate rates for discharge volumes of fluid and sediment for each inflowing stream over the time span represented in the experiment. These discharge rates can be defined at boundaries either in form of point sources such as rivers, or in form of line sources, as for example at marine boundaries. Fluid discharge is coupled with sediment discharge, usually expressed in form of suspended load within the water column and as bed-load transported along the stream bed, beach, or ocean floor.

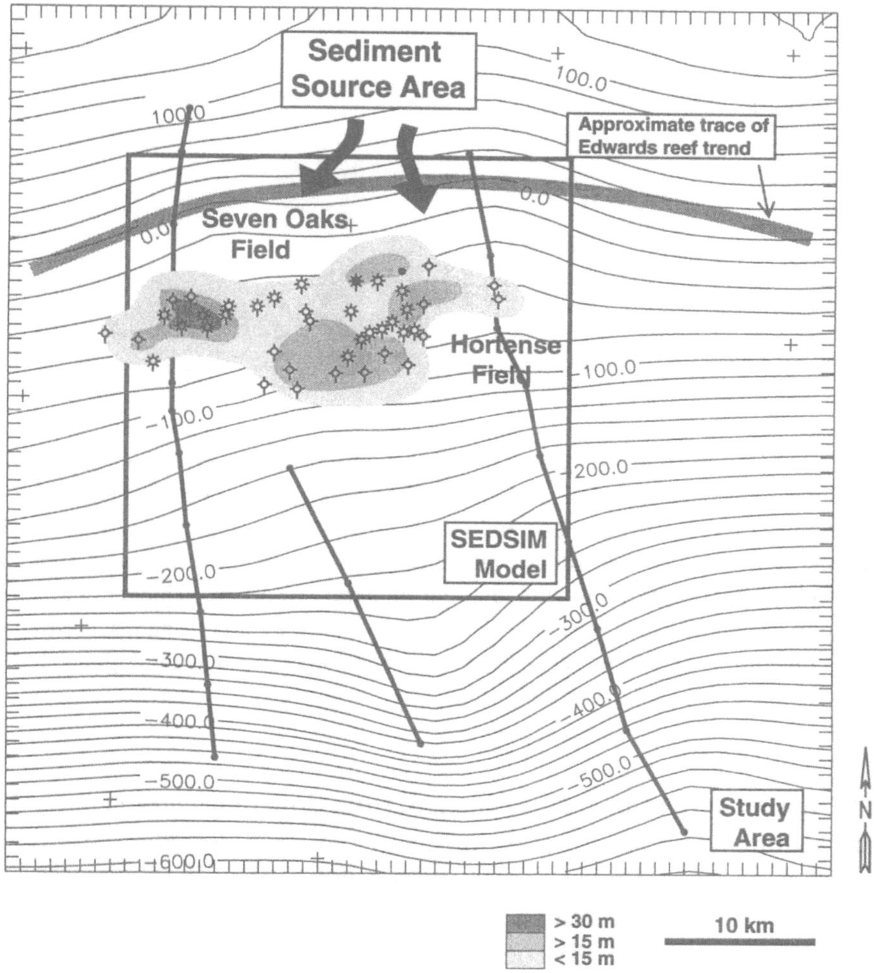

Figure 2. Structure contour map showing reconstructed ancient depositional topography at base of Cretaceous Woodbine Formation in Polk and Tyler Counties, East Texas. Map is based on three seismic lines whose traces are shown by solid lines. Contour interval is 20 m. Plus signs near edges of overall map denote places where additional data points were provided to constrain surface appropriately. Three shades of gray represent isopachs of local variations in thickness of sandstones that serve as oil reservoirs and are based on information from wells shown (Foss, 1979). Inset map whose outline is represented by bold lines denotes area of SEDSIM simulation experiment. Arrows denote main sediment transport directions and locations of sources of fluid and sediment in experiment. Results of experiment are shown in Figure 3 (from Wendebourg, 1994).

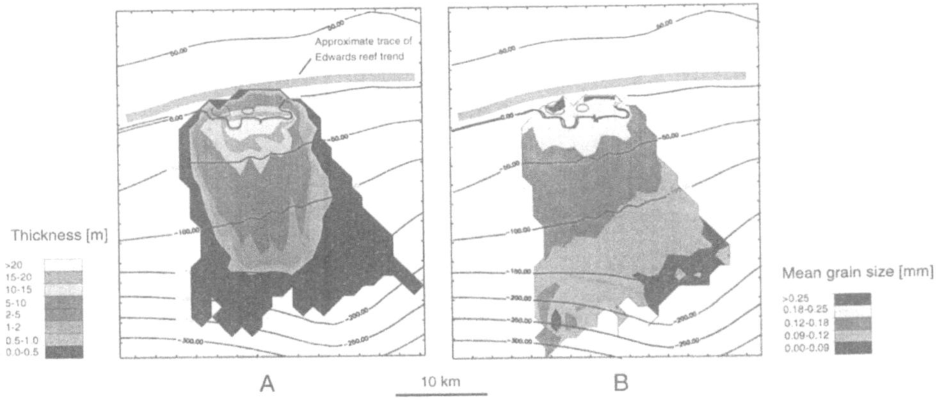

Figure 3. Maps produced in SEDSIM simulation experiment involving deposition of Woodbine sandstones that serve as oil reservoirs: A, isopachous map; B, and facies map showing areal distribution of mean grain sizes. Area of experiment is denoted with bold lines and labeled "SEDSIM Model" in Figure 2. Contour lines represent submerged topography after deposition, with contour interval of 50 m. Thicker and coarser sandstone deposits formed in simulation occur near Edwards reef trend, according with actual thickness variations of Woodbine sandstones shown in Figure 2 (from Wendebourg, 1994).

Although modern rates can be obtained from direct measurements of fluid and sediment discharge rates, such rates are not directly available for ancient deposits and must be determined indirectly.

Although there are several methods to determine paleodischarge rates (Koltermann, 1993), one specifically developed and applied to SEDSIM involves the "similar basin" approach (Koltermann and Gorelick, 1992) that assumes that variations between modern climatic zones can serve as approximations for climatic changes in the past. Koltermann and Gorelick compiled flow and sediment discharge data from modern topographic basins in California's Coast Ranges and utilized flood statistics to generate a climate history for the last 600,000 years for the Alameda Creek drainage basin in Central California (Fig. 6). Flood statistics in warm and dry interglacial periods of the past correspond with flood statistics of topographic basins today in southern California, whereas cooler and wetter glacial periods correspond with basins today

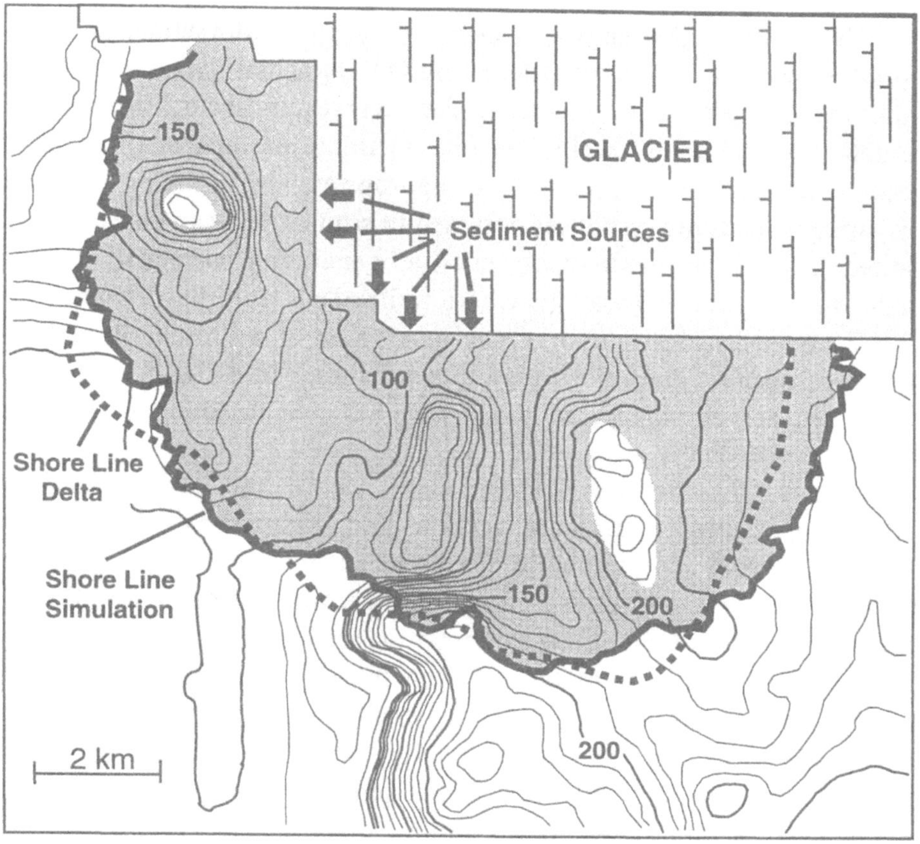

Figure 4. Map showing geographic features of ancient Gardermoen delta in southern Norway represented in SEDSIM simulation experiment involving ice-contact delta deposits. Streams emanating from glacier dumped sediment through five portals indicated by arrows labeled as sediment sources. Contours show bedrock topography based on borehole and seismic data that served as initial topography in simulation experiment. Contour lines are in m above present sealevel with interval of 10 m. Margin of actual deltaplain deposits is shown with heavy dashed line, and margin of simulated delta plain deposits is shown with heavy solid line. Light gray denotes area covered by glacier during creation of delta, whereas medium gray denotes simulated deltaic deposits. Hills not covered by sediment correspond with areas where bedrock crops out at present. Modified from Tuttle and others (1995).

in northern California. Such comparisons between present and past are feasible because of geomorphic similarities between the different basins of the Coast Ranges.

COMPARING RESULTS OF SEDIMENTARY SIMULATORS WITH ACTUAL DATA

The most challenging aspect of using sedimentary process simulators involves comparing their results with actual data. Because we cannot regulate process simulators so that their results accord closely with the details provided by wells, seismic sections, or outcrops, we must judge whether the responses are appropriate by making indirect comparisons. Graphic responses obtained in simulations can be especially adapted to facilitate comparison, such as generating simulated logs of grain-size variations at locations where wells occur (Fig. 5). Comparisons are influenced by the limited vertical resolution of a simulation and the number of discrete grain sizes used in simulations. In SEDSIM only four grain sizes are employed, so the degree of concordance needs to be evaluated accordingly.

Thus far, assessing the degree of accord has largely involved visual comparisons of maps and sections and logs, and is influenced by the level of detail represented in simulated sections and maps, which in turn is limited by the grid resolution both laterally and vertically. Details observed in sections based on outcrops or correlations between closely spaced wells are likely to be too fine to be represented in sections generated in simulations. For example, small channels or crevasse splays seen in the actual section shown in Figure 7A cannot be seen in the simulated section (Fig. 7B) because the actual processes that generated these details have not been simulated.

Although direct visual comparisons are straightforward, they do not enable us to measure the degree to which a simulation's results accord with actual data. Quantitative comparison can be based on statistical descriptions of simulation results (Tuttle and others, 1995). Furthermore, statistical descriptions can be used as input to stochastic simulators either directly in form of stationary parameters such as means and variogram statistics, or as nonstationary parameters involving trends of geometric forms and spatial distribution of facies (Beucher and others, 1992).

When comparing actual and simulated sequences, we can also make use of other relationships. In both simulated and actual sequences, subsurface fluid flow is constrained by the spatial distribution of petrophysical properties. In actual sequences these properties may be estimated from production tests, core analyses, and petrophysical logs

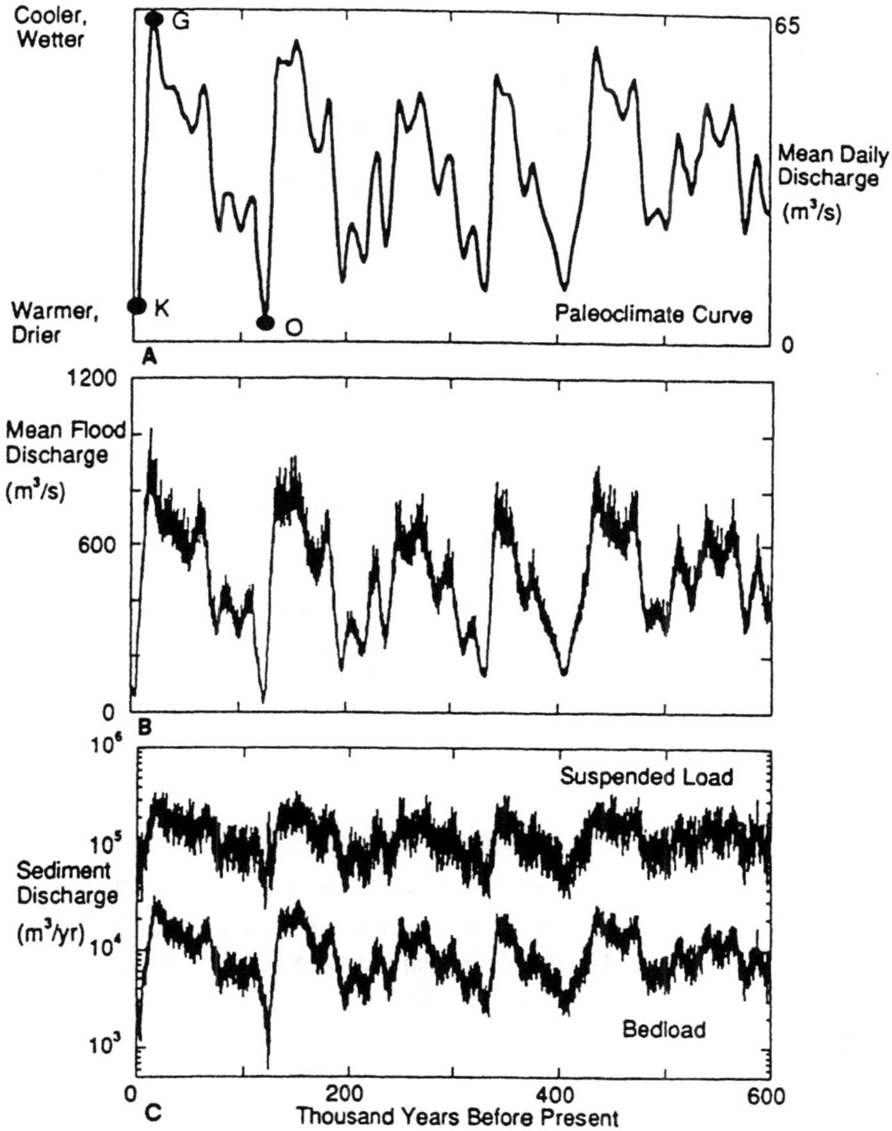

Figure 5. Graphs showing variations: A, in climate; B, fluid discharge rates; and C, sediment discharge rates for alluvial fan deposits formed by Alameda Creek (San Francisco Bay region, California) during last 600,000 years. Discharge rates for fluid and sediment shown in graphs B and C were used as input to SEDSIM simulation experiments whose results are shown in Figure 7B. Note that variations in both discharge rates include overall cyclic variations that reflect major changes in climate and smaller variations or "noise." representing stochastic variations in flood frequencies (from Koltermann and Gorelick, 1992, reproduced with permission of the American Association of the Advancement of Science).

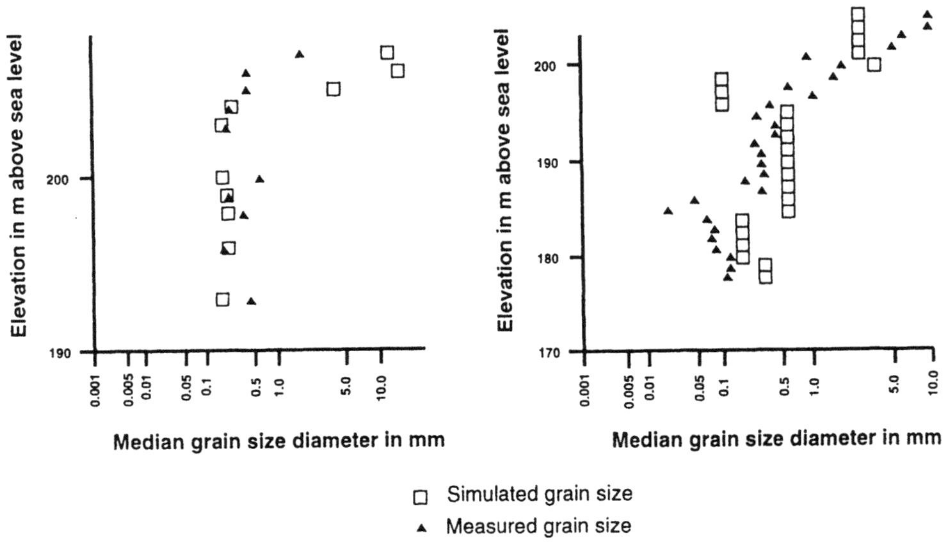

Figure 6. Example plots of two locations in Gardermoen delta of Figure 4 comparing variations in median grain size between actual and simulated sequences with respect to elevation. Note limited variability of simulated grain sizes because only four grain sizes are used in SEDSIM, as well as limited vertical resolution of simulated layers (from Tuttle and others, 1995)

whereas in simulated sequences, these properties can be derived from depositional processes in the form of grain-size distributions.

Two examples involving subsurface fluid flow can be cited: (1) Petrophysical properties derived from process simulation and used with groundwater flow simulations of the aquifer shown in Figure 7B have yielded in spatial distributions of variations in hydraulic head that compare favorably with actual data derived from water wells (Koltermann, 1993). (2) In an application involving oil migrating into reservoir beds at South Belridge field of the San Joaquin Valley of California, simulations helped determine the degree to which lithologic heterogeneities influenced migration, where facies interpretations, petrophysical logs, and oil saturations from cores of producing wells permitted simulations of depositional processes to be coupled with subsequent simulations of subsurface fluid flow (Wendebourg, 1994).

Details of a permeability-grain size function for South Belridge are shown in Figure 8, where a log-linear function has been fitted to scattered

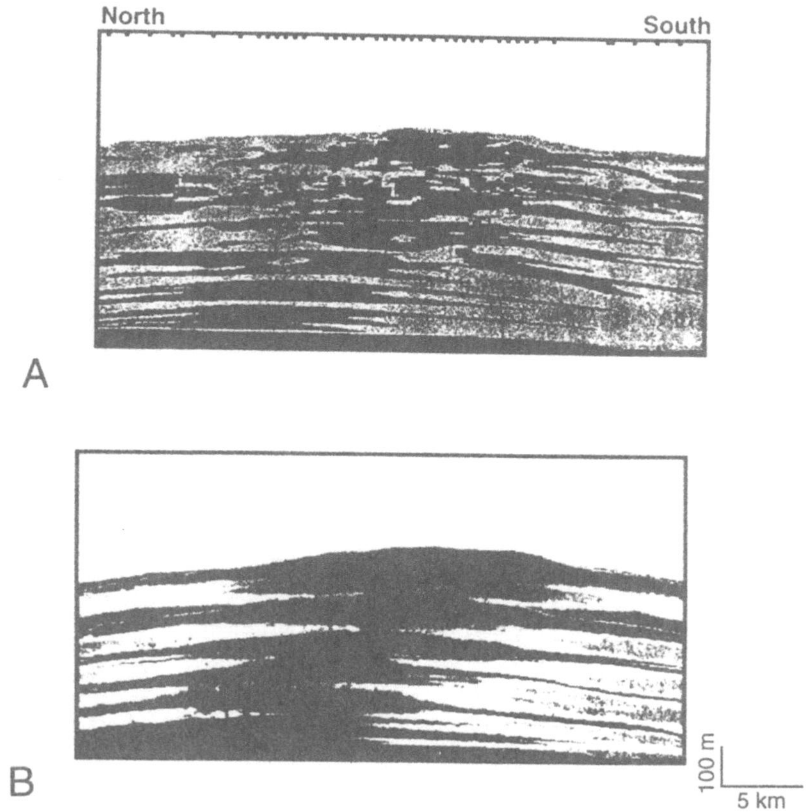

Figure 7. Cross sections showing: A, actual deposits interpreted from information from wells; and B, simulated deposits that form Alameda Creek alluvial fan. Length of section is 16 km and thickness of deposits in center is about 235 m. Gray shades denote variations in mean grain sizes, dark gray denoting coarse sediment and light gray denoting fine sediment. Ticks in upper cross section (A) indicate position of wells that provided information on which cross section is based. Note smooth representation of grain size variations in simulated cross section (B) as compared with complex detail of actual deposits in cross section (A) (from Koltermann and Gorelick, 1992, reproduced with permission of the American Association of the Advancement of Science)

data points for subsequent use in simulating flow. Figure 9 compares oil saturations with respect to permeability for measurements from cores versus those generated using a two-phase reservoir infilling simulator (Wendebourg, 1994). Although sediments with very high and very low permeabilities accord relatively well, the spread of simulated saturations for sediments with intermediate permeabilities is greater than those

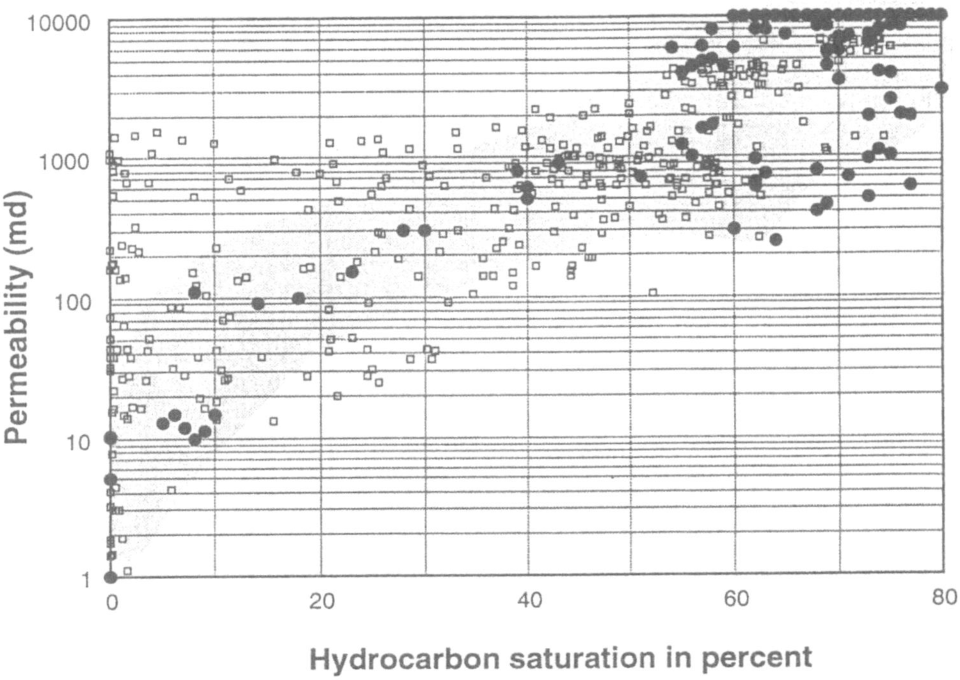

Figure 8. Permeability versus mean grain size in reservoir beds in South Belridge oil field. Larger solid circles denote averages based on measurements in cores (Miller and others, 1990) whereas small squares denote data values used in SEDSIM simulation experiment (from Wendebourg, 1994).

measured in cores. One explanation is that the representation of the reservoir in the simulation does not sufficiently reflect the true sedimentary heterogeneity and therefore grain sizes resulting from the simulator are too homogeneous and thereby overestimate the actual permeabilities. More realistic permeabilities would shift simulated permeability values downward and therefore improve the degree of accordance. Also it is possible that the simulation may not have reached an equilibrium state in its representation of the reservoir, as compared with present conditions in the actual reservoir (Wendebourg, 1994).

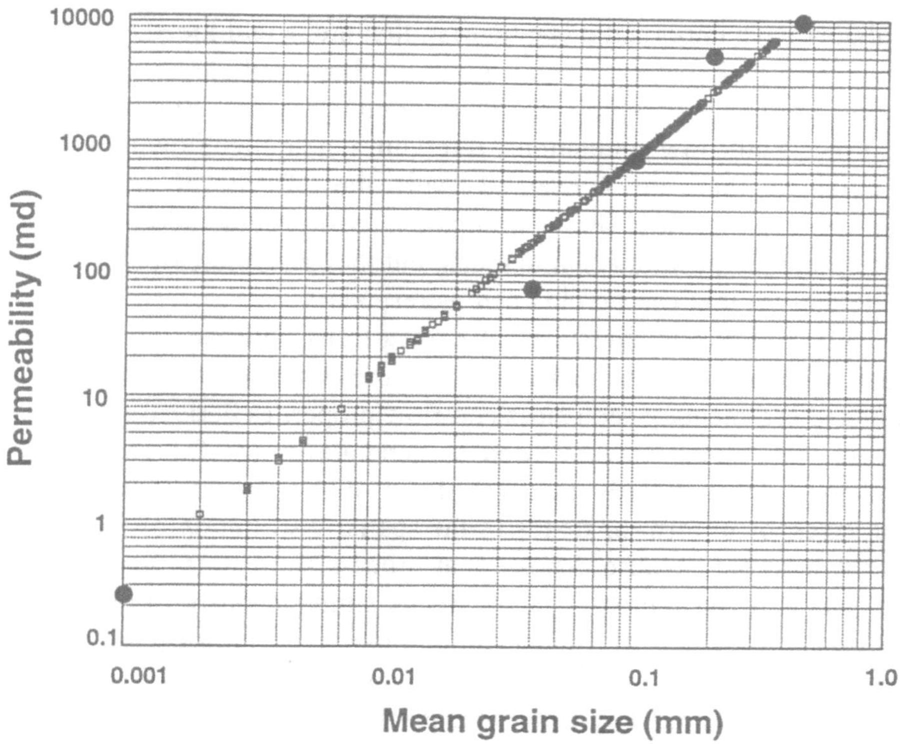

Figure 9. Permeability versus oil saturation in reservoir beds in South Belridge field involving reservoir infilling study conducted with SEDSIM. Solid circles denote measurements based on cores (Miller and others, 1990) whereas squares represent values obtained at individual grid cells generated by SEDSIM and subsequently used in simulaton of subsurface multiphase flow. Note that variations in simulated saturations in cells with high permeabilities are similar to those measured in cores, whereas variations in simulated saturations in grid cells with intermediate permeabilities are higher than those in cores (from Wendebourg, 1994).

CONCLUSIONS

Depositional process simulation is a procedure that can provide 3D descriptions of aquifers and oil and gas reservoirs in terms of the events that formed them. Because process simulators are irreversible and involve working forward through geologic time, it is difficult to constrain them directly to observations. Furthermore, they are regulated by initial and time-dependent boundary conditions involving rates of processes that must be defined a priori. In spite of these requirements, process simulators have important advantages in that they yield the 3D geometric forms of sedimentary deposits and their lithologic properties simultaneously. By combining depositional process simulators with subsurface fluid-flow simulations, reservoir and aquifer flow models can be obtained that are concerned with the effects of sedimentary heterogeneities on flow. Flow simulations also provide additional ways of calibrating depositional simulations, but they also increase the number of alternatives that can be obtained as responses.

In addition to direct visual comparisons between simulated and actual sequences, it is appropriate to use statistical descriptions of simulated sedimentary sequences and flow properties derived from them. Process simulations may be used to generate inputs for geostatistical simulators, either directly as descriptions of the spatial distribution of grain sizes, or in form of trends in the spatial distribution of facies. The latter is possible because depositional process simulators store histories of events and can represent facies trends simultaneously through space and time, properties of fundamental importance in geology. Finally, interpretations of simulation results require careful consideration of assumptions and boundary conditions that control the simulator, based on an understanding of the regional and local geological framework of the area and sequence of beds represented in the simulation.

ACKNOWLEDGMENTS

We thank our colleagues of Stanford's SEDSIM project who contributed to this paper, including Paul Martinez, Young-Hoon Lee, Christoph Ramshorn, Christine Koltermann, and Don Miller. In addition, Kevin Tuttle of the University of Oslo and Cedric Griffiths of the University of Adelaide contributed to our work. Finally, we thank Philippe Joseph of the French Petroleum Institute (IFP) for reviewing the manuscript.

REFERENCES

Beucher, H., Galli, A., LeLoch, A., Ravenne, C., and Heresim Group, 1992, Including a regional trend in reservoir modeling using the truncated Gaussian method, *in* Soares, A., ed., Geostatistics, Troia 1992, Vol. 2: Kluwer Academic Publ., Dordrecht, p. 555-566.

Chiocci, F.L., 1994, Very high-resolution seismics as a tool for sequence stratigraphy applied to outcrop scale - examples from Eastern Tyrrhenian margin Holocene/Pleistocene deposits: Am. Assoc. Petroleum Geologists Bull., v. 78, no. 3, p. 378-395.

Deutsch, C.V., and Journel, A.G., 1992, GSLib - Geostatistical Software Users Manual: Oxford University Press, New York, 340 p.

Doyen, P.M., Guidish, T.M., and de Buyl, M., 1991, Lithology prediction from seismic data, a Monte-Carlo approach, *in* Lake, L.W., Carroll, H.B., and Wesson, T.C., eds., Reservoir characterization II: Academic Press, London, p. 557-564.

Foss, D.C., 1979, Depositional environment of Woodbine sandstones, Polk County, Texas: Gulf Coast Assoc. Geol. Societies Trans., v. 29, p. 83-94.

Geehan, G.W., Lawton, T.F., Sakurai, S., Klob, H., Clifton, T.R., Inman, K.F., and Nitzberg, K.E., 1986, Geologic prediction of shale continuity, Prudhoe Bay field, *in* Lake, L.W., and Carroll, H.B., eds, Reservoir characterization: Academic Press, London, p. 63-82.

Griffiths, C., 1994a, SEDSIM's application in simulating the Brent lower-shoreface Rannoch Formation of the North Sea's Viking graben: Proc. SEDSIM Affiliates Meeting, Stanford University, Stanford, California, 27 p.

Griffiths, C., 1994b, SEDSIM's application in simulating a mid-Jurassic North Sea submarine apron fan for prospect risk analysis: Proc. SEDSIM Affiliates Meeting, Stanford University, Stanford, California, 21 p.

Haldorsen, H.H., and Damsleth, E., 1993, Challenges in reservoir characterization: Am. Assoc. Petroleum Geologists Bull., v. 77, no. 4, p. 541-551.

Harbaugh, J.W., and Bonham-Carter, G., 1970, Computer simulation in geology: Wiley-Interscience, New York, 575 p.

Isaaks, E.H., and Srivastava, R.M., 1989, An introduction to applied geostatistics: Oxford University Press, New York, 561 p.

Koltermann, C.E., and Gorelick, S.M., 1992, Paleoclimatic signature in terrestrial flood deposits: Science, v. 256, no. 5065, p. 1775-1782.

Koltermann, C.E., 1993, Geologic modeling of spatial variability in sedimentary environments for groundwater transport simulations: unpubl. doctoral dissertation, Stanford University, 250 p.

Lawrence, D., Doyle, M., and Aigner, T., 1990, Stratigraphic simulation of sedimentary basins: concepts and calibration: Am. Assoc. Petroleum Geologists Bull., v. 74, no. 3, p. 273-295.

Lee, Y.H., Haskell, N.L., Butler, M.L., and Harbaugh, J.W., 1991, 3-D modeling of Arkansas river sedimentation (abst.): Am. Assoc. Petroleum Geologists Bull., v. 75, no. 3, p. 619.

Lessenger, M., 1993, Forward and inverse simulation models of stratal architecture and facies distributions in marine shelf to coastal plain environments: unpubl. doctoral dissertation, Colorado School of Mines, 182 p.

Martinez, P.A., and Harbaugh, J.W., 1993, Simulating nearshore environments: Pergamon Press, Oxford, 265 p.

Miller, D.D., McPherson, J.G., and Covington, T., 1990, Fluviodeltaic reservoir, South Belridge field, San Joaquin Valley, California, *in* Barwis, J.H., McPherson, J.G., and Studlick, J.R., eds., Sandstone petroleum reservoirs: Springer, New York, p. 109-130

Murray, C.J., 1992, Geostatistical applications in petroleum geology and sedimentary geology: unpubl. doctoral dissertation, Stanford University, 230 p.

Ramshorn, C., Ottolini, R., and Klein, H., 1992, Interactive three-dimensional display of simulated sedimentary basins (abst.): Eurographics Workshop on Visualization in Scientific Computing (Clamart, France), 1 p.

Rivenaes, J.C., 1992, Application of a dual-lithology, depth-dependent diffusion equation in stratigraphic simulation: Basin Research v. 4, no. 2, p. 133-146.

Rudkiewicz, J.L., Guerillot, D., and Galli, A., 1990, An integrated software for stochastic modelling of reservoir lithology and property with an example from the Yorkshire Middle Jurassic, *in* Buller, A.T., Berg, E., Hjelmeland, O., Kleppe, J., Torsaeter, O., and Aalsen, J.O., eds., North Sea oil and gas reservoirs II: Graham and Trotman, London, p. 400-406.

Shuster, M.W., and Aigner, T., 1994, Two-dimensional synthetic seismic and log cross sections from stratigraphic forward models: Am. Assoc. Petroleum Geologists Bull., v. 78, no. 3, p. 409-431.

Slingerland, R., Harbaugh, J.W., and Furlong, K., 1994, Simulating clastic sedimentary basins: Prentice Hall, Englewood Cliffs, New Jersey, 220 p.

Strobel, J., Soewito, F., Kendall, C.G., Biswas, G., Bezedk, J., and Cannon, R., 1989, Interactive (SEDPAK) simulation of clastic and carbonate sediments in shelf to basin settings: Computers & Geosciences , v. 15, no. 8, p. 1279-1290.

Tetzlaff, M., and Harbaugh, J.W., 1989, Simulating clastic sedimentation: Van Nostrand Reinhold, New York, 202 p.

Tuttle, K., Wendebourg, J., and Aagaard, P., 1995, A process-based depositional simulation of the coarse-grained Gardermoen delta, Norway (abst.): Geol. Soc. London Conf. on Quantification and Modeling of Spatial Patterns in Permeable Rocks (Scarborough, England), 1 p.

Webb, E., and Anderson, M.P., 1991, Simulating the grain-size distribution of Wisconsian age glaciofluvial sediments: application to fluid transport (abst.): Am. Assoc. Petroleum Geologists Bull., v. 75, no., 3, p. 689.

Wendebourg, J. 1991, 3-D modeling of compaction and fluid flow of the Woodbine progradational sequence, Polk County, Texas (abst): Am. Assoc. Petroleum Geologists Bull., v. 75, no. 3, p. 691.

Wendebourg, J., Ramshorn, C., and Martinez, P.A., 1993, SEDSIM3.0 User's Manual, Proc. SEDSIM Affiliates Meeting, Stanford University, Stanford, California, 73 p.

Wendebourg, J., 1994, Simulating hydrocarbon migration and stratigraphic traps: unpubl. doctoral disseration, Stanford University, 258 p.

MODELING OF MULTICOMPONENT DIAGENETIC SYSTEMS

Robert Ondrak
GeoForschungsZentrum Potsdam, Potsdam, Germany

ABSTRACT

A computer model is presented, which permits the modeling of large-scale diagenetic evolution of sandstones resulting from mineral dissolution and precipitation. The model includes various mineralogical components and dissolved species which are major components of many sandstones. The model allows the simulation of temporal and spatial evolution of porosity and mineral distribution in a sandstone layer. Possible application to diagenetic problems is illustrated by several theoretical examples, representing different geological environments.

INTRODUCTION

Modeling diagenesis is a complex task because various processes may interact. In principle, however, two mechanisms determine the diagenetic evolution of rocks: mechanical compaction and chemical diagenesis. Several models have been published during the last decade, concerned with the numerical modeling of sandstone diagenesis (e.g. Wood and Hewett, 1982; Leder and Park, 1986; Chen and others, 1990; Nagy and others, 1990; Gouze and Coudrain-Ribstein, 1993). The models steadily increased in complexity from simple 0-dimensional quartz models (e.g. Leder and Park, 1986) to complex, multidimensional reaction-transport models (Chen and others, 1990). The model presented here focuses on regional diagenetic alterations, which are the result of the interaction of mineral reactions and fluid flow in sandstones. In a flow-driven diagenetic environment, the regional distribution of the flow system usually is inhomogeneous, varying in space as a function of the local porosity and permeability distribution. In addition, mineral dissolution or precipitation changes porosity and permeability with time, which, in turn, modify the flow system. The interaction between the flow system and the diagenetic alteration of the rock is illustrated by computer simulations.

THEORETICAL BACKGROUND OF THE MODEL

Modeling chemical diagenesis requires the numerical simulation of water-rock interactions. Depending on the mineral composition of the rock, a large number of minerals and dissolved species can be involved. The attempt to consider all possible components provides a model too complex to be handled numerically. The present example, therefore, is limited to the dominant constituents of many sandstones: quartz, calcite, albite, and kaolinite as mineral species. For each mineral a chemical reaction has to be defined describing the interaction between the pore fluid, the mineral, and the resulting dissolved species [Reaction (1)-(4)].

$$SiO_2 + 2H_2O \iff H_4SiO_4 \quad (1)$$
$$CaCO_3 + H_2CO_3 \iff 2HCO_3^- + Ca^{2+} \quad (2)$$
$$NaAlSi_3O_8 + 4H^+ + 4H_2O \iff Na^+ + Al^{3+} + 3H_4SiO_4 \quad (3)$$
$$Al_2Si_2O_5(OH)_4 + 6H^+ \iff 2Al^{3+} + 2H_4SiO_4 + H_2O \quad (4)$$

The interaction between a gaseous phase (CO_2) and the pore fluid also causes the formation of a dissolved species [Reaction (5)].

$$CO_2 + H_2O \iff H_2CO_3 \quad (5)$$

The reactions describe the mineral-fluid interaction, or in general the phase transition between the solid, the fluid, and the gas phase. In addition, a complete chemical description of the system requires the interaction of the dissolved species and the ions in the pore fluid [Reaction (7)-(8)]. In the presence of carbonic acid these are:

$$H_2CO_3 \quad <=> \quad HCO_3^- + H^+ \qquad (6)$$
$$HCO_3^- \quad <=> \quad CO_3^{2-} + H^+ \qquad (7)$$
$$H_4SiO_4 \quad <=> \quad H_3SiO_4^- + H^+ \qquad (8)$$
$$H_2O \quad <=> \quad OH^- + H^+ \qquad (9)$$

For each chemical reaction an equation can be formulated describing its equilibrium condition. The stability constants of the reactions differ nonlinearly as a function of temperature and pressure. Figure 1 shows the temperature dependence of the different stability constants along the vapor pressure line illustrating that temperature is one of the major controlling factors for water-rock interaction processes.

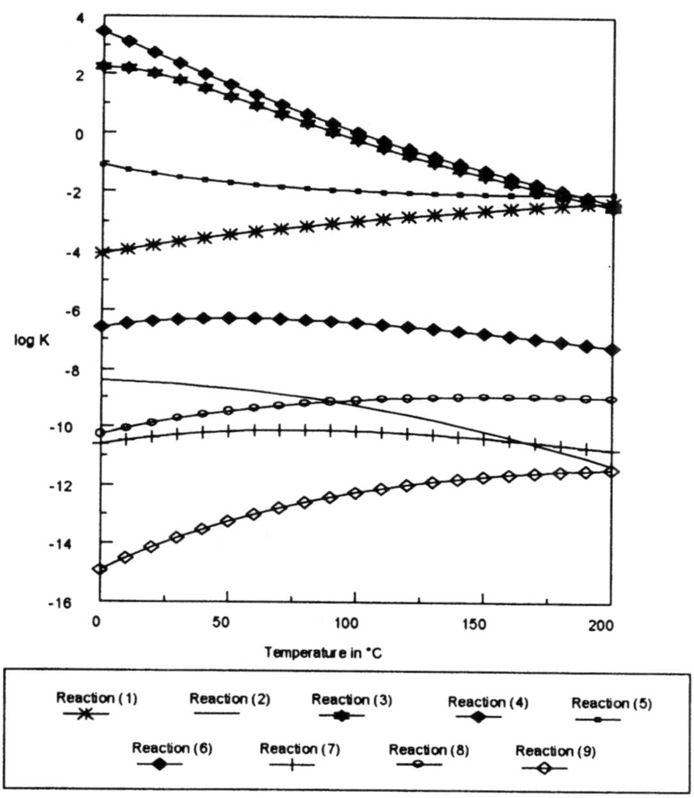

Figure 1. Temperature dependence of stability constants of reaction (1)-(9).

Lastly, it has to be considered whether the model should be regarded as an open or a closed system with respect to certain components. In general, the model describes an open system into which species can be imported or exported by fluid flow. But water-rock and gas-water interactions are phase transitions which only take place if the specific phase is present. Therefore, the system may become closed with respect to a particular component if this component is removed from the system resulting from dissolution. The diagenetic model must handle this modification of the physical system and it must ensure a correct mass balance for the entire system. The correct mass balance is achieved by logical separation of phase equilibria and phase transitions, which permits to model the time dependent diagenetic evolution of rocks (Ondrak, 1992, 1993; Ondrak and Bayer, 1993).

The dissolution and precipitation of minerals not only alters the mineralogical composition of the rock, but also the petrophysical properties, for example conductivity and permeability of the rock. The modification of the physical properties changes the flow regime of the pore fluid, which determines the transport rates of the dissolved species. Therefore, the flow system and the diagenetic evolution of the rock interact. To study the coupling between the two systems the chemical reaction model is combined with a two-dimensional flow model in which the flow rates vary spatially and temporally as a function of the evolving physical properties. The combination of the two modules allows to simulate the interaction of fluid flow, chemical evolution of the pore fluid, and mineral reactions with respect to the regional diagenetic evolution of a sandstone.

MODEL CALCULATIONS

The following section outlines the possible application of the model to an idealized sandstone. The following schematic geological scenario is assumed for the simulations. The thickness of the sandstone layer is small with respect to its lateral extension. Therefore, its thickness can be reduced to a two-dimensional surface (Fig. 2). The sandstone is bounded by almost impermeable claystone layers from above and below. Pore fluid, therefore, can only flow within the surface. Within the model a pressure gradient is defined between the left and right border of the layer causing a fluid flow from left to right (Fig. 2), whereas the lateral boundaries are sealed; the layer is 40 km long, 30 km wide, and is located at a depth of 3 to 2 km. Temperature along the layer changes from 110 °C at the deepest position to 65 °C at the highest. Both temperature and burial depth remain constant throughout the calculations. Permeability and porosity of the layer differs spatially, representing the heterogeneity of natural rocks, whereas the mineralogical composition is considered to be more uniform but changes from one example to the other. The three layers in Figures 3-10 represent the same layer at different time slices: at the beginning of the calculations (0 my), in the middle (5 or 2.5 my), and at the end (10 or 5 my).

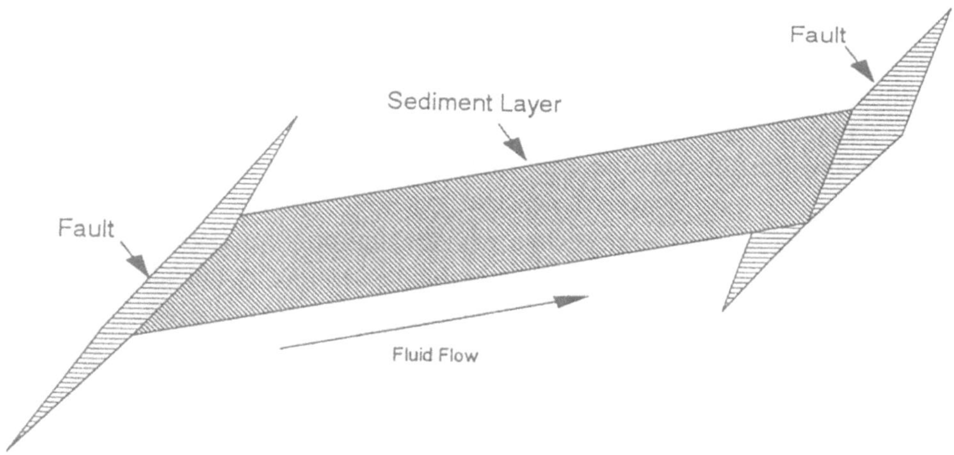

Figure 2. Schematic geological model for simulations. Pore water flow through sandstone layer is from left to right. For model, sandstone body is represented by 2D surface.

Example 1

In the first example the diagenetic evolution of a calcite cemented quartz sandstone is considered. The initial flux ranges between 15 and 20 m/y depending on the local permeability of the layer. The fluid entering the layer from the left side is in chemical equilibrium with respect to temperature, quartz, calcite, and a low CO_2-concentration. The conditions remain constant throughout the simulated time period. Within the layer, the chemical system is considered a closed system with respect to

CO_2. The amount of carbonic acid, therefore, can only be modified by calcite dissolution or precipitation [cf. Reaction (2)]. As the fluid flows updip along the layer it cools which changes the chemical equilibria of the dissolved species. The equilibrium concentration of quartz decreases with declining temperature (Fig. 1) causing the precipitation of silica cement. For the carbonate equilibrium the opposite trend is the situation. Calcite solubility increases with declining temperature. Therefore, calcite cement is dissolved continuously by the upward moving pore water. The effects on porosity are illustrated in Figure 3, which shows the initial porosity distribution and its temporal evolution after 5 and 10 million years. Most apparent is the overall porosity increase during the simulated time period. Near the left and right boundary of the model this trend is reversed towards the end of the simulations. For a detailed consideration of the interacting processes, the temporal evolution of the different mineral species is illustrated in Figures 4 and 5.

Figure 4 shows the changes of calcite cement resulting from dissolution and fluid flow. The moving fluid dissolves the pore filling cement because of its temperature dependent leaching capacity. Therefore, dissolution proceeds not homogeneously throughout the layer, but is most prominent where the highest temperature variations occur along the flow path. This is the situation, in particular near the right and left boundaries, because of the increased inclination of the layer. The calcite dissolution capacity is not only temperature controlled, but also depends on the chemical composition of the pore fluid. This effect can be seen in the central part of the layer, where calcite dissolution proceeds slowly because of the small temperature variations. In this part of the sandstone, calcite removal becomes important towards the end of the simulated time period. In this final phase, the fluid increasingly is undersaturated with respect to calcite because of the complete removal of calcite cement farther downstream and the dissolution proceeds into the central region of the layer following zones of high flow rates. In addition, the calcite dissolution increases the permeability of the layer, thereby, increasing flow rates and mass transport. The higher mass transport, in turn, enhances the dissolution process causing an interaction of fluid flow and mineral reactions.

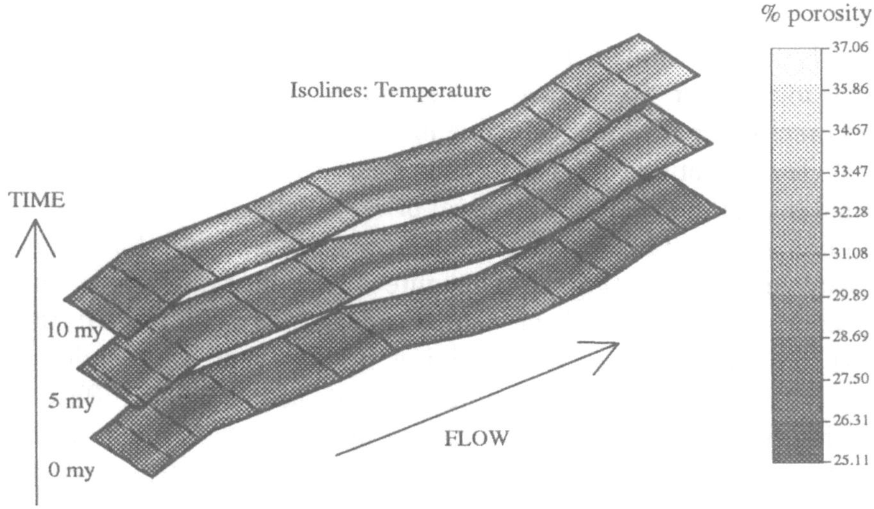

Figure 3. Porosity evolution of artificial calcite cemented sandstone for time period of 10 my. Bottom layer shows porosity distribution at beginning of simulation, whereas middle and top layer show porosity distribution after simulated time period of 5 and 10 my. Flow rates range between 15-20 m/y, and CO_2-concentration is low.

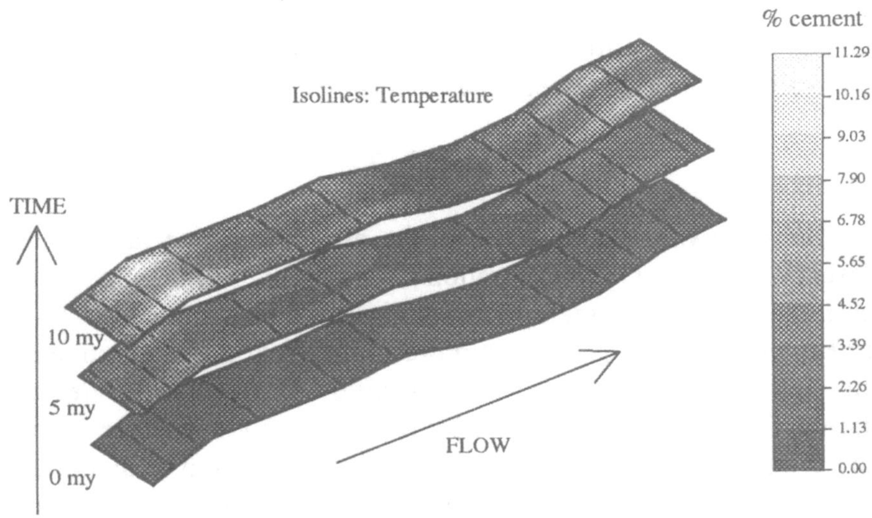

Figure 4. Temporal and spatial evolution of silica cement distribution of sandstone layer shown in Figure 3.

The interaction between fluid flow and diagenesis also can be illustrated by the development of the silica cement (Fig. 5). The large amount of newly formed silica cement in the left part of the layer, which corresponds to the zone of early calcite cement removal, is interesting. Here, silica precipitation is more effective in destroying porosity than is gained by calcite dissolution. The timing of increased silica precipitation (Fig. 5) coincides with the complete removal of the calcite cement (Fig. 4). The evolution of the calcite and the silica cement also is reflected in the porosity development (Fig. 3). In Figure 3 an initial porosity increase can be seen in this part of the layer, which is destroyed again towards the end of the simulated time period. The initial porosity increase is followed by porosity destruction, which indicates that the enhanced flow rates resulting from the permeability increase by calcite dissolution promoted the strong silica precipitation. The spatial variations in silica precipitation are connected to mass transport variations and varying temperature changes along the flow path. The results illustrate the interaction between fluid flow and diagenetic evolution of sandstones in a flow-dominated system. High-flow rates enhance the mass transport of dissolved components. The increased mass transport, in turn, promotes the dissolution or precipitation of minerals depending on the surrounding conditions. The mineral reactions alter the permeability of the rock which controls mass transport. The interaction and feed back of the processes results in interesting diagenetic patterns even in the situation of a relatively simple calcite cemented sandstone.

Example 2

The following example shows the same sandstone layer with the difference that the CO_2 concentration is two orders of magnitudes higher than in the previous example. The setting could be compared to an environment with active CO_2 production resulting from the maturation of organic matter or with an area where juvenile CO_2 occurs. In the present example, flow rates were reduced by approximately one order of magnitude having values of 1-2 m/y. The diagenetic evolution of the sandstone layer differs from the previous example because of the modified boundary condition. Again, an overall porosity increase can be observed (Fig. 6), which correlates with zones of high-flow rates.

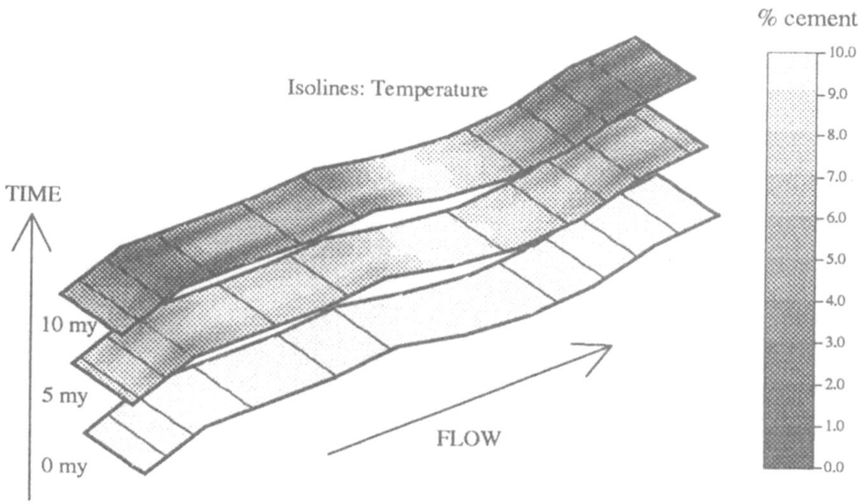

Figure 5. Temporal and spatial evolution of calcite cement distribution of sandstone layer shown in Figure 3.

With respect to the previous example, the main difference is the lacking porosity decrease in the left part of the layer towards the end of the simulated time period. Because of the low-flow rates of 1-5 m/y, silica cementation is too small to compensate the porosity increase caused by calcite dissolution. The diagenetic evolution is dominated by calcite dissolution, which is increased strongly by the high CO_2 concentration of the pore fluid. Figure 7 shows the temporal evolution of the calcite cement distribution corresponding to the porosity evolution of Figure 6. In general, the picture is comparable to the previous example with the main difference that changes occur at smaller flow rates than before. The distribution of the silica cement is not illustrated because the amount of newly generated silica cement is negligible. Silica precipitation reaches a maximum of 1% in some areas, which has little effect on the overall picture.

Example 2 illustrates that the diagenetic evolution of a sandstone is not controlled simply by temperature changes and mass transport but by all factors which alter the chemical composition of the pore fluid. In the present model, the high CO_2 concentration of the pore fluid is the dominant feature which enhances calcite dissolution. In the following example, the effect of feldspar dissolution on the model is studied as an other factor influencing the diagenetic evolution of the model sandstone.

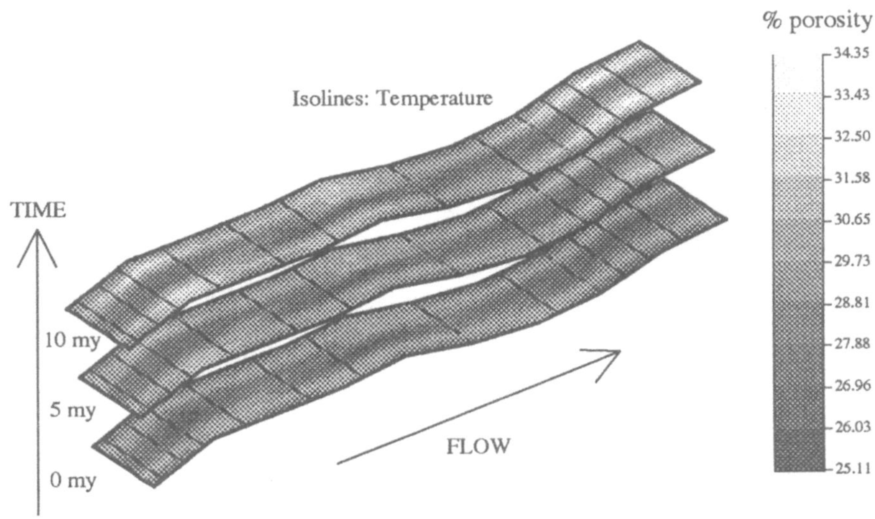

Figure 6. Porosity evolution of artificial calcite cemented sandstone for time period of 10 my. Bottom layer shows porosity distribution at beginning of simulation, whereas middle and top layer show porosity distribution after simulated time period of 5 and 10 my. Flow rates range between 1-2 m/y and CO_2 concentration is high.

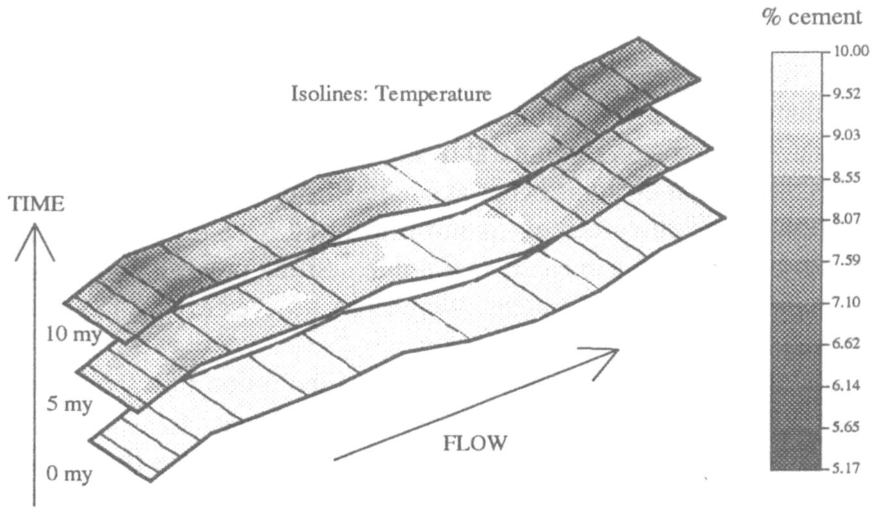

Figure 7. Temporal and spatial evolution of calcite cement distribution of sandstone layer shown in Figure 6.

Example 3

In the final example, the mineralogical composition of the sandstone layer differs from the previous ones, because albite is distributed homogeneously in the sandstone, in addition to calcite and quartz. Furthermore, it is postulated that the dissolution of albite will result in the formation of kaolinite as the new aluminium-bearing mineral phase. The fluid entering the layer is in equilibrium with calcite and with a relatively high CO_2 concentration but not with respect to albite. Figure 8 illustrates the temporal and spatial porosity evolution of the layer. The most prominent feature of this example is an overall porosity decrease of the sandstone. It is most pronounced along the zones of high initial porosity, and moves as a cementation front along the flow direction through the layer. In order to understand the processes causing this porosity evolution one has to look at the temporal and spatial distribution of the different mineral phases.

Figure 9 illustrates the temporal evolution of the albite distribution in the layer. It can be seen that albite is dissolved by the entering pore fluid which, although in chemical equilibrium with calcite, contains enough acidity to dissolve albite. A relatively sharp reaction front develops breaking up into different fingers as a result of the varying flow

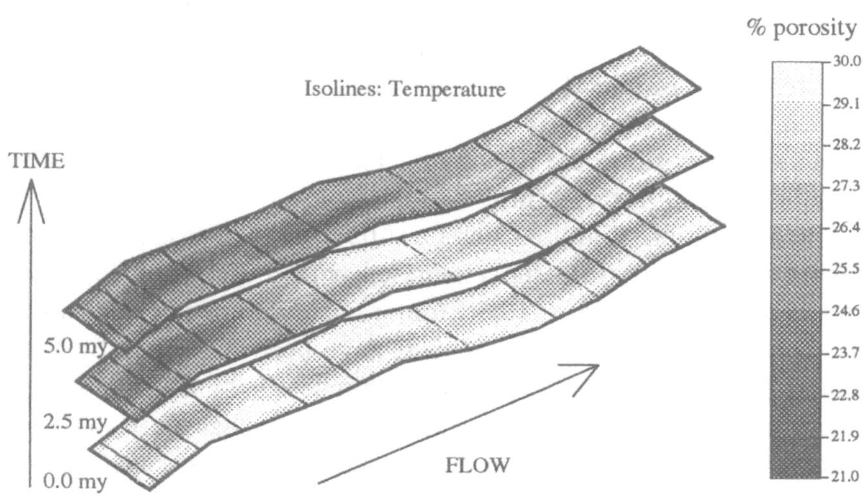

Figure 8. Porosity evolution of artificial sandstone for time period of 5 my. Bottom layer shows porosity distribution at beginning of simulation, whereas middle and top layer show porosity distribution after simulated time period of 2.5 and 5 my.

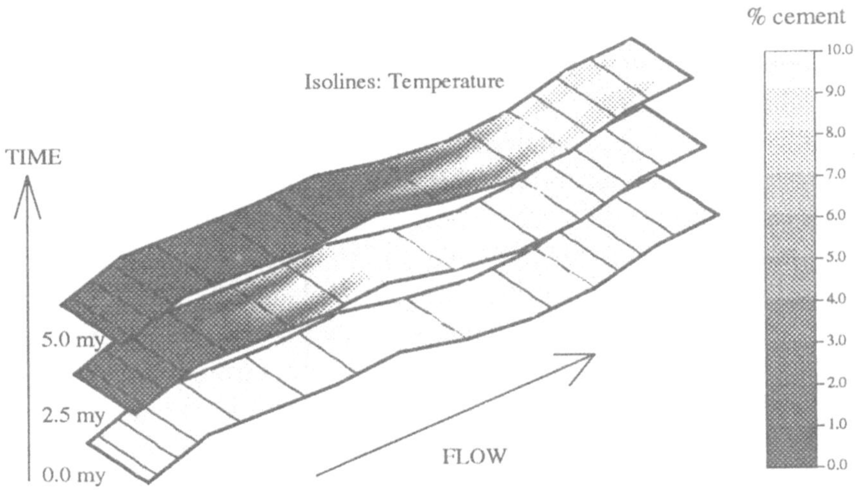

Figure 9. Temporal and spatial evolution of albite distribution of sandstone layer shown in Figure 8.

rates in the sandstone. At the dissolution front the leaching capacity of the pore fluid is consumed and, therefore, feldspar cannot be dissolved anymore. In the model, dissolution of albite buffers the fluid and inhibits mineral reactions farther down stream of the reaction front. This is a simplification, which neglects the small alterations resulting from temperature variations along the flow path. However, these changes are too small to be noticed in the figures.

The dissolution of albite causes a shift in pH towards an alkalic environment making the fluid supersaturated with respect to calcite which then starts to precipitate (Fig. 10). Similar observations of feldspar dissolution in the presence of calcite cement have been described by Siebert and others (1984) from investigations of thin sections. But the precipitation of calcite stops as soon as the albite has disappeared. Calcite starts to dissolve again, because of the increasing dissolution capacity of the cooling pore fluid.

According to Reaction (3) the dissolution of albite increases aluminium and silica concentration in the pore fluid. Following Reactions (1) and (4) an increased aluminium and silica concentration will result in

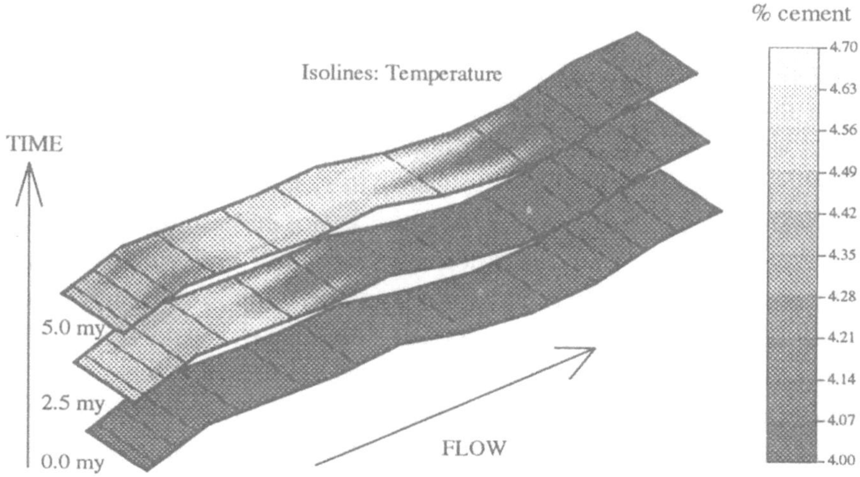

Figure 10. Temporal and spatial evolution of calcite cement of sandstone layer shown in Figure 8.

the formation of silica and kaolinite cement. The precipitation of these cements cause the overall porosity reduction shown in Figure 8. On the scale of the model the major amount of silica cement can be contributed to the dissolution of the feldspar present in the rock, whereas mass transport and temperature related changes are negligible. However, it must be kept in mind that this result is valid only for the special assumption of the model. For other geological settings the results may look different as shown by Examples 1 and 2.

CONCLUSIONS

The previous examples illustrated the interaction of mineral phases with the pore fluid and the flow system under various boundary conditions. The results should not be seen purely from their quantitative value, but also their qualitative aspect should be considered. A major problem for quantitative modeling is the quality of the thermodynamic data on which the calculations rely. Already little uncertainties in the data can modify drastically the results of such highly nonlinear dynamic system. In the present model the acidity of the pore fluid is related closely to the carbon dioxide content of the fluid. A main source of CO_2 is the maturation of organic material (Siebert and others, 1984). New models in organic geochemistry (e.g. Burnham and Sweeney, 1989) allow to quantify the amount and the time of CO_2 production. The next step to better understand the interaction of organic and inorganic diagenesis could be obtained by the combination of diagenetic modeling and maturation models. Another aspect which becomes interesting in systems with high-flow rates is the kinetics of the reactions. This point is not considered in the model because the flow rates are considered to be small enough for chemical equilibrium to be established. This consideration obviously must be modified when flow rates are increased. The model, therefore, cannot be applied to environments with flow rates of meters per day, where the equilibrium assumption is no longer justified. Therefore, the present model is limited to flow rates of centimeters to meters per year, which allow the establishment of chemical equilibrium (Ondrak, 1993). Within this framework it can be used as a tool to test diagenetic concepts derived from petrologic studies against chemical and physical first principles. But it should be kept in mind that the model is based on a number of assumptions and simplifications, which may have to be rethought if the simulation results do not provide plausible results with regard to geological and mineralogical observations. For this reason, the present model can be only one more step towards a better understanding of the complexity of diagenetic processes. It will evolve further as the comprehension of the processes grows. However, it provides a tool to obtain a better understanding of the complex geological system.

ACKNOWLEDGMENTS

Thanks are due to U. Bayer, T. McCann, and D.F. Merriam for critical reading of the manuscript.

REFERENCES

Burnham, A.K., and Sweeney, J.J., 1989, A chemical kinetic model of vitrinite maturation and reflectance: Geochim. Cosmochim. Acta, v. 53, no. 10, p. 2649-2657.

Chen, W., Ghaith, A., Park, A., and Ortoleva, P., 1990, Diagenesis through coupled processes: modeling approach, self-organization, and implications for exploration, *in* Meshri, I.D., and Ortoleva, P. J., eds., Prediction of reservoir quality through chemical modeling: Am. Assoc. Petroleum Geologists Mem. 49, p. 103-130.

Gouze, P., and Coudrain-Ribstein, A., 1993, Quantitative approach to geochemical processes in the Dogger aquifer (Paris Basin): from water analyses to the calculations of porosity evolution, *in* Doré, A. G., and others, eds., Basin modelling: advances and applications: NPF Spec. Publ., v. 3, Elsevier, Amsterdam, p. 343-351.

Leder, F., and Park, W.C., 1986, Porosity reduction in sandstone by quartz overgrowth: Am. Assoc. Petroleum Geologists Bull., v. 70, no. 11, p. 1713-1728.

Nagy, K.L., Steefel, C.F., Blum, A.E., and Lasaga, A.C., 1990, Dissolution and precipitation kinetics of kaolinite: initial results at 80°C with application to porosity evolution in a sandstone, *in* Meshri, I.D., and Ortoleva, P.J., eds., Prediction of reservoir quality through chemical modeling: Am. Assoc. Petroleum Geologists Mem. 49, p. 85-101.

Ondrak, R., 1992, Mathematical simulation of water-mineral interaction and fluid flow with respect to the diagenetic evolution of sandstones, *in* Kharaka, Y.K., and Maest, A.S., eds., Water-rock interaction: Proc. 7th Intern. Symp. Water-Rock Interaction: Balkema, Rotterdam, p. 1187-1191.

Ondrak, R., 1993, Untersuchungen zur mathematischen Simulation der Sandsteindiagenese und der damit verbundenen Veränderungen der Porosität und Permeabilität: Berichte des Forschungszentrum Jülich, v. 2770, 99 p.

Ondrak, R., and Bayer, U., 1993, Dissolution and cementation in basin simulation, *in* Harff, J., and Merriam, D.F., eds., Computerized basin analysis: the prognosis of energy and mineral resources: Plenum Press, New York, p. 59-82.

Siebert, R.M., Moncure, G.K., and Lahann, R.W., 1984, A theory of framework grain dissolution in sandstones, *in* McDonald, D.A., and Surdam, R.C., eds., Clasitc diagenesis: Am. Assoc. Petroleum Geologists Mem. 37., p. 163-175.

Wood, J.R., and Hewett, T.A, 1982, Fluid convection and mass transfer in porous sandstones-a theoretical model: Geochim. Cosmochim. Acta, v. 46, no. 10, p. 1707-1713.

MODELING PETROLEUM RESERVOIRS IN PENNSYLVANIAN STRATA OF THE MIDCONTINENT, USA

W. Lynn Watney, John A. French, and Willard J. Guy
University of Kansas, Lawrence, Kansas

ABSTRACT

Improved characterization of petroleum reservoirs must include better geologic models in order to predict quantitative attributes of reservoir units. By necessity, petroleum reservoir prediction and modeling must make both interpolations and extrapolations from limited data. Several approaches to modeling of sedimentary rocks include descriptive or conceptual (qualitative) geologic models and geostatistical, simulation (process), and visualization (quantitative) models. Each type of model has its advantages and limitations, including the appropriate scale of application, data requirements, and knowledge as to how the reservoir was formed. All of the models compliment one another, providing views of complex reservoirs from different perspectives. Quantitative modeling potentially can create a more coherent, integrated view of the reservoir than qualitative conceptual models. An optimum model probably includes a combination of approaches based on the extent and type of knowledge about the reservoir.

Geologic Modeling and Mapping
Edited by A. Förster and D. F. Merriam, Plenum Press, New York, 1996

The examples of stratigraphic simulation and 3D visualization models presented in this paper are based on multilayered oolite units developed in the Upper Pennsylvanian, carbonate-dominated Swope Formation depositional sequence in both the shallow subsurface of eastern Kansas and in the Victory Field in southwestern Kansas. The models discussed in this paper include local and regional scales and quantitative process and visualization types.

INTRODUCTION

Stratigraphic modeling is a powerful tool that can improve our ability to understand the complexities of petroleum reservoirs, and thereby assist in the recovery of oil and gas from those reservoirs. Types of models can be classified into (1) conceptual models -- a geologic interpretation or construct; (2) correlation models -- a manual interpretation of the spatial association of geologic units; (3) interpolation models -- a machine generated association of spatial data, that is a visualization; (4) forward simulations -- machine generation of geology based on input of processes; and (5) inverse models -- machine derivation of process parameters from geologic data. Two forms of modeling are presented in this paper: two-dimensional stratigraphic simulations and visualization modeling. These approaches, which include examples at different scales, are discussed in turn in the next sections.

Stratigraphic Simulation Modeling

Stratigraphic forward simulation modeling begins as an inductive exercise to ascertain the processes behind sedimentation and the resulting stratigraphic succession. Initially, a quantitative data set is obtained to infer and characterize the processes and develop a conceptual model such as the one for Upper Pennsylvanian carbonates in the upper part of the Midcontinent, U.S.A. (Fig. 1). The steps in database development include:

(a) identify, describe, correlate, map, and interpret 3D geometries of regional and locally significant temporally distinct genetic units (such as depositional sequences) (Table 1);
(b) characterize shelf paleotopography and subsidence history;

(c) conduct inverse modeling procedures to estimate timing and rates of sediment accumulation (e.g., time-series analysis of logs and lithologic data).

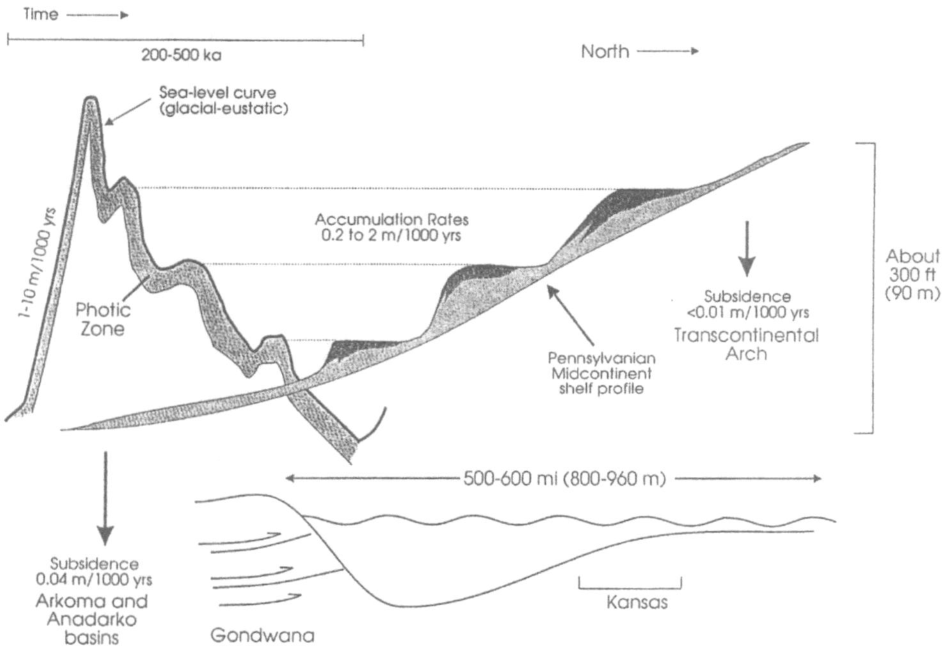

Figure 1. Conceptual model depicting processes that dynamically interact to produce 4th-order depositional sequences Upper Pennsylvanian cyclic strata of Midcontinent.

Sequence-stratigraphic analysis is used to define the temporally distinct packages needed in the simulation model. A nonscalar sequence stratigraphic terminology is used in which the duration and order classification of the cycles follows that of Goldhammer, Oswald, and Dunn (1991) (Table 1). These unconformity-bounded cyclothems (5-25 m thick) possess all of the components of Vail's seismic depositional sequences except thickness and duration (Youle, Watney, and Lambert, 1994). These are classified as 4th-order units based on estimated durations of 0.25 to 0.5 Ma (Heckel, 1986; Klein, 1990). Cyclothemic 4th-order depositional sequences are fundamental mapping units in the Pennsylvanian of the Midcontinent, as evidenced by their designation as formations in formal stratigraphic nomenclature (Zeller, 1968). Fourth-order cycles can be correlated easily with wireline logs. Thicknesses and contrasting lithologies of component units that comprise these cycles generally can be distinguished on well logs (Fig. 2). Biostratigraphic information derived mainly from conodonts and ammonoids is being used to resolve independently individual 4th-order depositional sequences as well. These sequences are correlatable regionally, and because they are bounded by laterally extensive unconformities, they can be considered as temporally distinct units (i.e. depositional sequences) for purposes of paleogeographic reconstructions, interpretation of detailed facies relationships, and for modeling.

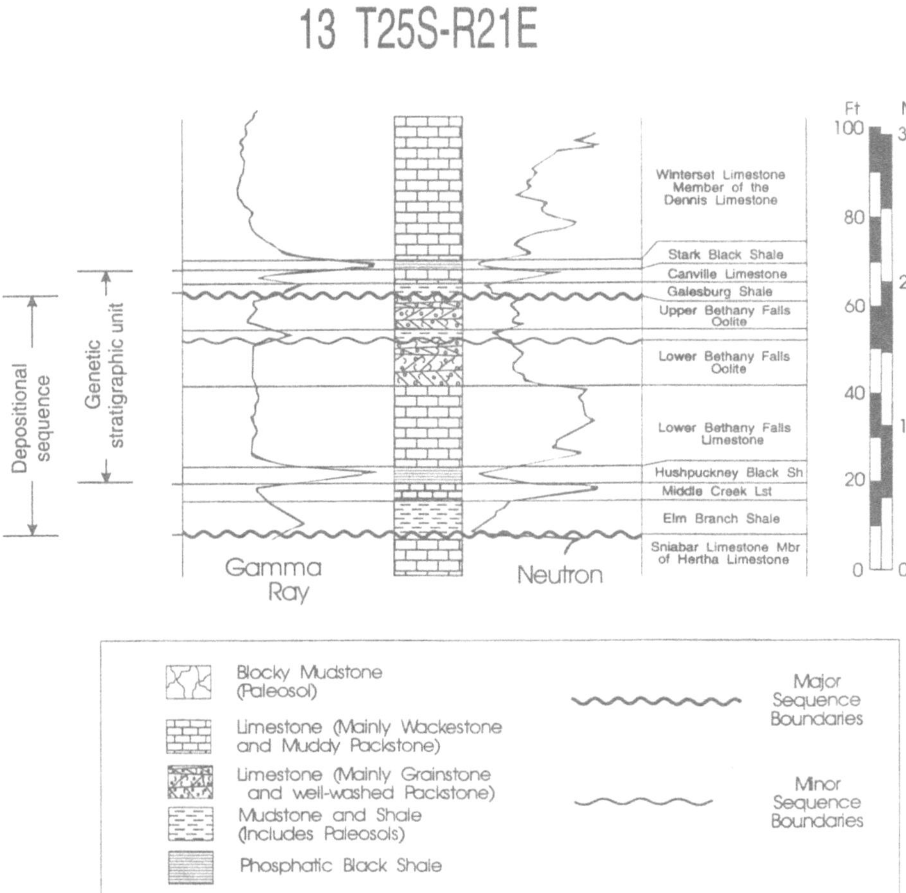

Figure 2. Unscaled gamma ray-neutron wireline log annotated with Swope sequence and its component genetic units: Middle Creek Limestone = transgressive unit; Hushpuckney Shale = condensed section; Bethany Falls Limestone = regressive (forced) unit; Galesburg Shale = capping paleosol in the location of this well. Wavy lines are subaerial unconformities.

Other scales of unconformity-bounded sequences are present, including 3rd-order cycles consisting of sets of five to seven 4th-order depositional sequences. In some situations where 4th order cycles are thick, 5th-order unconformity-bounded minor cycles are recognized, thus the stratal packaging is complex. Any model must resolve units down to the fifth order, because these are flow units fundamental to reservoir performance and reservoir prediction.

The stack of unconformity-bounded oolite and wackestone packages developed within the Bethany Falls Limestone of the 4th-order Swope sequence modeled in this paper are classified as 5th-order minor cycles. These cycles exhibit transgressive bases and regressive tops, with basinward shifts leading to subaerial exposure interpreted as evidence of forced regression by eustacy. These minor cycles are therefore not parasequences even though the thicknesses of these units might suggest such an interpretation.

Table 1. Cycle orders (modified after Goldhammer, Oswald, Dunn, 1993)

Cycle Order	Duration (million yrs., Ma)	Amplitude (meters)	Rise/Fall rates (m/ka)	Sequence terminology	This study
1st order	>100		<.01		
2nd order	10-100	50-100	0.01-0.03	Supersequence	
3rd order	1-10	50-100	0.01-0.1	Sequence	Sequence Set
4th order	0.1-1	1-150	0.4-5	Sequence, Cycle	Cycle, Cyclothem
5th order	0.01-0.1	1-150	0.6-7	Parasequence, Cycle	Minor Cycle

The correlatable cyclicity observed in these strata is attributed to changes in accommodation space effected by changes in relative sealevel. These changes in accommodation are interpreted as a response to an interplay of tectonism, eustasy, and sediment supply. Simulation modeling is used to examine the relative contributions of these causal mechanisms.

Parameter Definition

Specific procedures to derive parameters used in the simulation modeling are divided into six categories (Table 2). These parameters were discussed in Watney, Wong, and French (1991).

Table 2. Parameters used in modeling Pennsylvanian cyclothems.

(I)	Characterization and interpretation of temporally distinct stratigraphic units;
(II)	Estimates of subsidence across Kansas shelf during the Upper Pennsylvanian;
(III)	Estimates of elevation across the Pennsylvanian shelf and magnitude of relative sealevel change;
(IV)	Characteristics of Pennsylvanian eustacy;
(V)	Sealevel history estimated from Quaternary analog;
(VI)	Sediment accumulation rates from Recent and Quaternary analogs.

The values of the parameters used to build the simulations include:

(1) 4th-order sealevel rise and fall ranging to maximum of 100 m characterized by rapid rise and gradual fall, patterned after late Quaternary eustatic fluctuations;
(2) elevation differences across the depositional surface of up to 100 m based on study of regional isopach, facies, and decompaction (135%) of the Pleasanton clastic platform;
(3) sediment accumulation rates determined by water depth and slope of depositional surface (energy level) ranging under 2 m/thousand years (m/ka) and based on Quaternary analogs (Fig. 3);
(4) subsidence ranging between 0.02 m/ka and 0.2 m/ka determined through backstripping;
(5) the maximum depth at which carbonates accumulated (euphotic zone) was set at 35 meters, using modern data to help constrain this value.

2D Stratigraphic Simulation Modeling (KANMOD)

Our initial geologic simulation model of carbonate sedimentation was developed for a personal computer using one-dimensional output generating a single stratigraphic succession at one location. This 1D model incorporated parameters derived from characterization of Upper Pennsylvanian carbonate-dominated depositional sequences as well as data on sedimentation and stratigraphy derived from Quaternary and Recent analogs (Watney, Wong, and French, 1991) (Fig. 3). The 1D model has user defined input of eustatic sealevel curve, subsidence rate, and carbonate sediment accumulation rate. The output is a time vs. elevation plot of the sealevel curve, initial depositional surface, and new sediment surface. Rapid sealevel rise leads to attenuation of sediment accumulation rate. A maximum depth for carbonate accumulation is given. No erosion occurs during subaerial exposure.

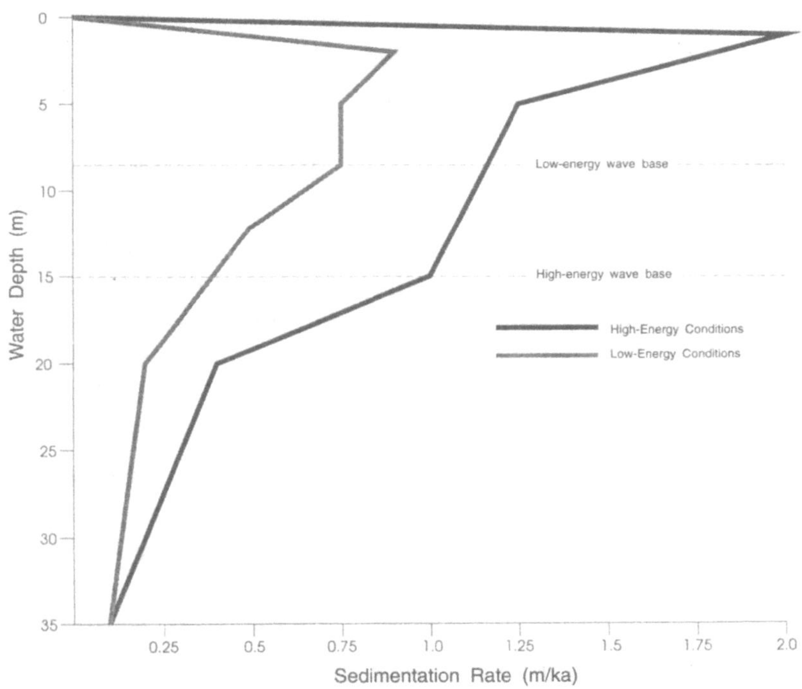

Figure 3. Carbonate sediment accumulation rate vs. water depth used in 2-D model.

The second phase of simulation model development was an expansion of this preliminary one-dimensional simulation into a two-dimensional model (French and Watney, 1990). Programming was done by J. French using the Pascal language; the program is tentatively titled KANMOD. This stratigraphic simulation model generates carbonate sediments according to water depth and energy levels related to depositional slope. This model is used to simulate the development of a set of oolite shoals that occur as part of individual 5th-order minor cycles developed in the 4th-order Swope sequence in southeastern Kansas (Fig. 2). This study area is serving as a shallow subsurface to outcrop analog site to oil and gas fields occurring in south-central and southwestern Kansas. Shelf conditions, elevation, slope, and general deposition setting were apparently similar in these two areas even though they are separated by distances of 200 to 400 mi.

KANMOD generates up to 11 different carbonate facies, including carbonate skeletal grainstone and oolite, which is the most abundant reservoir type in the Missourian carbonate-dominated depositional sequences being studied. Depositional rates are significantly reduced by rapid relative sealevel rise (subsidence + eustasy) such as those inferred for the modeled Pennsylvanian strata. Essentially, sedimentation can not keep up with rapid rising sealevel. Also, sediments are slowly eroded during subaerial exposure.

The model generates a color 2D image of the resulting stratigraphy, as well as more detailed individual stratigraphic profiles at selected locations defined by the user in the modeled profile. The computer screen output can be captured to obtain a hard copy using commercial screen capture programs. Input parameters include subsidence rates, eustatic sealevel changes, and initial depositional topography. Sediment accumulation rates are inferred from modern sediment analogs defined via depth-rate relationships. Subsidence rate can be changed across the area that is being modeled. A eustatic sealevel curve can be drawn using the keyboard, read from a file, or generated through the use of convolved sinusoids such as Milankovitch orbital parameters. The initial shelf configuration on which deposition takes place can be drawn on the screen or input from a file based on information interpreted from mapping and facies analysis.

The horizontal and vertical scales of the graphical output can be changed for successive runs to maximize the image. Parameters that can be varied include the duration of individual runs and the vertical exaggeration of the output. Smaller areas of the output can be selected and expanded to full screen by drawing a box around the area of interest and repeating the run. The second screen then is filled with the boxed-in area.

Run length of the 2D model can be differed within two basic scales: (1) 100,000 to approximately 2 million years; or (2) longer term, 5+ million years for 3rd-order sequence modeling. Time steps differ with run length; shorter runs are composed of shorter time steps. At each time step during a model run, a pixel of a given "facies" is "deposited."

Model output is comprised of a set of these pixels. Model runs typically range from 3.5 minutes to 5 minutes on a 386-33 IBM-compatible computer.

The basic stratigraphic unit that is generated by the model is the depositional sequence. All of the primary types of genetic units that comprise these depositional sequences can be produced, including flooding units, condensed sections, late-highstand (or forced-regressive) units containing complex internal stratigraphy, minor cycles within the depositional sequence, high-energy grainstone units, and subaerial-exposure surfaces. Lowstand, transgressive, and highstand systems tracts can be developed under the proper conditions. Geometries usually observed are produced including progradation of late highstand, regressive genetic units with toplap and downlap. Time lines can be included at user-selected intervals to identify where the depositional surface was at various times during development of the deposit.

A high-resolution data file also can be generated during the model run capturing pixel information. The size of the file is limited by the available RAM. These files include information from each time step across the entire two-dimensional model "area," including facies and thickness information as well as temporal data, such that the duration of the hiatuses and the relative ages of unconformities can be captured.

TWO-DIMENSIONAL SIMULATION OF STACKED OOLITES

Overview

The southeastern flank of the Pleasanton siliciclastic platform in eastern Kansas is a pronounced depositional slope (Fig. 4). At least three temporally distinct sets of strike-oriented oolite buildups were deposited along this basin-facing slope in the Bethany Falls Limestone Member (regressive phase exhibiting forced regression) of the 4th-order scale Swope depositional sequence. Interpretation of data suggests that the oolites onlap landward and downlap basinward. In some locations several of the oolites form what is interpreted as a shingled stack of oolites representing separate episodes of accumulation during falling sealevel. Some of the bounding surfaces separating the oolites are subaerially exposed. Subaerial exposure of these surfaces are suggested by thin blocky claystones with rhizoliths along some of these surfaces and light

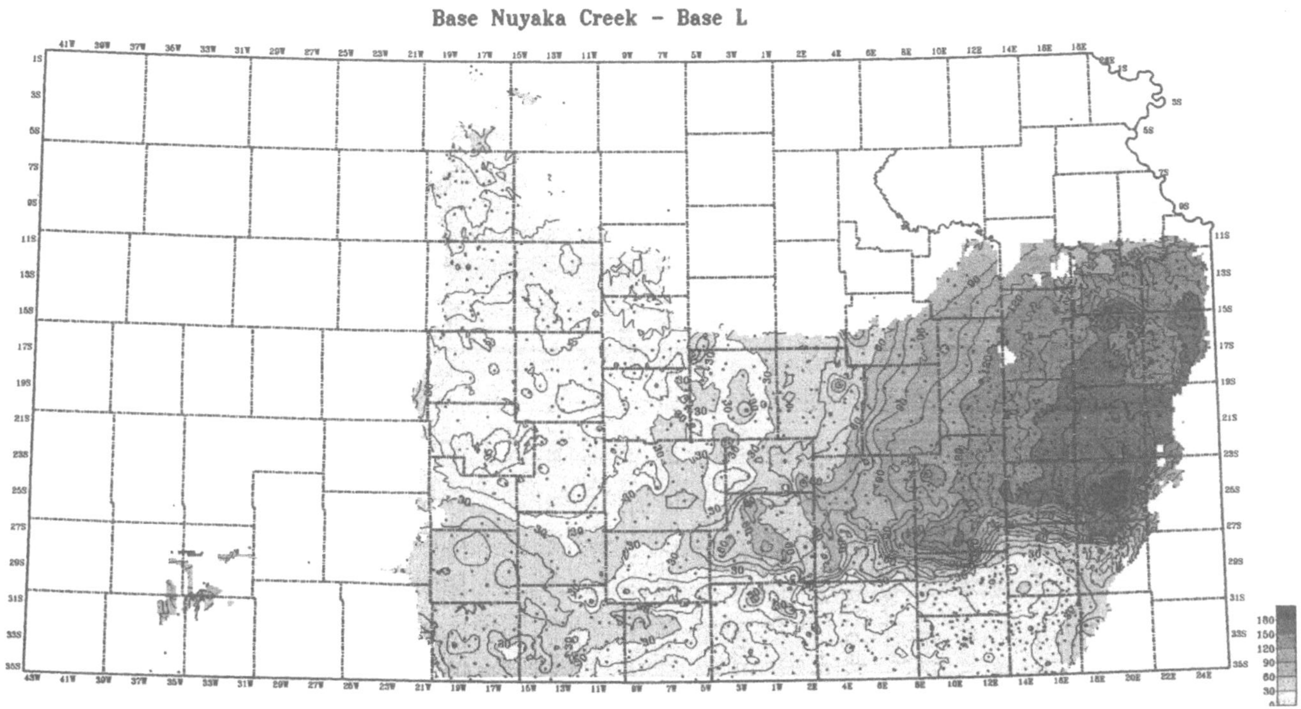

Figure 4. Isopachous map of Pleasanton Group showing large clastic wedge in eastern Kansas. 10-ft contours with gray levels distinguishing 30 ft (9 m) increments. Thicknesses up to 200 ft (61 m) in area of southward prograding deltaic platform. White bar is location of cross section in Figure 5. Platform thins gradually to west against Nemaha Uplift. Abrupt southern margin is interpreted as a depositional slope. Southward beyond this margin is time equivalent sediment-starved interval reflecting lower shelf adjoining deeper actively subsiding foreland basin (Arkoma Basin). Several subsequent cycles are draped over this topography including Swope sequence, the focus of this paper.

isotopic excursions of carbon and oxygen profiles. These isotopic profiles shows depletion of heavier carbon at the position immediately below several of the apparent exposure surfaces, suggesting meteoric exchange with the limestone. In some situations, the base of the oolites is marked by thin zones of probably slightly deeper water subtidal carbonate, suggesting that deposition of the oolitic units was punctuated by minor marine transgressions.

A similar stack of oolitic units in the Bethany Falls Limestone is developed in the Victory Oil Field, which is located 480 km to the west in Haskell County, Kansas. Victory Field is located roughly along depositional strike of the near-surface occurrences, based on regional mapping of the Swope sequence. Results of 3D visualization modeling of this sequence in Victory Field in southwestern Kansas are presented in a later section of this paper.

The 2D simulation modeling program, KANMOD, was used to generate a dip-oriented cross section of the Swope sequence in the shallow-subsurface analog area in southeastern Kansas. The cross section is datumed on a persistent marker that occurs below the Pleasanton clastic wedge. The Pleasanton thins southward toward the basin along this line of section. The slope shown here is interpreted as a minimum approximation of depositional slope at the time the Swope sequence was deposited.

An initial examination of the oolite might suggest that the oolite represents the culmination of a simple regressive, shallowing-upward succession. However, multiple cores taken near this cross section and up and down the slope reveal an oolitic grainstone that is punctuated by multiple thin skeletal packstones and subaerial exposure surfaces resulting in a stacking of these units (French and Watney, 1993). The correlations of the oolites shown on the cross section between the various shelf positions is conjectural, but suggests the development of different oolite-bearing minor cycles as accommodation is gained or lost. Internal correlation of these high-frequency minor cycles is not definitive.

The resulting stratal pattern is an en echelon or imbricate set of oolitic units. Although the oolitic facies occupies the same general stratigraphic position in the Bethany Falls Limestone, it is interpreted to be comprised of multiple, temporally distinct packages that were deposited as the result of relative changes in sealevel. This has significant implications for reservoir characterization and estimating fluid flow in these strata because the bounding surfaces and intervening lithologies usually lack permeability. Each stratal package exhibits genetically different oolites with differing textures, adding further to the heterogeneity. The stratigraphic simulation is an attempt to generate this stratal pattern via inferred geologic processes.

Results of Simulation

The Swope sequence was simulated based on a dip-oriented cross section (Fig. 5). The input parameters were derived from this local setting including slope angle and relief. A sealevel curve was defined based on interpretation of the facies successions of the Swope sequence and adjoining units in core, outcrop, and wireline-log data obtained in the area (French and Watney, 1993).

Three separate oolitic units were generated from three separate stillstands of sealevel that were part of an overall regression (caused by falling relative sealevel) that occurred near the end of deposition of the Swope sequence (Fig. 6). The latter portion of the sealevel curve that generated the entire Swope sequence is shown in the figure. The distribution of oolite in the computer model resembles that depicted in the cross section, with shingled oolites stepping down the depositional slope. This model output includes time lines for every 30,000 yrs. This is instructive to show the time separation of the oolites and the downlap relationships of the oolite with contemporaneous deeper water facies. Also, while an oolite unit is being deposited on a low shelf elevation, the upper shelf is undergoing subaerial exposure. The convergence of time lines in a basinward direction (downlap) occurs in the black, organic-rich marine shale in this sequence (the Hushpuckney Shale) (Fig. 2). This convergence of time lines in this shale reflects a condensed section. Water depth during the accumulation of this black shale is interpreted to be below the effective photic zone. The depth of the photic zone in this model run in Figure 2 is set at 35 m. Thus, limited or no carbonate accumulation occurred, and because of the lack of evidence of significant clastic influx accumulation rates were set low in the model (.01 m/ka). The simulation results suggest that carbonate deposition occurred along updip shelf locations during equivalent intervals of time when black shale accumulated on the lower shelf and basin. The updip carbonate section essentially downlaps onto the black shale.

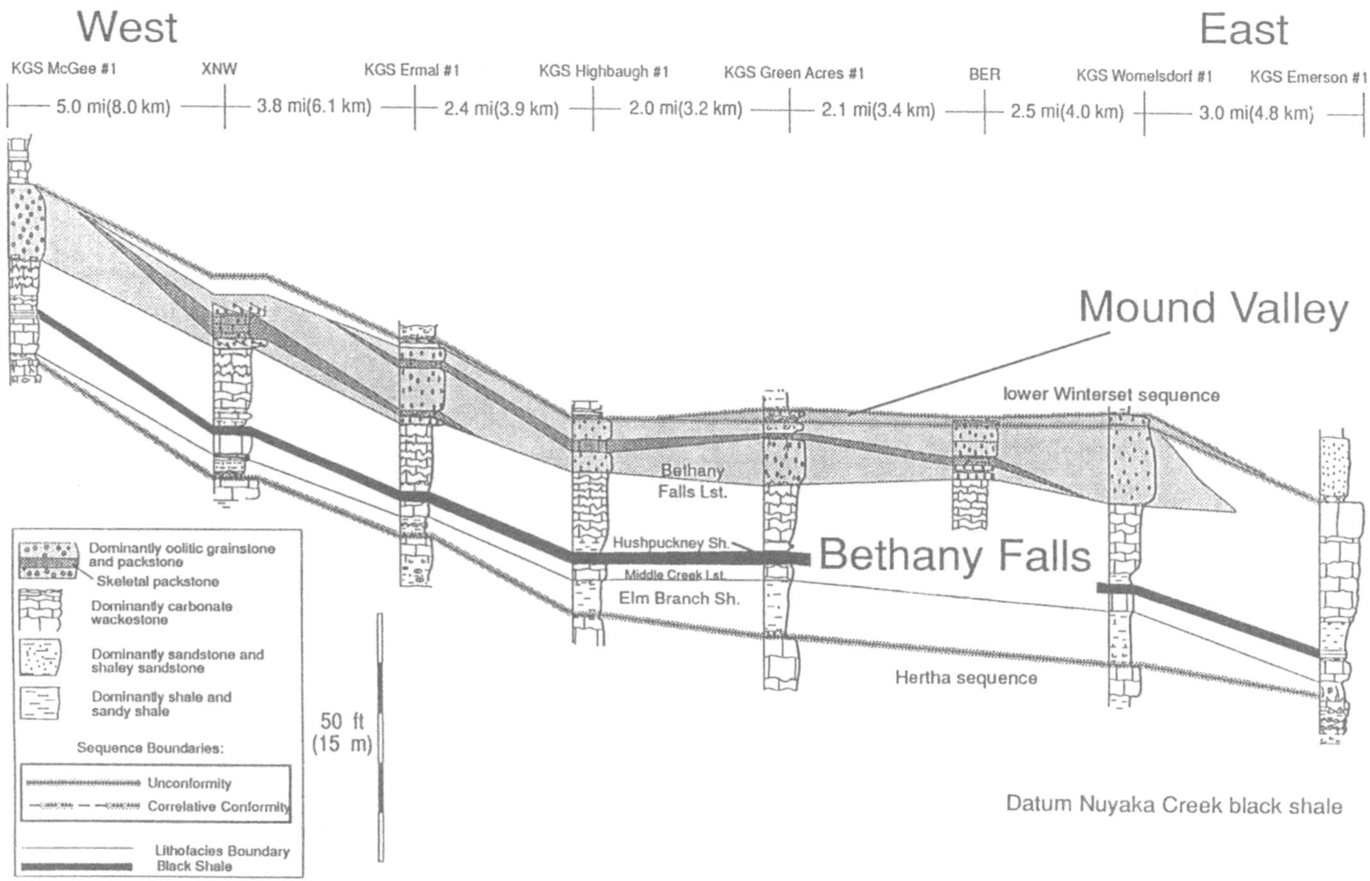

Figure 5. Diagrammatic core and outcrop cross section Swope sequence comprised of Bethany Falls and Mound Valley limestones. Dip-oriented cross section identifies three discrete, shingled oolite layers in Bethany Falls Limestone and additional oolite Mound Valley Limestone.

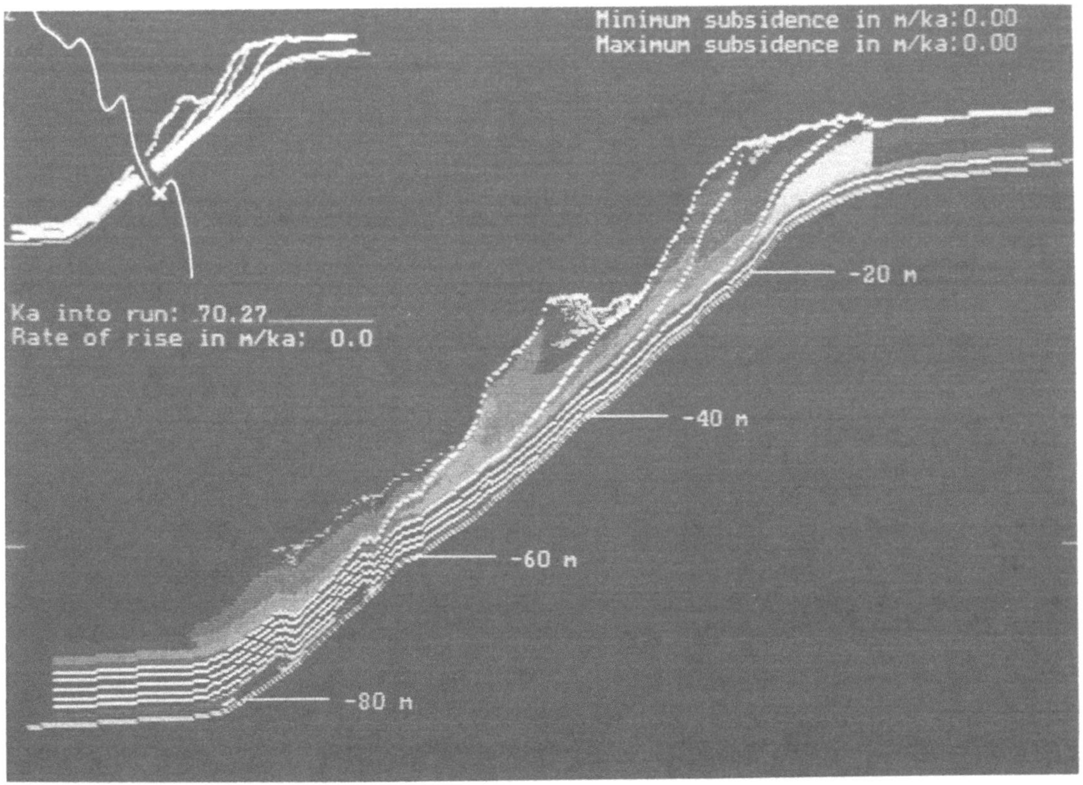

Figure 6. Stratigraphic simulation of three oolitic grainstones developed along slope as result of three minor sealevel oscillations (stillstands) during overall sealevel fall. Time lines are white dashed lines cutting through facies indicating contemporaneous facies.

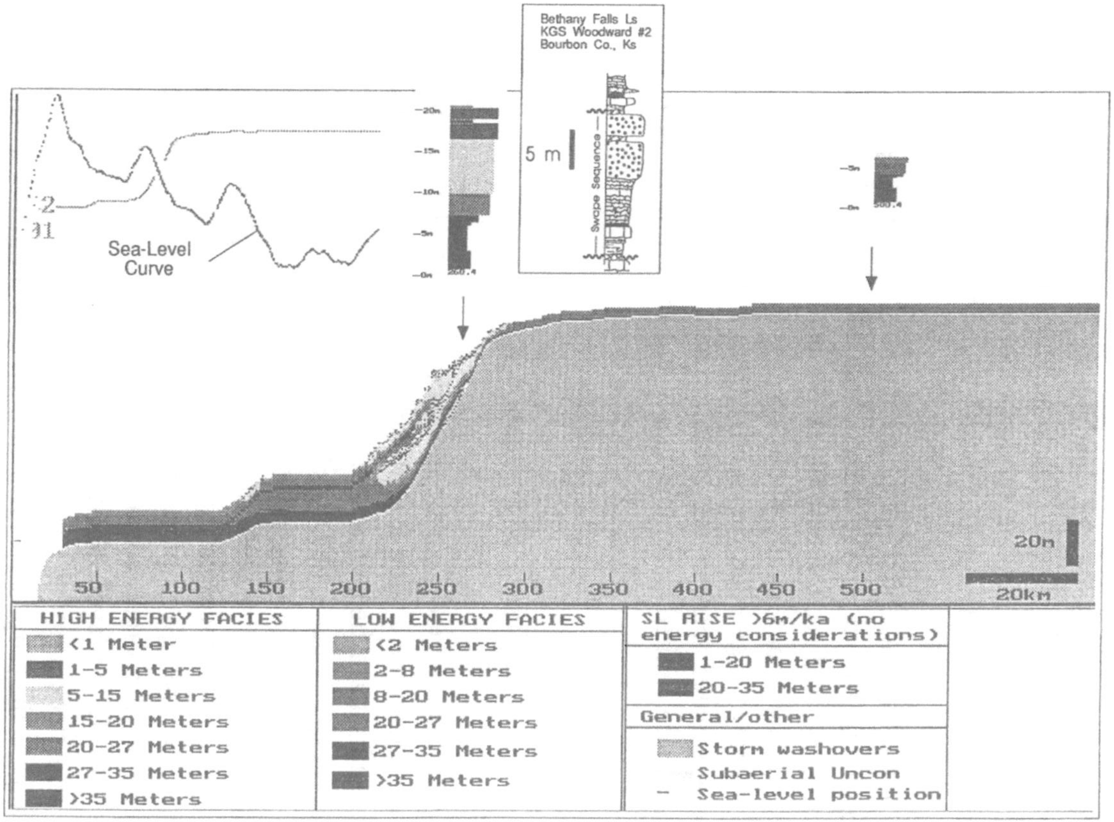

Figure 7. Regional 2D stratigraphic simulation of KANMOD. Swope 4th-order carbonate-dominated sequence depicting deposition across depositional topography created by Pleasanton clastic wedge (underlying uniform gray interval). Section extends north-south across northern Midcontinent (500 km, 310 mi).

A regional stratigraphic simulation of the Swope sequence in Figure 7 depicts the sequence along a 500 km-long south-to-north cross section spanning the carbonate-dominated shelf across the northern Midcontinent. The model includes the relatively steep Pleasanton slope. The model run includes only two stillstands in sealevel during late-stage regression of the Swope sequence, leading to formation of only two oolitic units on the break in slope. The simulation includes an expanded stratigraphic profile of the Swope sequence containing the two oolites. The location of this profile on the model section is identified by the arrow.

An actual measured stratigraphic section obtained from a core on the shelf break is shown in the boxed area alongside the simulated section. The simulated section resembles the actual section in terms of facies type, thickness, and stratigraphic succession. Another simulated succession is shown higher on the shelf. Here the Swope sequence is much more attenuated than to the south. The landward, higher shelf locations had more limited accommodation and thus thinner succession in general.

Conclusion of Simulation Results

The simulation model corroborated the conceptual model inferred from observations in the analog area by adjusting input parameters. The simulation results have helped to constrain the a priori concept. The model has successfully generated the observed stratal patterns via inferred geological processes. The procedure aids in prediction of the oolite petroleum reservoirs in the sense that the stratal template to be likely caused by geological processes that affect other analogous parts of the Kansas shelf. Accordingly, a more intelligent and focused search can be made to locate similar stratal geometries and lithofacies. The timing and magnitudes of the sealevel stillstands seem strongly to control the occurrence and stacking of the oolites. The calibration of this relative sealevel change is the purpose of the near-surface analog site, where cores can be taken to resolve the distribution of units and reconstruct a stacking geometry. Regional studies also are needed to ascertain the shelf configuration and relative elevations of other locations and to develop a robust relative sealevel curve.

Relative sealevel change involves both tectonics and eustacy. Thus, sealevel histories interpreted from two widely separated areas are expected to differ depending on the local tectonic history and relative changes in elevation and subsidence history during accumulation of the stratigraphic interval of interest. This is the challenge that can be addressed through integrated, multiscaled studies, and sensitivity analyses of the critical parameters.

THREE-DIMENSIONAL VISUALIZATION OF OOLITIC RESERVOIR IN VICTORY FIELD

Overview

Three-dimensional visualization is a form of modeling, whereby a framework of attributes of a rock volume is viewed using computer graphic displays. Correlations of attributes between wellbores as in these examples are simply linear interpolations within predefined layers. The definition of this layering determines how the attributes are interpolated between the wells. The primary layers are the depositional sequences, with internal layers assigned to best approximate how the rock was deposited. Geostatistical models are available to make extrapolations conditioned to the available well data. Computer power and graphic capabilities provide real-time 3D viewing and manipulation of the rock volume. Stratamodel (version 3.1), running on an SGI Personal Iris, is the software used in this example. Stratamodel is a stratigraphic geocellular modeling program that can display various well attribute information interpolated into a 3D volume displayed as surfaces, panels, cross sections, and solids. Key stratigraphic surfaces such as correlatable bounding surfaces of temporally distinct 4th-order cycles are used to provide an initial framework.

Wireline-log data in Victory Field contain critical information for recognizing bypassed oil and evaluating improved oil recovery potential. Visualization techniques are shown to help in identifying and characterizing the reservoir heterogeneity.

The dominant reservoir rock in Victory Field is oolitic grainstone. The 4th-order Swope sequence is the focus of this example. It is comprised of apparent 5th-order cycles analogous to those described in the near-surface analog site in southeastern Kansas. The visualization model indicates subtle packaging within a thick (up to 18 m) oolitic unit, with internal layers intricately stacked and shingled. Porosity is dominantly oomoldic, and is interpreted to have resulted from freshwater dissolution associated with early subaerial exposure. The porosity apparently was enhanced by later dissolution and crushing. The crushing of the fabric is inferred to represent the product of early burial compaction of the more weakened areas of the dissolved oolite. Locally, these pores are occluded by late-stage calcite and dolomite cements which further adds to the complexity. The porous and permeable rock is spatially limited as isolated pods in the thicker lobes of oolite. These pods are perhaps related to sites where greater volumes of undersaturated meteoric water percolated through the unit (Watney and others, 1993; 1994).

Effective pay is recognized in this complex pore system using "Super" Pickett plots. Typically, the porosity cutoff is high (>16%) and water saturation is low (<20%). Together these parameters indicate a low bulk volume water (porosity times water saturation) in the pay intervals. Stratigraphic intervals with these cutoffs and reservoir thickness exceeding 1 m are considered to be effective pay. Visualization of these calculated parameters was accomplished using Stratamodel. The information together is used to assess the potential for bypassed oil and gas accumulations. Utilizing these techniques, an estimated 3 million barrels of oil are believed to be behind pipe in Victory Field (Watney and others, 1994).

A 7.5 sq mi area was selected from Victory Field for 3D visualization modeling (Figs. 8 and 9). Forty-seven wells were digitized from this area representing about 15,000 m of log traces from an interval extending between several cycles above and below the Bethany Falls Limestone. Log type types used included gamma ray, sonic, neutron and density porosity, and induction resistivity. The area is located in the northern part of Victory Field, encompassing one of the two large oolite lobes developed in the Bethany Falls Limestone. The particular objectives of the field study were to:

(1) establish best-fit stratigraphic layering and compartmentalization of the oolitic reservoir using visualization modeling;

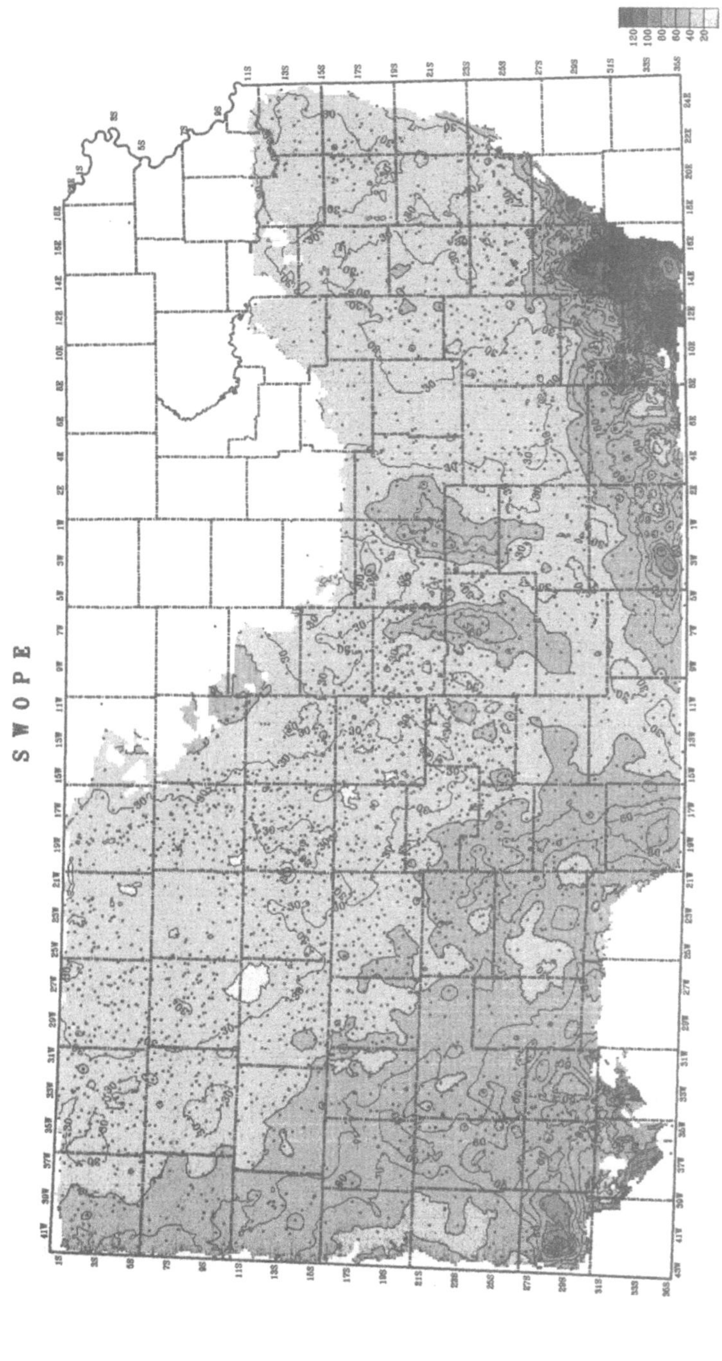

Figure 8. Isopachous map of Swope sequence in Kansas. Contour interval is 10 ft (3 m). Shading is every 20 ft (6 m). Medium gray area is location of thicker oolitic grainstones. Width of this region is narrower in southeastern Kansas because of steepness of shelf edge. Thick black area in southeastern Kansas includes clastics added by southerly source, lapping against depositional topography. Analog site and Victory Field are identified.

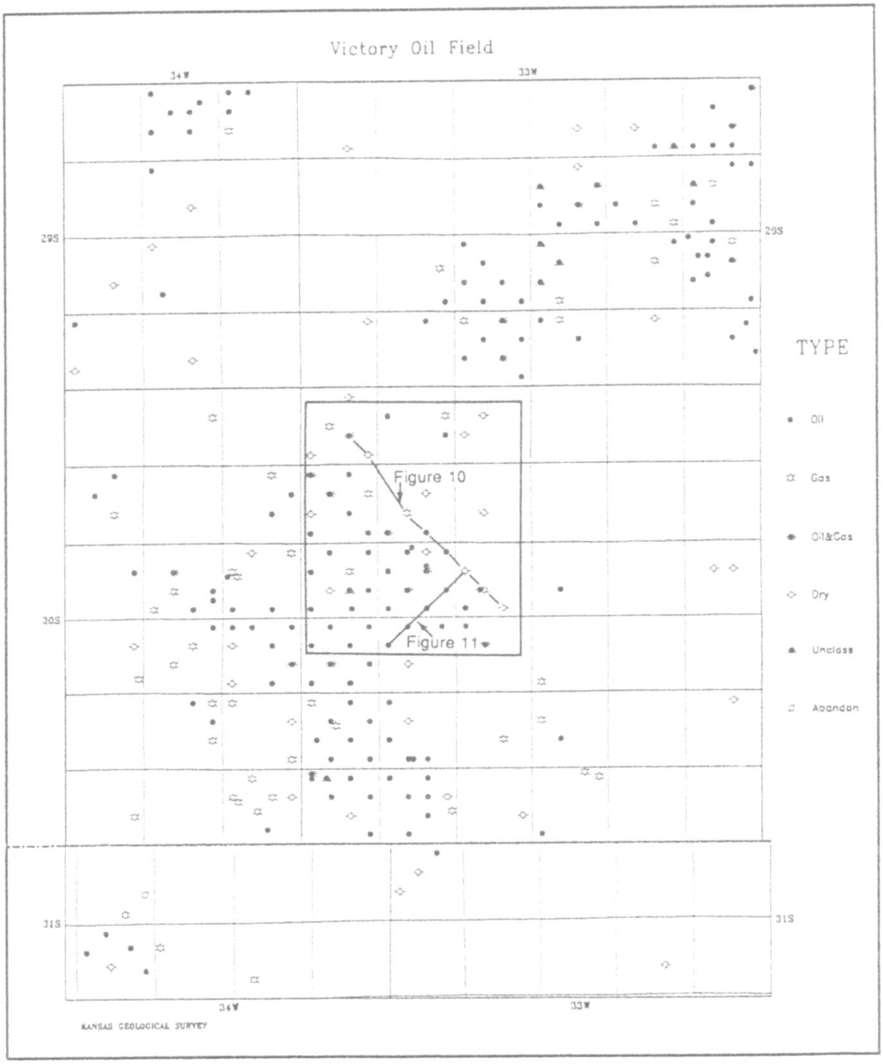

Figure 9. Index map to Victory Field identifies wells used to make 3D visualization with Stratamodel software.

(2) calculate and display variables (such as water saturation and sonic-minus-density porosity) as color attributes to show interparticle and total porosity, bulk volume water (S_w x porosity), and visual displays of an empirical estimate of separate vug porosity developed by Lucia and Conti (1987);

(3) apply filters to target areas of the field with specific reservoir properties based on selected porosity-water saturation cutoffs;
(4) systematically examine variables in the modeled reservoir volume to assess controls and perhaps enhanced views of areas to be targeted for additional oil recovery;
(5) compare this visual, empirically derived model of a reservoir to the simulations of a temporally equivalent oolite in southeastern Kansas.

Computer elevation grids of key stratigraphic surfaces from Victory Field were read into the program. Each datum serves as an event surface, which in this study are divisions between depositional sequences and the genetic units that comprise them. Additional layers are generated within these temporally distinct packages for use in interpolating attributes. The layering template can be selected that most closely fits the internal stratigraphic framework:

(1) onlap -- layers parallel the upper event boundary;
(2) offlap -- layers parallel to an inclined internal surface that is assigned to approximate progradation;
(3) truncation -- layers of equal thickness parallel the lower event boundary where thickening and thinning of interval is accomplished by adding or truncating layers at the top event surface, that is the upper surface is an unconformity;
(4) proportional -- layers are a constant number in interval, each varying in thickness in proportion to overall thickness

The regressive limestones were best treated by truncation whereas the shales and flooding units were assigned layers that onlap.

The layers define the correlation template for interpolation between the wells. The grid consists of square cells with a spacing of 180 ft (90 m). The cells have the same thickness as the layers, set at a constant 2 ft (0.7 m) for the Bethany Falls Limestone. As the unit thickens, layers are added to the top of the unit. This is referred to as truncation layering. Each cell retains the values of the attributes assigned by interpolation from the wells. The grid size (60 x 60 cells) and number of layers (739) determine the size of the model that is manipulated by Stratamodel. The total number of cells in the Victory Field model is approximately 2.6 million.

Digitized well data, sampled every 0.5 ft (0.15 m), were read into Stratamodel. The program reassigns (averages) the vertical well data to values for each layer, for example, into 2 ft (0.61 m) layers in the oolite example. The values of the attributes in the layers then are interpolated using a weighted distance function, analogous to the surface gridding procedure. The search radius and other parameters can be selected to conform with that of the imported grids.

An attribute model was built from Victory Field log data transforming the raw log values to new composite variables, for example, water saturation and bulk volume water. An option termed geobody also was used to determine the amount of connectedness between cells that met certain criteria, for example, cutoff values of porosity and water saturation.

Results of Visualization

The visualization modeling conducted for Victory Field well-log data included construction of a series of cross sections between wells. These sections facilitated both relating the model back to the original well data and comparison with other analyses such as petrography (Figs. 10 and 11). Panels of multiple, regularly intersecting cross sections were generated that serve as the primary method for presentation of 3D properties (Figs. 12 through 15).

Working with the computer, models can be rotated and zoomed in and out. Attributes are selected for display and color values assigned (scaled or given default values). This tool is particularly powerful when attributes are filtered to isolate specific intervals, layers, areas, or attributes in a certain range.

The fine color resolution and distinction provides an enhanced stratigraphic view of the layering of the oolitic reservoir within the Bethany Falls. The truncation method provides the most coherent layering of the interpolated model. Attempts of other types of layering such as parallel to base and proportional exhibit more lateral heterogeneity within the layers. The truncation method builds layers from the base of the limestone upward with layers parallel to the base as previously described. This truncation layering appears to be the most coherent because it most closely approximates the stratigraphic layering

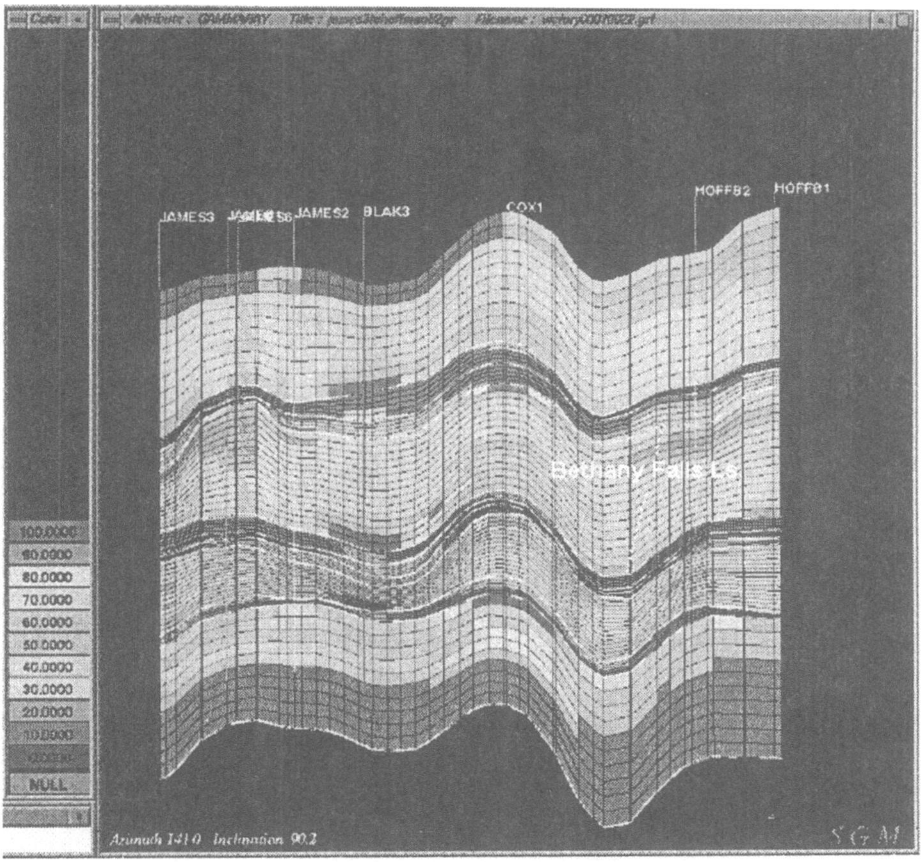

Figure 10. Southeast-to-west cross section generated using Stratamodel software showing interpolated gamma ray between wells. Section is indexed in Figure 9. Even-toned gray units are shales separating cycles of limestone. Swope sequence is highlighted.

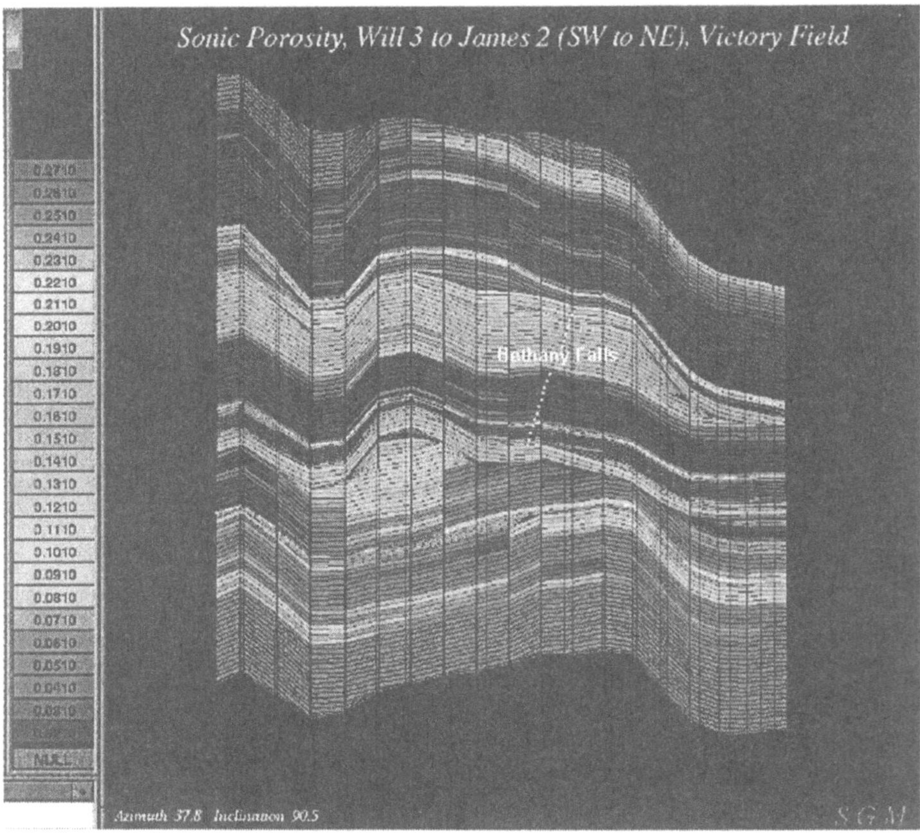

Figure 11. Southwest-to-northeast structural cross section showing interpolated sonic porosity. Section indexed on map in Figure 9. Bethany Falls Limestone is identified. 2-ft thick cells involved in making interpolations are highlighted in this presentation. Truncation model for generating cells made best fit, approximating how limestone was deposited. Higher porosity (lighter shade of gray) thins markedly to northeast.

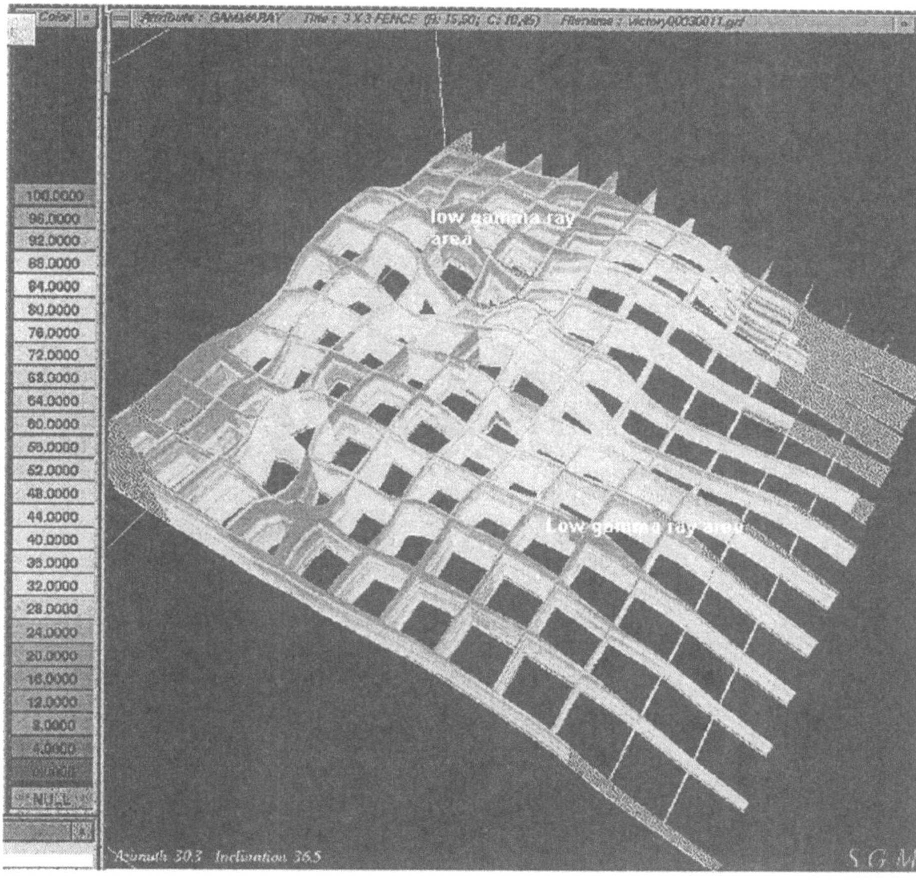

Figure 12. 3D panel, looking northwest, showing interpolated gamma ray for Bethany Falls Limestone. Lowest gamma-ray limestone corresponds to medium gray at top of interval. This is location of oolitic grainstone. Notice how it thickens in southeastern area.

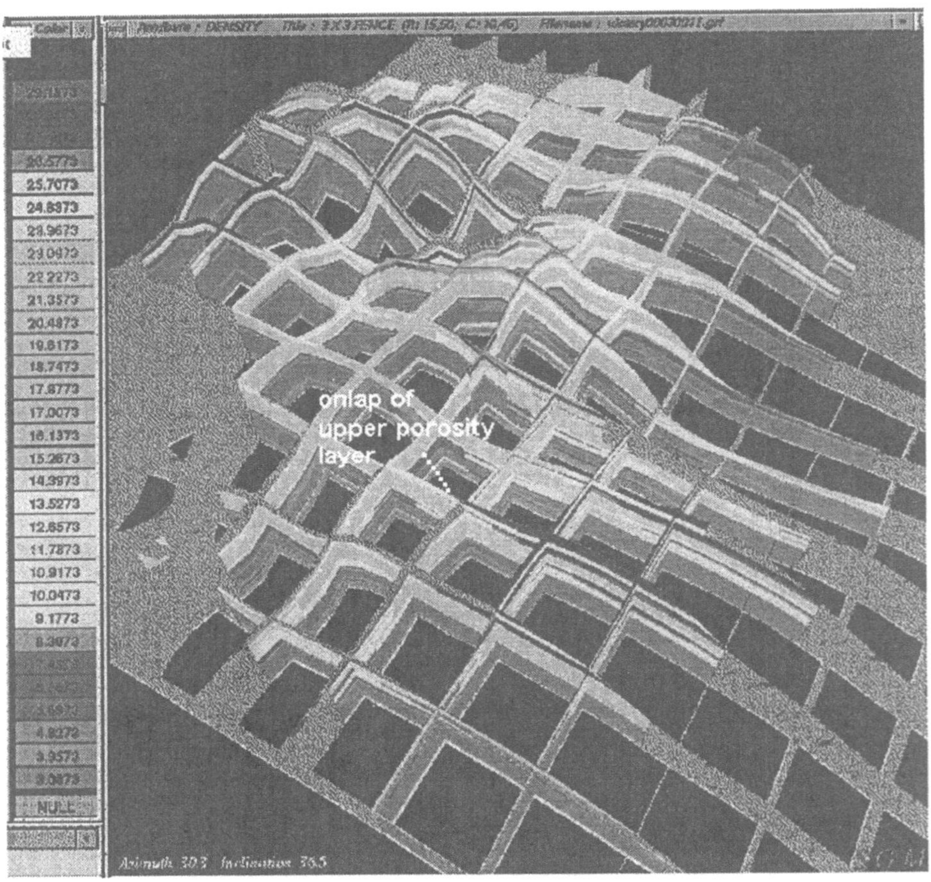

Figure 13. 3D panel looking northwest showing interpolated density porosity in Bethany Falls Limestone. Light shaded layers are in upper Bethany Falls and correspond to oolitic grainstone intervals. Even gray on edges is no data where density logs are lacking. Some of layers appear to lap out into structure.

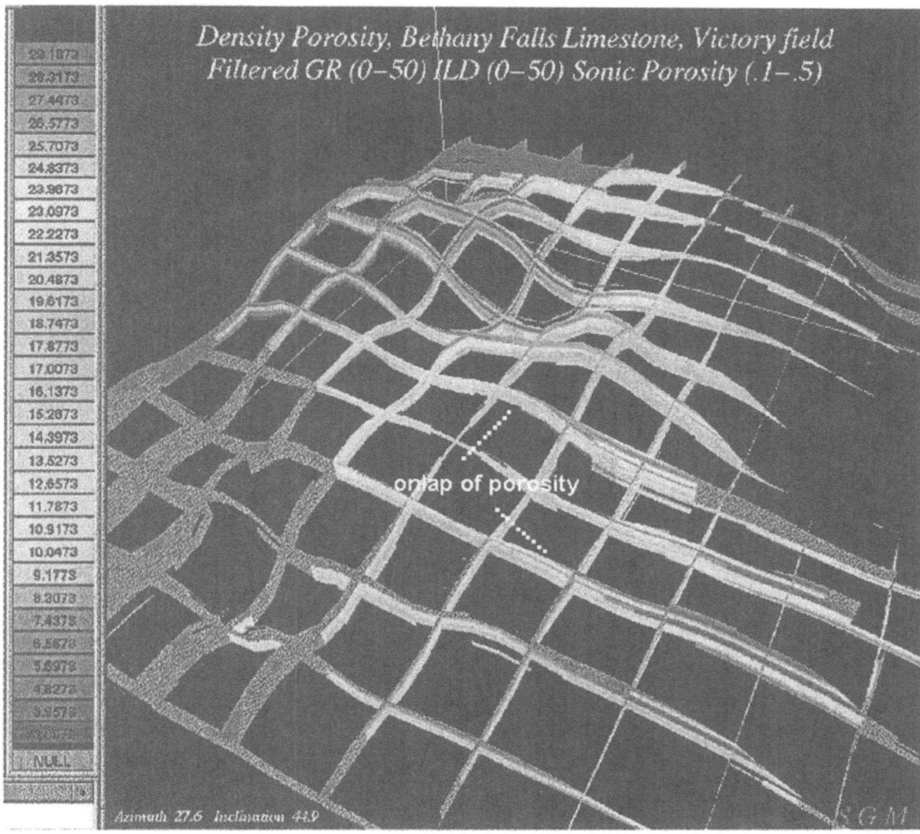

Figure 14. 3D panel of filtered (>20%) interpolated density porosity for Bethany Falls Limestone looking to northwest. Onlap of porous carbonate layers to northwest is indicated.

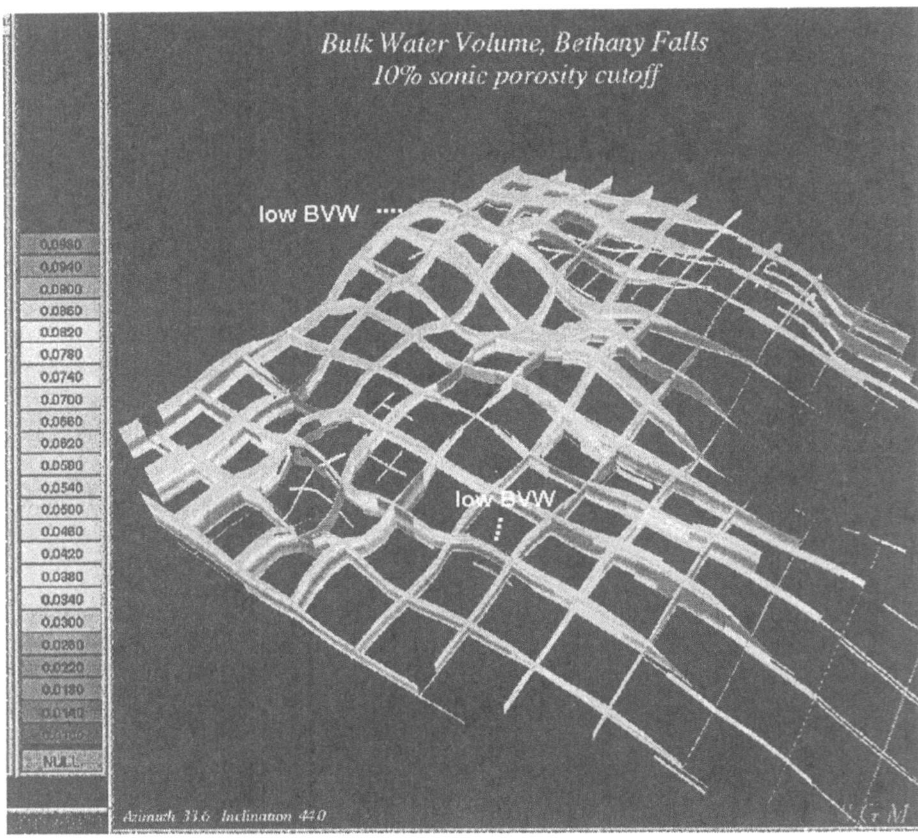

Figure 15. 3D panel of calculated bulk volume water (BVW) for Bethany Falls Limestone using sonic porosity, filtered at 10% cutoff. Areas with low BVW (medium gray) are thought to be prospective reservoir rock with significant oil or gas saturation.

that is represented by the well logs. Thus, the selection of layering procedure is necessarily an iterative procedure.

The filtering option is used to isolate portions of the reservoir that appear to contain effective pay. Filtering can be used to isolate individual sequence, layers, or range of attribute values. Using this approach, a previously undetected isolated porosity pod was located in Victory Field, located slightly down dip of the structural crest on the southeast flank of the modeled area (Fig. 14). The fact that the area is isolated from porous intervals updip suggests that hydrocarbon may be trapped there.

Three separate porous layers in the Bethany Falls Limestone are recognized in the model (Fig. 14). The porous layers are oolite as confirmed by cuttings and core. The lower porous layer laps out updip onto the structure, the middle layer extends over the crest of the structure, and the top porous layer onlaps out on the flank of the structure similar to the lower layer. It therefore seems that the current structure was a paleohigh. Other evidence such as general stratigraphic thinning suggest that Victory Field was a localized positive structure.

Base and top of the oolite succession are subaerial exposure surfaces and suggest a higher order cyclicity than the general 4th-order cycle. Other internal surfaces that define the suggested layering are not expressed clearly as exposure surfaces. Isotopic analyses are needed to examine if these surfaces were indeed exposed. A relative rise and fall of sealevel may have lead to development of the stacked oolites.

The onlap/offlap relationships of the oolites in the Swope sequence at Victory Field resemble the shingled offlapping oolites modeled in the analog sites in southeastern Kansas. The implications of the simulation modeling in the analog area suggested that this type of minor cyclicity could express itself elsewhere where a similar shelf configuration presented itself (slope, shelf elevation that is appropriate for sealevel history to play out, and depositional conditions). The Victory Field area is such as site. The correlation of temporal equivalency of these minor cycles between the areas is currently not possible because the minor cycles are currently below biostratigraphic resolution. Regardless, a subdued basin-facing slope is suggested to have influenced the formation and distribution of several small-scale oolite units in Victory Field in a manner similar to the outcrop analog site in eastern Kansas. Other downdip and overlooked "sweet spots" in other fields along this shelf trend may be located and extend the life of these fields.

The areas of low bulk volume water shown in a visualization (Fig. 15) compare closely with favorable (effective reservoir) bulk volume water cutoffs for oomoldic porosity as defined from crossplot analysis. This off-structure pay is analogous to the development of flanking effective pay in the underlying Sniabar Limestone reservoir in nearby Northeast Lemon Field. The implications for similar traps and compartments in other fields or wildcat locations beyond existing fields and in new pays is significant. In general, stratigraphic compartmentalization coupled with diagenetic overprinting is important in defining the location of remaining reservoirs and establishing reservoir continuity.

Slices through the individual layers of the reservoir further illustrate the lateral onlap/offlap relationships of the porosity in the oolite layers (Fig. 16). Cells that were connected were used to define a geobody. The surface of this resulting volume is displayed in Figure 17. It reveals the basic tripartite division of the oolite reservoir created by the layering.

Lateral and vertical heterogeneities are many in the Lansing-Kansas City Groups including multiple cycles, subzones, and layers within the cycles. Each cycle and possibly each subcycle can have different reservoir properties due to separate, temporally distinct depositional environments, and differences in diagenesis. Three-dimensional modeling initially is expensive and time consuming, especially with regard to data preparation, but the ease in examining the reservoir, drawing inferences, and communicating with other collaborators is facilitated through the use of such techniques. The modeling provides an excellent vehicle to communicate the nature of reservoir heterogeneity and can be helpful in later design of recovery methods. Also, this software permits the geologic model to be scaled up and exported to a reservoir simulator.

CONCLUSIONS

Quantitative geologic modeling is an integral component in improving the understanding and prediction of petroleum reservoirs. Different modeling approaches have inherent advantages and disadvantages, but can complement each other when used judiciously. Future modeling efforts probably will combine components of several

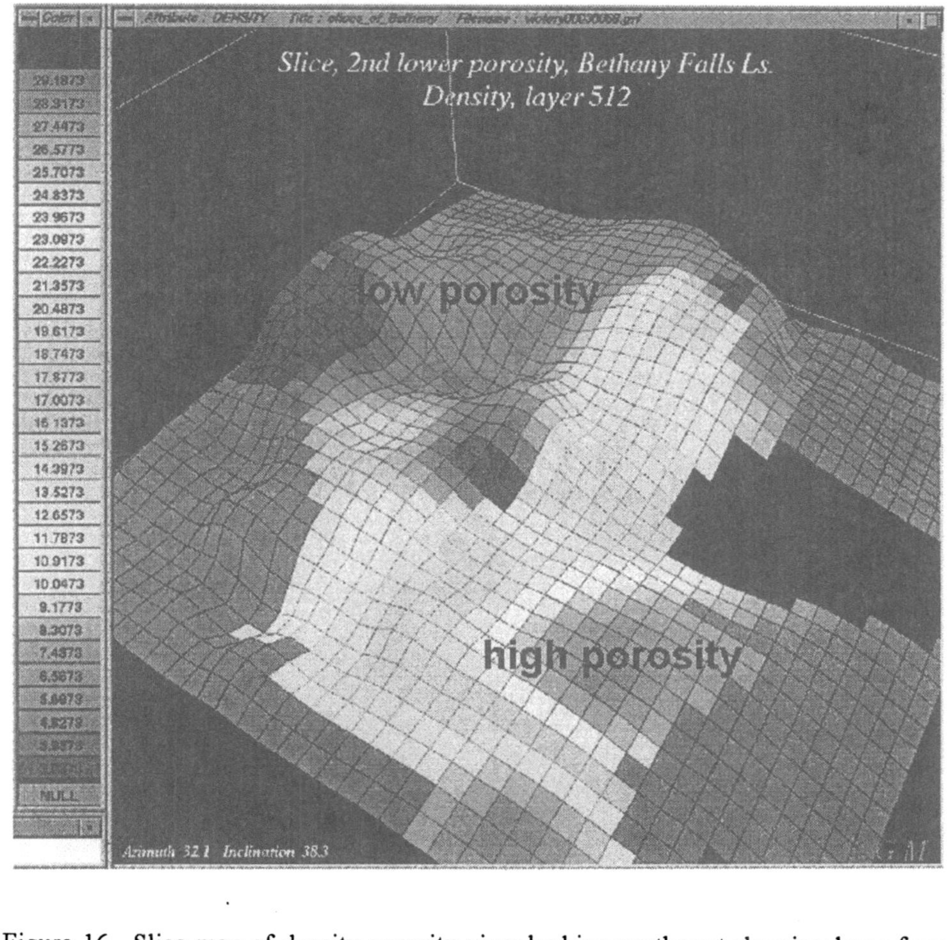

Figure 16. Slice map of density porosity view looking northwest showing layer from lowest porous grainstone. Porosity laps out westward onto structure.

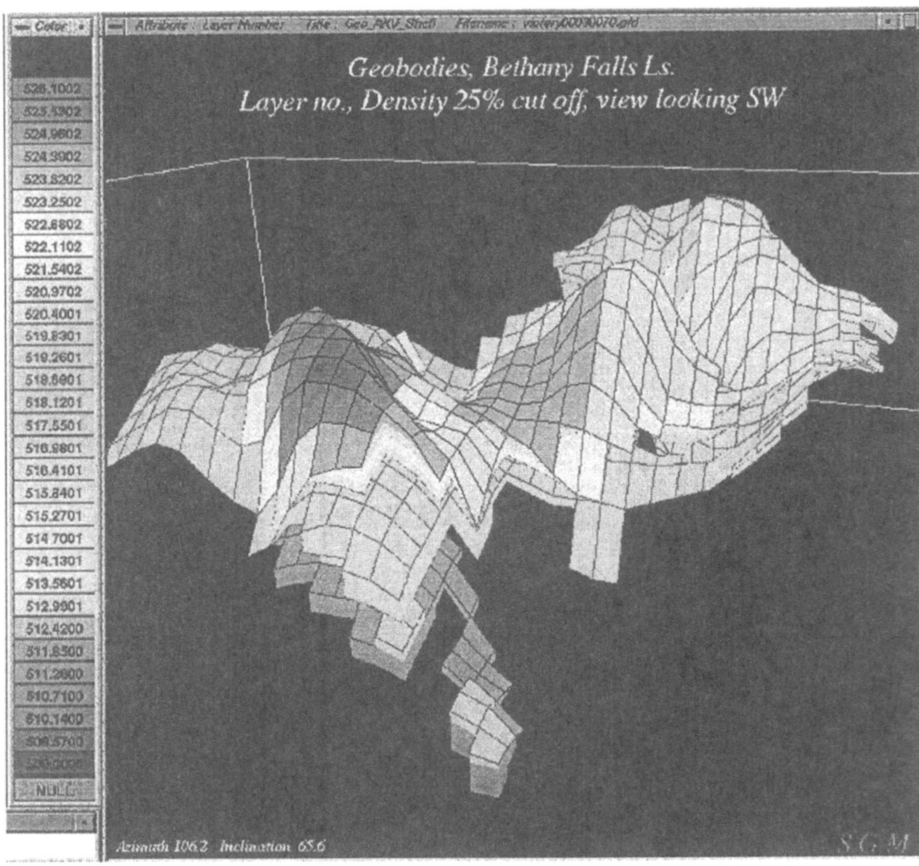

Figure 17. Geobody showing connected cells within Bethany Falls using density porosity with 25% cutoff. This view looks southwest showing layer numbers. Lateral discontinuities are noted on this gray-scale presentation.

types of modeling to obtain the desired ranges of scale of application and optimization of the available data.

Stratigraphic simulation modeling operates on estimates of process variables, producing output that resembles actual stratigraphic and facies data. Experiments using this modeling produces insights as to how the strata were deposited and aids in improving predictions utilizing sequence stratigraphic analysis. The nature and distribution of temporally distinct depositional sequences are fundamental to successful utilization of this type of modeling because timing of events is of paramount importance.

Visualization modeling as described in this paper utilizes 3D presentation and manipulation of the stratigraphy to help reconstruct complex reservoir geometries. Again, the approach is based on an established framework of temporally distinct stratigraphic units in which quantitative reservoir attributes are interpolated thorough various methods. The interpolation is an iterative, empirical approach. The derived model can provide additional insight and understanding that is useful in reservoir prediction and compliments the simulation models.

ACKNOWLEDGMENTS

Johannes Wendebourg and John Davis are thanked for their reviews of this paper, their comments were very helpful and improved the paper significantly. J.C. Wong is thanked for implementing Stratmodel. Ajit Verma and Tracy Gerhard are thanked for their assistance in digitizing well logs and formatting information for Stratamodel.

REFERENCES

French, J.A., and Watney, W.L., 1990, Computer modeling of Mid-Continent cyclothems and its application in the prediction of hydrocarbon reservoirs (abst.): Am. Assoc. Petroleum Geologists Bull., v. 74, no. 5, p. 657.

French, J.A., and Watney, W.L., 1993, Integrated field, analog, and shelf-scale geologic modeling of oolitic grainstone reservoirs in the upper Pennsylvanian Kansas City Group in Kansas, *in* Lineville, B., ed., Reservoir characterization III: PennWell Books, Tulsa, Oklahoma, p. 983-993.

Goldhammer, R.K., Oswald, E.J., and Dunn, P.A., 1991, Hierarchy of stratigraphic forcing: Example from Middle Pennsylvanian shelf carbonates of the Paradox basin, *in* Franseen, E.K., Watney, W.L., Kendall, C.G.St.C., and Ross, W., eds., Sedimentary modeling: computer simulations and methods for improved parameter definition: Kansas Geol. Survey Bull. 233, p. 361-413.

Goldhammer, R.K., Lehmann, P.J., and Dunn, P.A., 1993, The origin of high-frequency platform carbonate cycles and third-order sequences (Lower Ordovician El Paso Group, West Texas): Constraints from outcrop data and stratigraphic modeling: Jour. Sed. Pet., v. 63, no. 3, p. 318-359.

Heckel, P.H., 1986, Sea-level curve for Pennsylvanian eustatic marine transgressive-regressive depositional cycles along midcontinent outcrop belt, North America: Geology, v. 14, no. 4, p. 330-334.

Klein, G.deV., 1990, Pennsylvanian time scales and cycle periods: Geology, v. 18, no. 5, p. 455-457.

Lucia, F.J., and Conti, R.D., 1987, Rock fabric, permeability, and log relationships in an upward shoaling, vuggy carbonate sequence: Texas Bur. Economic Geology, Geol. Circ. 87-5, 22 p.

Watney, W.L., Wong, J.C., and French, J.A., 1991, Computer simulation of Upper Pennsylvanian (Missourian) carbonate-dominated cycles in western Kansas (United States), *in* Franseen, E.K., Watney, W.L., Kendall, C.G.St.C., and Ross, W., eds.: Sedimentary modeling: computer simulations and methods for improved parameter definition: Kansas Geol. Survey Bull. 233, p. 415-430.

Watney, W.L., French, J.A., Guy, W.J., Carlson, R., and Wong, J.C., 1993, Geological characterization of oolitic grainstone reservoirs in the Upper Pennsylvanian Lansing-Kansas City groups in Victory Field and near-surface analog in Kansas, *in* Reservoir description workshop: Energy Research Technology Transfer Series 93-1, p. 159-190.

Watney, W.L, and others, 1994, Depositional sequence analysis and sedimentologic modeling for improved prediction of Pennsylvanian reservoirs, Final Report, U.S. DOE Contract DOE/BC/14434-13, 212 p. + attachments, 212 p. +figs.

Youle, J.C., Watney, W.L., and Lambert, L.L., 1994, Stratal hierarchy and sequence stratigraphy -- Middle Pennsylvanian, southwestern Kansas, U.S.A., *in*, Klein, G.D., ed., Pangea: paleoclimate, tectonics, and sedimentation during accretion, zenith, and breakup of a supercontinent: Geol. Soc. America Spec. Paper 288, p. 267-285.

Zeller, D.E., ed., 1968, The stratigraphic succession in Kansas: Kansas Geol. Survey Bull. 189, 81 p.

THERMAL MODELING AT AN ANCIENT OROGENETIC FRONT WITH SPECIAL REGARD TO THE UNCERTAINTY OF HEAT-FLOW PREDICTIONS

U. Bayer, B. Lünenschloß, J. Springer, and C. v. Winterfeld
GeoForschungsZentrum Potsdam, Potsdam, Germany

ABSTRACT

The geothermal field of the upper crust results from a variety of interacting processes in combination with the transport properties of rocks. The study attempts to illustrate how the various factors may interact and how they may modify the thermal field. The main result focusses on the uncertainties related to the extrapolation from the near-surface field into greater depth. It is shown that a unique solution may not be guaranteed. However, the modeling approach allows to weight different scenarios, and perhaps may help to localize exploration wells at the most informative point.

INTRODUCTION

The thermal field frequently is considered conductive, sometimes with the addition of heat sources within the crust, which are the result of radioactive decay. Toth (1962) illustrated that hydrostatic forces related to topography may induce a deep-reaching pore fluid flow. The importance of convective heat transfer has been discussed repeatedly,

Geologic Modeling and Mapping
Edited by A. Förster and D. F. Merriam, Plenum Press, New York, 1996

especially with regard to sedimentary basins (Bethke, 1985; Bjoerlykke, Mo, and Palm, 1988; Clauser and Villinger, 1990; Smith and Chapman, 1983). The examples usually are related to basin and graben situations with a sedimentary filling. The anisotropy of the rocks (permeability, thermal conductivity) is frequently ignored, a feature, which becomes of special interest in tectonically deformed areas, such as orogenetic fronts.

We will illustrate by example how rock properties and especially greater permeability zones may modify the thermal field down to several thousands of meters at a regional level. In terms of a conductive heat flow, the anisotropy of rocks is of minor importance as long as the temperature distribution is considered. However, if one considers the vertical heat flow, then the models indicate that the fluxes are affected by the inhomogeneity of subsurface structures. Because the vertical heat flow is one of the "measurables" in boreholes, the models can provide an important tool to understand observed subsurface "thermal anomalies". Models may allow the reduction of the uncertainty of predictions of the thermal field, a problem that has occurred in the context of the location of wells, which have been positioned with regard to anomalies of the geothermal field, for example the KTB, Urach, and Soultz (Haenel, 1982; Jobmann and Schulz, 1992; Kappelmeyer and others, 1991).

The paper resulted from an interdisciplinary approach. U.B. designed the project, B.L. is responsible for the conceptual model, J.S. did the numerical modeling and designed appropriate programs, C.v.W. provided the basic geological model.

THE GEOLOGICAL MODEL

The area under consideration is located near Aachen (Germany), and all modeling is based on a cross section through the northern border of the ancient Variscian orogenetic front that extends into the ancient "molasse" basin. This geological cross section (Fig. 1A) is part of a balanced cross section, which was studied in detail by v.Winterfeld (1994). It summarizes all the currently available surface and subsurface data. Nevertheless, the cross section itself provides a geological model which may change as our knowledge and our understanding of orogenetic processes increase. A somewhat different view of the geological setting was given by Knapp (1980). Within this study we take the geological model of v. Winterfeld (1994) as granted because any future modifications may modify only the regional aspects, but not the principal considerations of this paper.

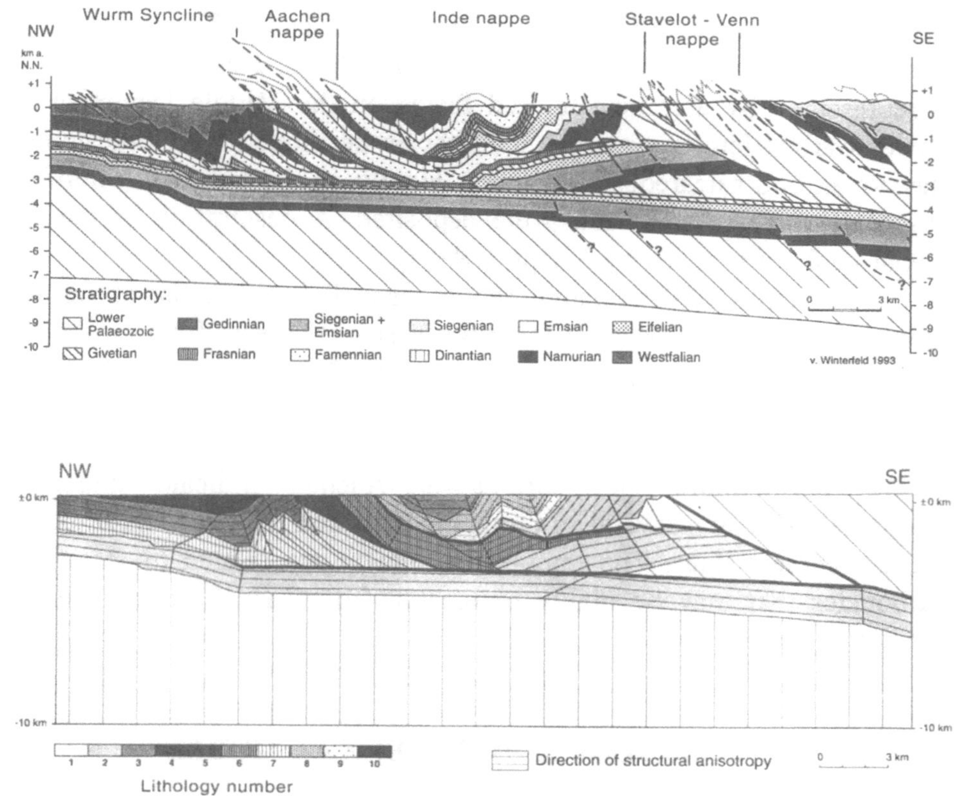

Figure 1. A (top), Geological structure of Variscian front near Aachen (v. Winterfeld, 1994); B (bottom), conceptual model of Variscian front. Different main lithotypes are indicated. Physical properties are summarized in Table 1 corresponding to material number. Hatching indicates direction of first principal axis of conductivity tensors.

The cross section illustrates the main features of the Northwestern Rhenish Massif with its thin-skinned tectonic setting of nappes, which are related to the Aachen detachment and extend into the northern foreland transition zone. The main tectonic structures have been formed by late Variscian compressional deformation. Remarkable structural elements are the low-grade metamorphic parts of the Eifel nappe in the southeast, the repetitions and tectonic discontinuations of the stratigraphic units by blind thrust imbrications, and the overthrusting and nappe implacement. Further structural elements are folding, cleavage, and the depression of the marginal molasse area (for details see v. Winterfeld, 1994).

The lowermost strata consist of Cambrian and Ordovician units of predominantly slates and quarzites that have been folded by the Caledonian orogeny. Above these strata, a sequence of Devonian and Carboniferous sedimentary units follows. The lower Devonian sequence is represented by arenitic Old Red facies. In the Middle Devonian to Lower Carboniferous a marine shelf facies developed with limestones, claystones, and calcareous sandstones. Finally, in the Upper Carboniferous coaly molasse sediments were deposited. A description of the regional geology is given by Knapp (1980).

The repeated stratigraphic sequences include alterations of various lithotypes as well as tectonic foliation in several areas. Both structures induce anisotropies of the physical properties. The transport coefficients especially have to be considered as tensors with different values parallel and perpendicular to layering and foliation (Ondrak, Bayer, and Kahle, 1994). Furthermore, hot springs at Aachen indicate increased permeabilities along several stratigraphic units and fault zones, for example detachment lines. The fault zones locally are related to Devonian carbonates that may be highly permeable from karstification, even at greater depth.

CONCEPTUAL AND NUMERICAL MODEL

The geological model discussed previously provides many stratigraphic and tectonic details. However, in order to model transport processes, geological information has to be translated into physical properties, and the scales have to be adapted to the software and hardware capabilities, that is a conceptual model has to be developed (Welte and Yalcin, 1987; Welte,1989). This transformation process introduces a certain "roughness" to the geological model, which is dictated by the modeling environment. In our special situation, we selected the main tectonic units as the principal structural components (Fig. 1B). On this level, all the information about stratification and foliation is lost. To overcome this problem "effective transport properties" have been introduced that account for the modeling scale and the anisotropy of the rocks. The direction of the first principal axis of the permeability and conductivity tensors is indicated in Figure 1B.

The values for the transport coefficients parallel and perpendicular to the stratification have been estimated as the weighted arithmetic and harmonic means of the lithotypes involved. The weights are simply the thicknesses of the strata. A more detailed discussion of such estimations is given by Ondrak, Bayer, and Kahle (1994). For foliated units, laboratory data have been used to determine the transport tensors. The physical property data used for all the models considered here are summarized in Table 1. They are compiled from various sources and include data specific to the area as far as available (Schürmeyer and Wohlenberg, 1985; Bless and others, 1981; Haenel, 1983; Heitfeld and others, 1985), and average data for the typical rock types as present in the literature (e.g. Angenheister, 1982; Freeze and Cherry, 1979; IES, 1991; Somerton, 1992). These estimated physical properties certainly are a sensible point of the entire model. However, a model is a simplification of nature, and has to be modified if the specific knowledge increases. Within the context of this study, the true values of the physical properties are of secondary importance.

Table 1. Physical property data for lithotypes used in model calculation. Distribution of mateiral parameters in model is indicated by lithology number in Figure 1B.

Lithology number		1	2	3	4	5	6	7	8	9	10
$\lambda\perp$	[W/Km]	3.469	2.049	1.981	1.772	2.17	2.294	2.325	2.518	2.45	2.05
$\lambda\parallel$	[W/Km]	3.565	2.08	2.065	2.041	2.22	2.344	2.371	2.538	2.475	2.05
$k\perp$	[mDarcy]	0.001	0.012	0.05	0.053	0.065	0.096	0.109	0.324	0.244	2.344*
$k\parallel$	[mDarcy]	0.01	0.228	0.252	0.305	0.464	1.056	1.299	1.754	2.366	2.344*
A	[$\mu W/m^3$]	1.306	1.404	1.389	1.296	1.278	1.03	0.938	0.719	0.537	—

* variable 20, 200, and 2000 mDarcy

The maximum thermal conductivity (3.6 W/Km) and the minimal permeabilty (0.001 mDarcy) are present in the Cambrian and Ordovician units. The lowest thermal conductivity (1.8 W/Km) occurs in the Westfalian strata of the foredeep due to interbedded coaly layers. The highest permeabilities are related to the carstic and dolomitic limestones of the Devonian (2.4 mDarcy), and to the fault systems.

The control parameters of the models are the mean thermal gradients known from well data (Schürmeyer and Wohlenberg, 1985; Dornstädter and Sattel, 1985; Graulich, 1969; Haenel and Staroste, 1988;

Hurtig, 1992), and data from the hot springs at Aachen that have been studied in detail by Pommerening (1992) and Langguth and Plum (1984). The geochemical studies indicate an origin of these springs at 120-130°C, a control parameter which has to be approached by the model.

The calculations have been accomplished by finite element methods using commercial software as well as especially designed and developed software which accounts for the coupling of heat conduction and forced convection. At the present state of the modeling approach we ignore possible effects of free convections, which could be induced by density variations of the pore fluid under the variable depth dependent P/T-conditions. Finally, it is assumed that heat conduction and pore fluid flow are in a stationary state. In this situation the coupled system of partial differential equations has the following form:

Darcy flow $$\mathbf{q} = -\frac{k}{\mu}\left(grad\ p + \rho g \mathbf{e}_z\right)$$

Mass balance $$div\ \mathbf{q} = -div\left(\frac{k}{\mu}\left(grad\ p + \rho g \mathbf{e}_z\right)\right) = 0$$

Energy balance $$div(\lambda\ grad\ T) - \rho\ c\ \mathbf{q}\ grad\ T + Q = 0$$

where

q	:	Darcy flow rate,
k	:	permeability tensor,
μ	:	viscosity,
p	:	pore fluid pressure,
ρ	:	pore fluid density,
g	:	acceleration of gravity,
e_z	:	vertical unit vector,
λ	:	thermal conductivity tensor,
T	:	temperature,
c	:	specific heat of pore fluid,
Q	:	heat production.

Another problem in a two-dimensional finite scenario is the boundary conditions. They effectively determine the special solution of the partial differential equations, and the models may be biased by boundary effects. For our models, we select a constant average annual surface temperature of 10°C and a zero hydrostatic pressure along the surface. At the northern and southern faces, the boundaries are closed for both types of fluxes. These closed boundary conditions provide some disturbances in their vicinity, however in the southern region, the permeability of the rocks, in generally, is extremely low so that the boundary effects remain small. At the northern end, the section actually approaches a fault that may be considered closed and impermeable. Towards the lower boundary at 10 km depth we assume that the permeability approaches zero while a heat flux is entering from beneath. This heat flux has to be determined for different variants of the models in order to approach the control data.

MODELING RESULTS

Figures 2 to 6 illustrate the modeling results within the area with regard to effects of heat conduction, heat production, and hydrostatically forced convection.

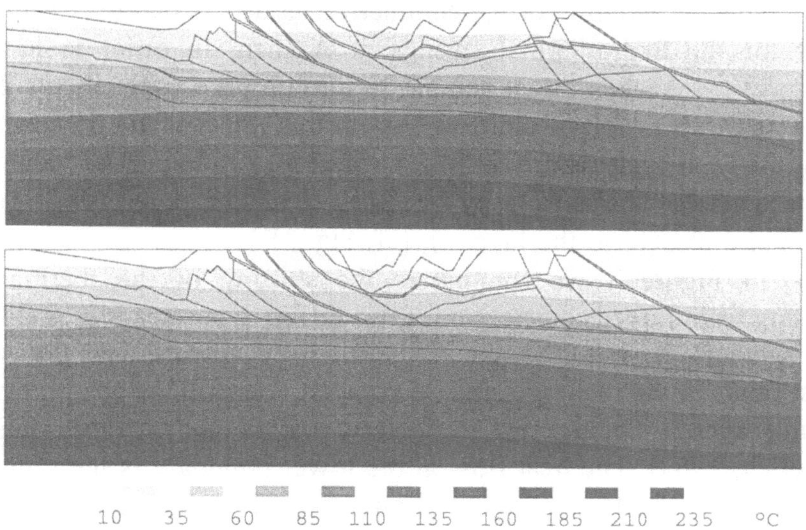

Figure 2. Two models with conductive heat transport. A (top), Simple conductive heat transport with basal heat-flow density of 60 mW/m^2; B (bottom), heat conduction combined with heat production corresponding to data given in Figure1B and Table 1 and basal heat-flow density of 47.5 mW/m^2.

Figure 2A illustrates the classical approach of basin analyses (e.g. Yükler, Cornford, and Welte, 1978) with a heat flux entering the area from the lower boundary (60mW/m^2). The isotherms resulting from this model reflect the variable conductivities of the different rock types. However, in this simple setting, a series of one-dimensional models also would provide acceptable results as long as we consider the temperature field only. But the anisotropies become of importance if the heat fluxes are considered, as will be discussed next.

The model of Figure 2A implies a nearly linear thermal gradient, which is slightly modified by the thermal conductivities of the various rock types. However, its extrapolation to greater depths would collide with our understanding of the temperature at the lithosphere/astenosphere boundary. The classical method to overcome this discrepancy is to assume heat production within the crust. Figure 2B illustrates how the thermal field is modified if radioactive decay is assumed to be a source for heat production. In order to adapt the near-surface heat-flow field to the control data, the heat flow at the lower boundary has to be reduced to about 47.5 mW/m^2. The temperatures increase now less rapidly with depth, the deeper parts are cooler, and the gradient declines with depth as required by geophysical reasoning.

The shortcoming of the conductive models is that they cannot account for the hydrothermal springs at Aachen. In order to have hot springs at the Aachen thrust one has to turn on pore water fluxes. Figure 3 gives an impression of the flow system that arises if the permeabilities along the fault systems are set to 2.4 mDarcy, a rather small value. Here the influence of the anisotropy becomes clear, the pore flow directions correspond closely to the layering (cf. Fig. 1B).

In Figure 4 we summarize the shifting of the thermal field resulting from forced convection for two different permeabilities along the fault systems. As required, the isotherms bend upward near the Aachen thrust fault. In the recharge area, in contrast, the isolines bend downward causing a cool upper crust, a feature that is known from this area (Graulich, 1969). The heat flow at the lower boundary is the same as in the simple conductive model of Figure 2A. In parts, it may have to be adjusted as our knowledge about surface heat fluxes increases.

A critical review of these modeling results shows that they do not yet reflect the details of the hydrothermal conditions at Aachen. Geochemical thermometers indicate that the water of the springs was heated originally to about 120-130°C. In our models the best results are achieved with permeabilities between 20 to 200 mDarcy along the fault system. In this situation the permeable zone is located near the 100 to 110°C

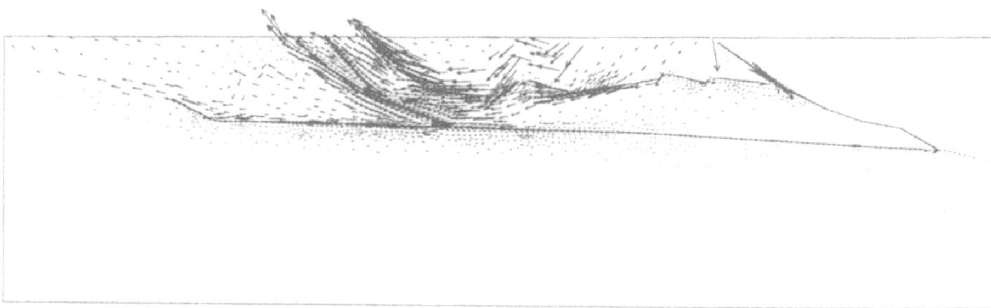

Figure 3. Flow field of hydrostatically forced pore water flow. Permeability along faults is set to 2 mDarcy.

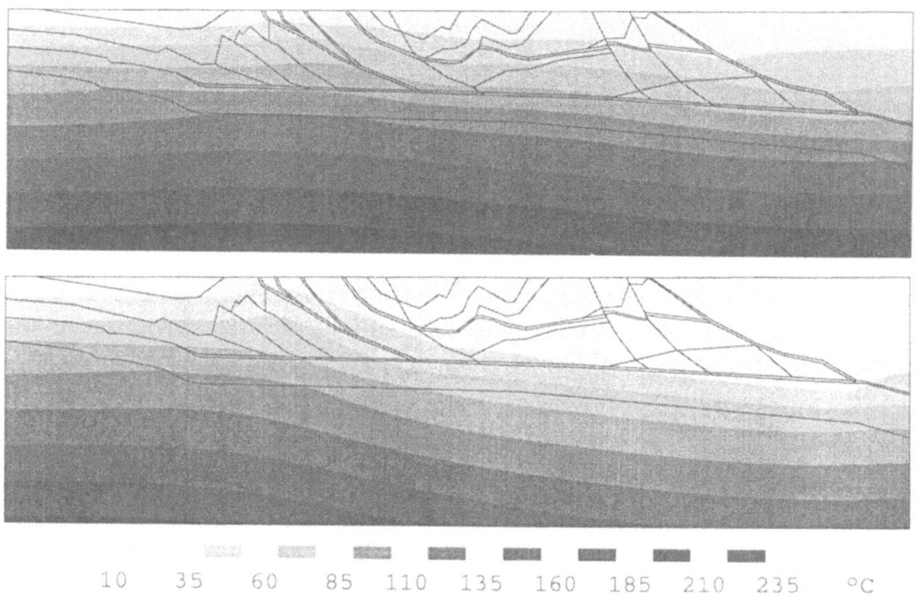

Figure 4. Thermal field under influence of hydrostatically forced pore fluid flow. Thermal boundary conditions are same as in Figure 2A. Permeabilities at fault zones are 200 mDarcy (top) and 2 Darcy (bottom), respectively.

isotherm. The model can be adapted further to the control data by introducing a variable heat flux at the lower boundary of the model. In Figure 5 the basal heat flow increases from 60mW/m^2 at the northern boundary to 85 mW/m^2 at the southern boundary. This boundary condition lifts the thermal field towards the surface and the possible source area of the springs now is located near the 120°C isotherm, whereas the surface heat fluxes are still near the observed values. The trend in basal heat flux now parallels the trend in near-surface heat fluxes, however, it is contrary to the near-surface mean thermal gradient, which in the model as in nature decreases towards the south. These complex patterns result from the complex interaction of basal heat flux, thermal conductivity, and pore fluid flow. Although the model may not approach reality, it approaches the complexity of reality and indicates that control data from a deep borehole (Walter and Wohlenberg, 1985) would be necessary to proof or improve the model.

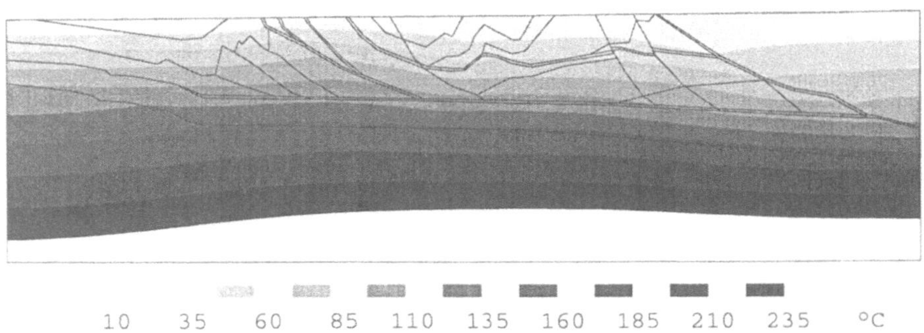

Figure 5. Thermal field under influence of forced pore fluid flow (as in Fig. 4A, 200 mDarcy along faults) and variable basal heat-flow density (60 mW/m^2 at northern boundary, increasing to 85 mW/m^2 at southern end).

UNCERTAINTIES IN PREDICTING THE GEOTHERMAL FIELD

At the end of the last section we were faced with the fact, that there is no definitive test for even a "best fit" model if nonlinear interactions are considered. In this section we will consider the uncertainties associated with extrapolations to depth in some detail.

One of the main control parameters is the near-surface field of heat fluxes. One could assume that the degrees of freedom can be reduced as the knowledge of the near-surface field increases. One possibility would be to locate a sequence of shallow wells along the profile, which then are studied in detail with regard to the heat flux by separating it into its conductive and convective components (Clauser and Villinger, 1990). Next, we will illustrate that this approach may fail if the medium is inhomogeneous and anisotropic.

Figure 6 shows the vertical heat flux for the two conductive models with and without heat production as given in Figure 2. The heat-flow representation corresponds to a high-pass filter that is applied to the thermal field, thus it enhances small variations. In our example, these variations are the result of the different thermal conductivities of the rocks, although there may be numerical errors that confuse the patterns - an aspect which will not be considered within this context.

Figure 6 illustrates that the two models may be calibrated so that the near-surface heat-flow pattern agrees almost perfectly. Correspondence was achieved simply by modifying the heat input at the lower boundary, as previously discussed. Down to about 1000 m the heat-flow patterns are so similar, that they actually cannot be distinguished. The same is true for the near-surface temperature field down to about the same depth as can be illustrated by taking the difference between the temperatures of the two models. Based on current knowledge of near-surface heat fluxes, the two models are indistinguishable. This result reflects our current uncertainty about the temperature distribution in the upper crust.

More important is that in a one-dimensional well that would be drilled into the model's scenario the inhomogeneities caused by the anisotropic rock properties easily could be misinterpreted as convective effects so that the attempt to separate conductive and convective components could fail. Studies and interpretations of the thermal field, therefore, should be accompanied by numerical models. The modeling approach allows to test the sensitivity of the system with regard to various types of disturbances and allows to locate areas with the highest potential for relevant additional information.

CONCLUSIONS

Geothermal studies usually rely on one-dimensional models and on the concept of heat conduction. Most geothermal maps are compiled on this basis (Haenel and Staroste, 1988; Hurtig, 1992). In our study, modeled conductive heat transport results in variations of temperature gradients and heat-flow densitiy. In addition, forced convective and heat-generating processes can transform strongly the temperature distribution calculated from thermal conduction. The models allow to separate and investigate the effects of thermal conduction, forced convection and

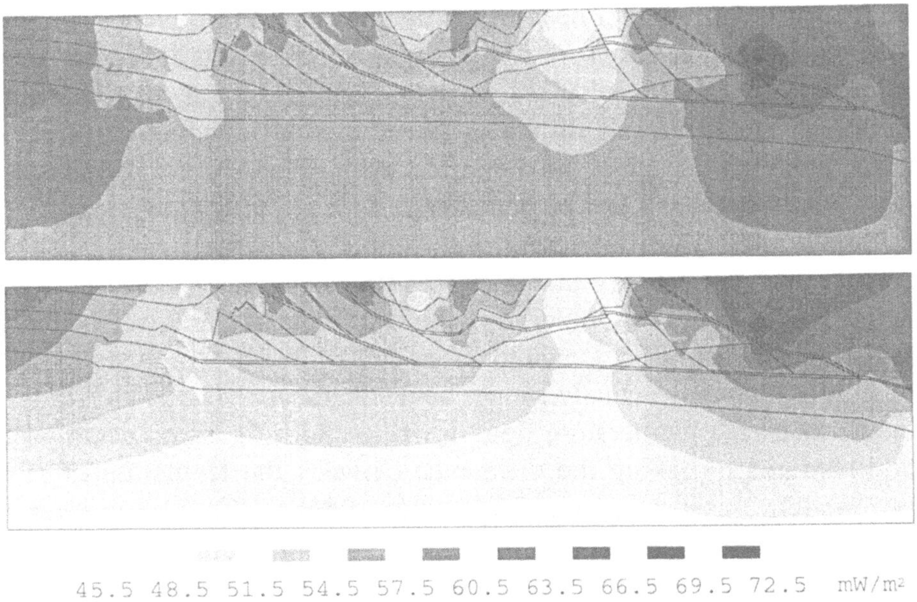

Figure 6. Vertical heat-flow densities for conductive models of Figure 2. A (top), No heat production, basal heat-flow density 60 mW/m^2; B (bottom), heat production according to Table 1, basal heat-flow density 47.5 mW/m^2.

radiogenetic heat generation, which occur coupled in nature. On the other hand, the attempt to model the geothermal field at an ancient orogenetic front indicates that near-surface data are not sufficient to understand the thermal structure of deeper parts of the upper crust. Identical surface heat-flow density can be derived from different temperature fields.

Heat conduction and forced convection provide an interacting system. The complexity of the models can be increased if the transport coefficients depend on the P/T conditions, and if free convection is introduced as an additional force into the system. In such highly nonlinear systems, we may enter a "new world" of possible phenomena.

Besides the uncertainties the models allow to study various scenarios in advance to the location of a well. Although the models do not allow a definite prediction of the thermal subsurface structure, they enhance the differences between various alternative scenarios. These differences may help to localize exploration wells at points with the highest information content.

ACKNOWLEDGMENTS

The study is part of a project funded by the "Bundesministerium für Forschung und Technologie" (BMFT), Projekt No. 0326761A.

REFERENCES

Angenheister, G., ed., 1982, Landolt-Börnstein. Numerical data and functional relationships in science and technology. Physical properties of rocks, v. 1a: Springer-Verlag, Berlin, 373 p.

Bethke, C.M., 1985, A numerical model of compaction-driven groundwater flow and heat transfer and its application to the paleohydrology of intracratonic sedimentary basins: Jour. Geophys. Res., v. 90, no. B7, p. 6817-6828.

Bjoerlykke, K., Mo, A., and Palm, E., 1988, Modeling of thermal convection in sedimentary basins and its relevance to diagenetic reactions: Marine Petrol. Geology, v. 5, no. 4, p. 338-351.

Bless, M.J.M., and others, 1981, Preliminary report on Lower Tertiary - Upper Cretaceous and Dinantian-Famennian rocks in the boreholes Heugem-1/1a and Kastanjelaan-2 (Maastricht, the Netherlands): Meded. Rijks Geol. Dienst, N.S. 35-15, p. 333-415.

Clauser, C., and Villinger, H., 1990, Analysis of conductive and convective heat transfer in a sedimentary basin, demonstrated for the Rheingraben: Geophys. Jour. Intern., v. 100, no.3, p. 393-414.

Dornstädter, J., and Sattel, G. ,1985, Thermal measurements in the research borehole Konzen, Hohes Venn (West Germany): N. Jb. Geol. Paläont. Abh., v. 171, no. 1-3, p. 117-130.

Freeze, R.A., and Cherry, J.A., 1979, Groundwater: Prentice-Hall, Inc., Englewood Cliffs, New Jersey, 604 p.

Graulich, J. M., 1969, Eaux minerales et thermales de Belgique: XXIII Intern. Geol. Congress, v.18, p. 9-15.

Haenel, R. 1982, The Urach geothermal project (Swabian Alb, Germany): E. Schweizerbart'sche Verlagsbuchhandlung, Stuttgart, 419 p.

Haenel, R., 1983, Geothermal investigations in the Rhenish Massif, *in* Fuchs, K., Gehlen, K. von, Mälzer, H., Murawski, H., and Semmel, A. eds., Plateau uplift. The Rhenish shield - a case history: Springer Verlag, Berlin, p. 228-246.

Haenel, R., and Staroste, E., eds., 1988, Atlas of geothermal resources in the European Community, Austria and Switzerland: Verlag Th. Schaefer, Hannover, 74 p.

Heitfeld, K.-H., Völtz, H., Yü, S., Kopf, M., and Krapp, L., 1985, Geotechnical rock tests research borehole Konzen, Hohes Venn (West Germany): N. Jb. Geol. Paläont. Abh., v. 171, no. 1-3, p. 195-205.

Hurtig, E., ed., 1992, Geothermal atlas of Europe: Haack Verlag, Gotha, 165 p.

IES (Integrated Exploration Systems) GmbH, 1991, PDI/PC 1-Dimensional Simulation Program, User Manual Version 2.2: Jülich, Germany, 264 p.

Jobmann, M., and Schulz, R., 1992, Temperatures, fractures and heat flow densitiy in the pilot-hole of the Continental Deep Drilling Project (Oberpfalz, Germany): Scientific Drilling, v. 3, no. 1-3, p. 83-88.

Kappelmeyer, O, and others, 1991, European HDR project at Soultz-sous. Forets, general presentation: Geothermal Science and Technology, v. 2, no. 4, p. 263-289.

Knapp, G., 1980, Erläuterungen zur geologischen Karte der nördlichen Eifel 1 : 100 000: GLA Nordrhein-Westfalen, Krefeld, 155 p.

Langguth, H.-R., and Plum, H., 1984, Untersuchung der Mineral- und Thermalquellen der Eifel auf geothermische Indikationen: BMFT-Bericht T84-019, Nichtnukleare Energietechnik, 176 p.

Ondrak, R., Bayer, U., and Kahle, O.,1994, Characteristics and evolution of anisotropic geologic media, *in* Kruhl, J., ed., Fractals and dynamic systems in geosciences: Springer Verlag, Berlin, p. 355-367.

Pommerening, J., 1992, Hydrogeologie, Hydrogeochemie und Genese der Aachener Thermalquellen: unpubl. doctoral dissertation, RWTH Aachen, 169 p.

Schürmeyer, J., and Wohlenberg, J., 1985, The anisotropy of the thermal conductivity of Ordovician rocks from the research borehole Konzen, Hohes Venn (West Germany): N. Jb. Geol. Paläont. Abh., v. 171, no. 1-3, p. 131-143.

Smith, L., and Chapman, D.S., 1983, On the thermal effects of groundwater flow, 1. Regional scale systems: Jour. Geophys. Res., v. 88, no. B1, p. 593-608.

Somerton, W.H., 1992, Thermal properties and temperature-related behavior of rock/fluid systems: Elsevier Science Publishers B.V., Amsterdam, 257 p.

Toth, J., 1962, A theory of groundwater flow in small basins in central Alberta, Canada: Jour. Geophys. Res., v. 67, no. 11, p. 4375-4387.

Walter, R., and Wohlenberg, J., 1985, Proposal for an ultra-deep research borehole in the Hohes Venn Area (West Germany): N. Jb. Geol. Paläont. Abh., v. 171, no. 1-3, p. 1-16.

Welte, D.H., 1989, The changing face of geology and further needs: Geologische Rundschau, v. 78, no. 1, p. 7-20.

Welte, D.H., and Yalcin, M.N., 1987, Basin modeling - a new comprehensive method in petroleum geology, *in* Mattavelli, L., and Novelli, L., eds., Advances in organic geochemistry: Organic Geochem., v. 13, p. 141-151.

Winterfeld, C. von, 1994, Variszische Deckentektonik und devonische Beckengeometrie der Nordeifel - Ein quantitatives Modell (Profilbilanzierung und Strain-Analyse im Linksrheinischen Schiefergebirge): unpubl. doctoral dissertation, RWTH Aachen, 309 p.

Yükler, M.A., Cornford, C., and Welte, D., 1978, One-dimensional model to simulate geologic, hydrodynamic and thermodynamic development of a sedimentary basin: Geologische Rundschau, v. 67, no. 3, p. 960-979.

EFFECTIVE TRANSPORT PROPERTIES OF ARTIFICIAL ROCKS – MEANS, POWER LAWS, AND PERCOLATION

O. Kahle and U. Bayer
GeoForschungsZentrum Potsdam, Potsdam, Germany

ABSTRACT

Two-component systems provide simple idealized models for complex anisotropic geological structures. Such artificial media are studied in terms of their "effective" transport coefficients. It is shown that means and power laws provide empirical descriptions for the transport tensor and relate it to statistical measurements of the composition and structure of the medium. General anisotropic media are related to statistically isotropic media. The isotropic medium is well described by an iterative mean for low contrasts of conductivities. This iterative mean is derived from the arithmetic and harmonic means. For high conductivity contrasts rules of percolation theory come to application which expand the classical concepts of percolation theory. At the percolation threshold both accesses fall together, as is illustrated by appropriate power laws for the effective transport coefficients.

INTRODUCTION

The estimation of "effective" transport properties is a classical problem associated with modeling of transport processes in porous rocks. In nature, transport properties differ within a wide range of length scales which cannot be incorporated in numerical models. In addition, most

Geologic Modeling and Mapping
Edited by A. Förster and D. F. Merriam, Plenum Press, New York, 1996

rocks are anisotropic on a large scale, an aspect which should be included into models because it modifies the flow path. Thereby, the dimension of the model is of secondary order, it occurs in microscopic pore models as well as in megascopic basin models.

Generally, there are two ways of modeling phenomena in rocks. The first approach is based on the classical equations of transport or on the continuum model. The second one is based on network models of pore space and fractured rocks, that is the models are at the smallest scales with regard to the physical properties. Based on these microscopic models modern concepts of the statistical physics of disordered systems can be used to derive the macroscopic properties of the system. In this context the concepts of fractals and percolation play important roles (Sahimi, 1993).

A typical way to represent the average effects of small-scale flow in coarse grid reservoir or basin models has been termed pseudo relative permeabilities (Kyte and Berry, 1975; Stone, 1991). These "effective" permeabilities usually are generated by a stepwise coarsening process. The results derived from fine grid models then are used as input for coarser grid models. A variety of techniques has been used in the past to achieve the coarsening process. These techniques range from simple statistic averages which are inaccurate in real applications to detailed numerical simulations that are consuming much computation time. An alternative strategy is real-space renormalization, where the effective property is calculated recursively for increasingly coarser grids from small groups of cells. This process continues until the entire grid is reduced to a single block (King, Muggeridge, and Price, 1993).

The aim of our work was to determine simple rules from detailed simulations which allow an easy calculation/estimation of the effective transport property derived from two-component models which resemble sedimentary sequences. The property under consideration can be different as long as the basic transport equations are the same, for example permeability, heat conductivity, and diffusion coefficients are all related to the Laplace equation under stationary transport conditions.

The paper provides power laws for the effective transport coefficients which are derived from a pseudoization method which is applied to a domain in space consisting of two materials with distinct properties. Special attention is drawn on anisotropic distributions of the materials within the domain, and a simple algorithm allows a change in the internal structure of the distribution from an isotropic to a perfectly layered system.

GEOMETRIC MODELS OF ARTIFICIAL ROCKS

Figure 1 illustrates a set of artificial, computer-generated sedimentological patterns or rocks. These artificial rocks form the base for the modeling approach. The frequency of both components is identical within all blocks, however, the ordering (stratification, lamination, or layering) increases from the upper left to the lower right corner. The generation process is rather simple: square planes (bricks of height 1 and side lengths l) are distributed randomly within the modeling area, a 3-dimensional cube of size 60^3 'pixels'. If the 'bricks' have a side length l = 1 (1 pixel), then the medium is statistically homogeneous and equivalent to site percolation models (Stauffer, 1985). In this special situation the concepts of percolation theory such as percolation threshold, cluster-size distribution, and fractal properties can be applied.

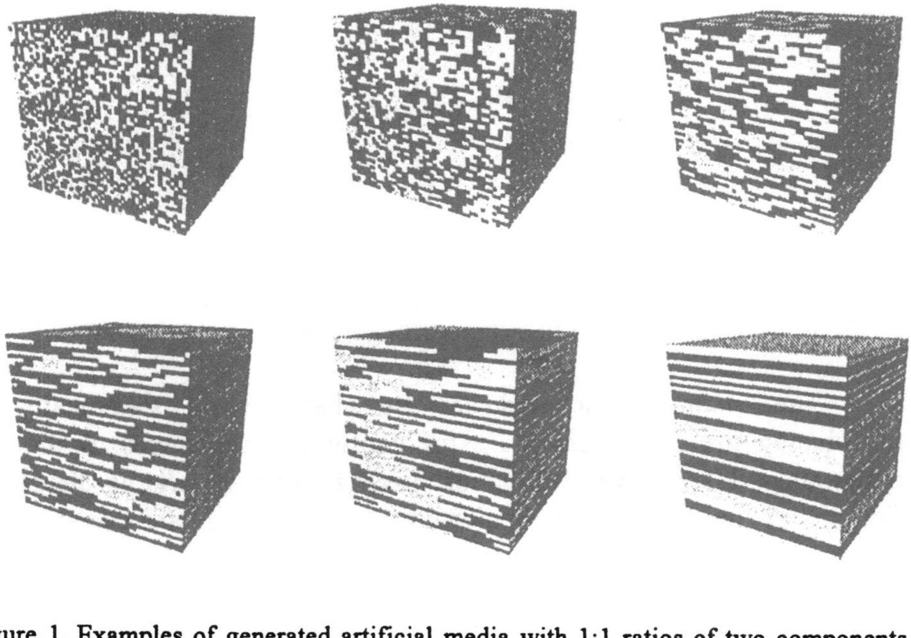

Figure 1. Examples of generated artificial media with 1:1 ratios of two components. Structural anisotropy increases from upper left to lower right corner, generating lengths 1, 2, 5, 10, 20, and 50 pixels.

However, if planes of sizes $l > 1$ are used, the properties of the artificial media change. The main reason is that some of the randomly placed objects may overlap. These overlappings are counted once in determining the density distribution. The "events", therefore, are not independent, and the filling mechanism is somewhat modified. Within this study we will consider only fillings with squared plane elements, a process which provides structures similar to sedimentological patterns. The two components within the model area should not be considered as pore space and rock matrix, but rather as the distribution of two lithological types such as clay and sandstone, or as two mineral phases with different conductivities. In natural systems the pore systems remain connected at porosities much below the percolation threshold of our model. In the situation of pore models, other approaches, such as the grain consolidation model (Schwartz and Kimminau, 1987), are more successful. Another way would be to replace our site oriented approach by a bond-oriented one (Wong, Koplik, and Tomanic, 1984).

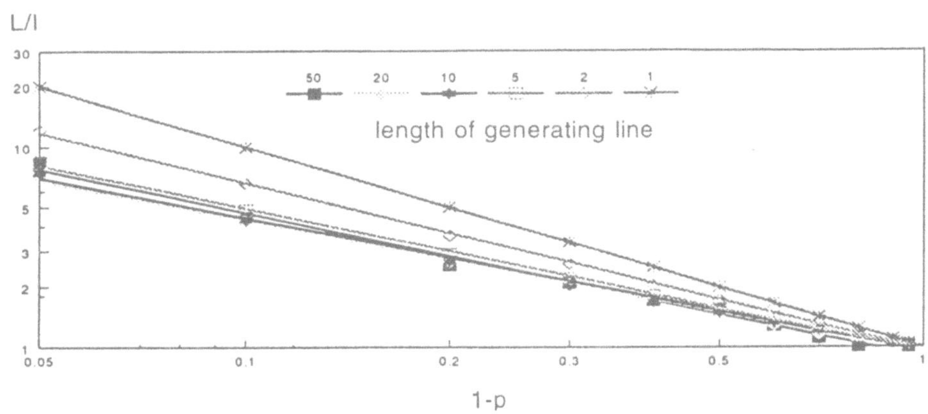

Figure 2. Power law dependence of characteristic length L from frequency of higher conductive material (p_2) and different generating lengths l.

The medium discussed can be characterized statistically by introducing the mean length $L_{j,i}$ for continuous sequences of the two materials j=1,2 in every direction i=x,y,z. The generating lengths l_i (lengths of the square planes) and the mean lengths L_i are correlated by a power law (Fig. 2):

$$L_{2,i} = l_i(1-p_2)^{-(\alpha+(1-\alpha)\frac{l_i}{l_0})} \quad , \text{ for } l_i > l_0 \tag{1}$$

whereby

$$\frac{p_1}{L_{1,i}} = \frac{p_2}{L_{2,i}} \quad , (p_1+p_2=1) \tag{2}$$

The parameter a is a characteristic constant (a=0.58..) for this type of model. The parameter l_0 is the basic unit for the grid, for example one pixel. If the generation length equals l_0 in all directions, then the equation simplifies to

$$L_{2,\mathrm{i}} = l_\mathrm{i}(1 - p_2)^{-1} \text{ , for } l_\mathrm{i} \leq l_\mathrm{o} \tag{3}$$

The power in Equation (1) expresses the fact that the medium may be rescaled to the statistically isotropic medium [Eq. (3)] if it has been generated by statistically independent events (no overlapping), independently of the size of the objects used. Note that there is no cross dependence between the different directions and the mean lengths. Both components are distributed symmetrically concerning the frequencies of the components [Eq. (2)], but they are not symmetrical with regard to the distribution of the lengths itself. The geometry, in addition, can be characterized by some invariant parameters such as the inner surface S and various measurements of anisotropy which have been discussed in Ondrak, Bayer, and Kahle, (1994). These parameters can be helpful in order to relate the model parameters to measurable data.

CALCULATION OF EFFECTIVE TRANSPORT COEFFICIENTS

In the situation of a stationary transport in porous media, the mathematical description may simplify to the Laplace equation which accounts for fluid flow, thermal conductivity, and diffusion processes in homogeneous media. For our heterogeneous medium we will consider fluid flow as a special situation, and we assume that Darcy's laws holds in absence of gravity effects (Fick's law for the diffusion processes):

$$\vec{q} = -K\nabla p \tag{4}$$

where q is the Darcy velocity [m/s], K the coefficient of permeability ($K = k/h$), k the intrinsic permeability of the rock [cm²], h the dynamic viscosity of fluid [Poise, kg/ms], and p the pressure potential [Pa, kg/ms²]. The equation for conservation of mass (continuity equation) is

$$(\varphi\dot{\rho}) = \nabla \cdot \rho\vec{q} + e \tag{5}$$

with ρ density of fluid [kg/m³], ϕ porosity of rock, e generation density of fluid [kg/m³s]. For incompressible fluid and constant temperature in the domain, that is $\rho =$ const. and for the absence of sources, $e = 0$, this equation reduces to

$$\nabla \cdot \vec{q} = 0 \tag{6}$$

Together with (4) we get the elliptical differential equation

$$\nabla \cdot (K \nabla p) = 0 \tag{7}$$

which has to be solved together with appropriate boundary conditions. Given a solution of Equation (7), the effective permeability can be computed by use of Equation (4). These modeling results resemble laboratory measurements and provide the database for the approximate methods discussed next.

If K is a known function of the space variable, the usual theory of linear elliptic equations can be applied. However, the geometrical generation process (Fig. 1) leads to a random field of K-values. The permeability at each site corresponds to the material component at that site. This progress initially provides two delta-functions describing the frequency distribution of K in space. For further investigations it is easy to generalize the delta-distributions to log normal distributions in order to incorporate fluctuations of the conductivity within each material component. On the other hand, it is known that for broad distributions of conductances the transport is dominated by those conductances with magnitudes greater than some characteristic value, at which the set of conductances forms an infinite connected cluster. Transport in such systems reduces to a percolation problem with this characteristic value and, therefore, resembles the approach by delta functions.

We use a finite difference method to solve (7) and to calculate the average effective value of $\overline{K}$ for the entire modeling domain. The flow is driven by a constant pressure gradient between two opposite faces of the cube. The other boundaries are unpermeable. This definition of the boundary conditions allows an easy calculation of $\overline{K}$ for the stationary situation, and Equation (4) provides a tool, in addition, to prove the quality of the numerical solution. The total flux has to be constant through any cross-section perpendicular to the flow direction.

The initial conductivities along a bond between two sites is taken as the harmonic mean of the values at the sites, an approach used in flow models for porous media (Kinzelbach, 1986). Figure 3 gives an example of the calculated pressure potential for such a stationary solution and illustrates how the anisotropic properties of the rock modifies the flow pattern.

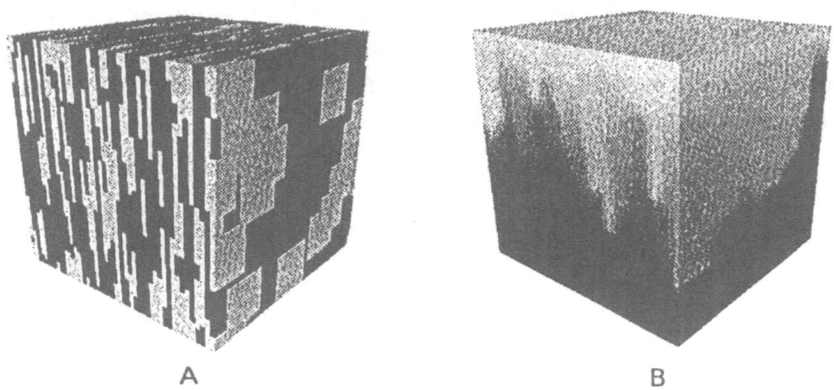

Figure 3. Example for stationary solution of transport problem. Distribution of two materials is given in A, potential is elucidated in B. Flow shows typical features of fingering. Ratio of conductivities is 100:1, and black areas in A indicate component of higher conductivity.

The calculations were made for different frequencies of the components, a wide range of conductivity contrasts (several orders), and various structural anisotropy. The principle axes of the transport tensor $\overline{K}$ are determined by calculating the flow normal and tangential to the geometrical structure. These calculations are sufficient because of the symmetry of the planes used to generate the layering. Some calculation results are given in Ondrak, Bayer, and Kahle (1994), by example, which elucidate the influence of anisotropy.

REPRESENTATIVE VOLUMES

In order to achieve accurate results with a minimum computation effort the modeling domain has to be sufficiently large to keep the statistical fluctuations small enough. The error related to the size of the domain should be of equal order similar to the error of the numerical flow calculations. Figure 4 shows frequency distributions of the calculated effective conductivities for modeling domains which range from size 60^3 down to a single site (size 1^3) for a statistically isotropic and an anisotropic medium. It becomes clear that a cube size of 1^3 produces two delta-functions located at the two conductivities associated with the two materials. For larger domains the distribution broadens and then fuses into a single peaked distribution around a well defined mean value. Finally, the distribution transforms into a single peak with small variance. But the isotropic and anisotropic situations are distinguished by a significant difference in the representative domain size for which the deviations from the effective value are negligible.

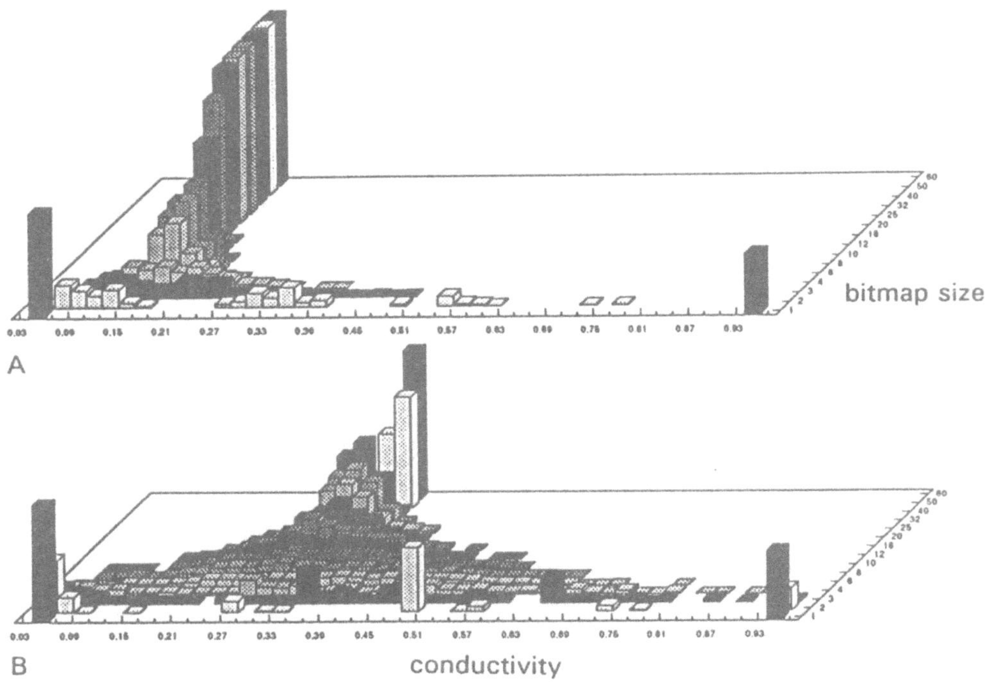

Figure 4. Two sets of frequency distributions of effective conductivity for isotropic medium A and anisotropic one B based on different domain sizes. Conductivities of two components are $k_1 = 0.05$ and $k_2 = 0.95$; frequency of component 2 is $p_2 = 0.4$. For anisotropic medium generating length is $l = 10$, and flow is parallel to structure (compare to Fig. 3).

In the isotropic situation a domain size of 32^3 is sufficient to equalize the random fluctuations of the generation process. An anisotropic structure requires a larger domain size to reach the same balancing. For the example shown in Figure 4B, the standard deviation of the effective value will be lower than 1% if the domain sizes are greater 50^3 (Fig. 5). It can be expected that the required domain sizes are proportional to $L_{2,i}$. The flow calculations and the determinations of $\overline{K}$ used below were all done with a domain size of 60^3. This is a good compromise between good statistics and calculation time.

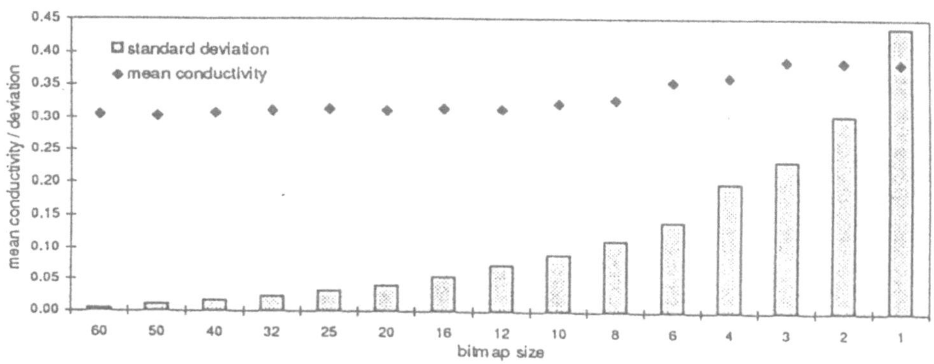

Figure 5. Standard deviation and effective conductivity for anisotropic medium of Figure 4B for different representative volumes.

MEANS AND POWER LAWS FOR EFFICIENT TRANSPORT COEFFICIENTS

A general aim is to describe the effective mean conductivity by simple relations in terms of measurable features of the medium. The effective conductivity, per definition, is some type of a mean taken over the different subdomains under consideration. Here, we will illustrate that this concept together with basic measurables, similar to the frequency of materials and simple measures of structural anisotropy (L), is sufficient for an approximate description of the effective mean conductivities.

First, we consider the most simple situation of a medium of infinite structural anisotropy, that is the medium is completely layered. This situation is understood fully, the medium can be imagined as consisting of resistors all in series for flow perpendicular to the layering and in parallel for flow in direction of the structure. The thickness of the layers is unimportant, and the effective conductivity $\bar{K}$ can be approximated by the weighted arithmetic and harmonic means:

$$\bar{K} = p_1 k_1 + p_2 k_2 \equiv A \text{ (for flow } \parallel \text{ structure)}, \tag{8}$$

$$\bar{K} = \frac{k_1 k_2}{p_2 k_1 + p_1 k_2} \equiv H \quad \text{(for flow } \perp \text{ structure)}. \tag{9}$$

These values provide upper and lower bounds for the effective conductivity.

Another situation considered is the statistically isotropic medium which will be discussed in some detail. A number of formulations have been inventoried for the effective conductivity (Prakouras and others, 1978) with contrary results in several instances. These equations incorporate many effects and corrections which are not suitable to describe our artificial medium. We propose the following access: An isotropic medium can be imagined as a mixture of areas of complete layering where flow is either normal or tangential to the structure. Thus, we expect that the effective conductivity is a mean bounded by Equations (8) and (9). Such a effective conductivity can be constructed by the following iteration process (Beckenbach and Bellmann, 1965; Schoenberg, 1982):

$$k_1^n = f(k_1^{n-1}, k_2^{n-1}, p_2) \equiv H \tag{10}$$
$$k_2^n = g(k_1^{n-1}, k_2^{n-1}, p_2) \equiv A \tag{11}$$

where $k_1^0 = k_1$ and $k_2^0 = k_2$, and

$$k_1^{n-1} < k_1^n < k_2^n < k_2^{n-1}, \tag{12}$$

with the limit

$$\lim_{n\to\infty} k_1^n = \lim_{n\to\infty} k_2^n \equiv M(k_1, k_2, p_2) \tag{13}$$

This iteration process provides some analogy to real space normalization, however, the initial means are taken over the entire domain.

For equal frequencies of the two components ($p_1 = p_2 = 0.5$) the iterative mean M is equal to the geometric mean $G = k_1^{p1} k_2^{p2}$ (see Schoenberg, 1982, for a theoretical derivation). However, for different ratios of the components there are small, but significant differences between M and G, whereby M provides a better approximation for the effective conductivity than the usual proposed geometric mean. This averaging technique provides sufficient approximations only at low conductivity contrasts between the two components. Figure 6 illustrates the relative error between the computed effective conductivity and the approximation by the iterative mean. Obviously, the conductivity contrasts should be lower than 20:1 in order to keep the error below 2%. Surprisingly, the relative error vanishes at the percolation threshold for all

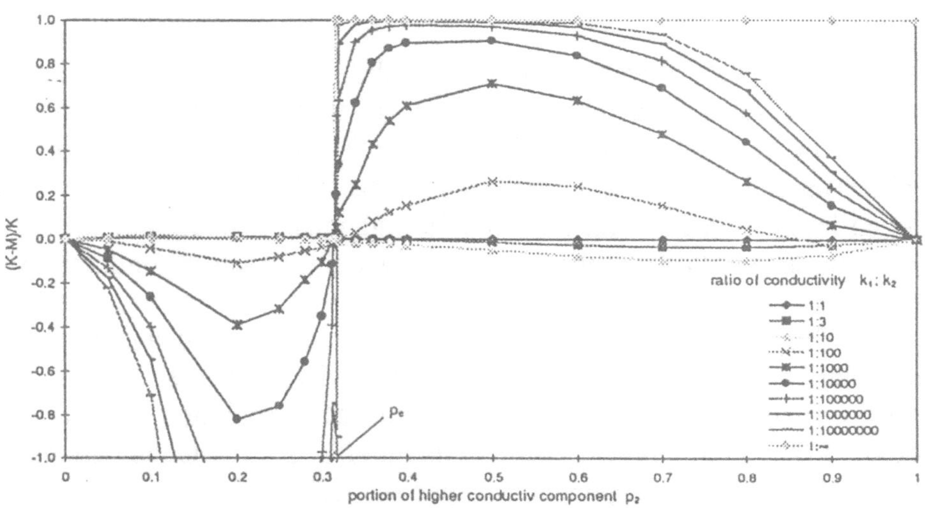

Figure 6. Relative difference between calculated conductivity and iterative mean for different ratios of conductivity.

conductivity contrasts. At this point the effective conductivity $\overline{K}$ is exactly described by M (Fig. 7)

$$\overline{K}(p_c) = M(k_1, k_2, p_c). \tag{14}$$

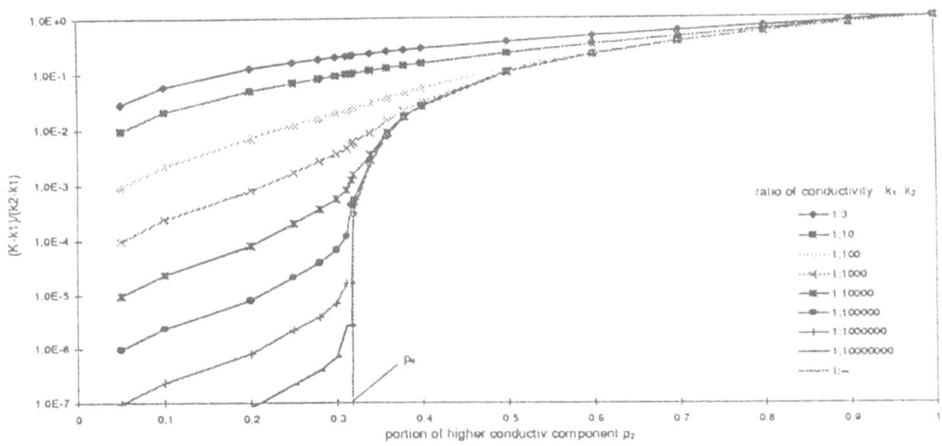

Figure 7. Normalized effective conductivity calculated from flow model for different ratios of conductivities.

At the percolation threshold the statistically isotropic medium is known to have a fractal structure, and the correlation length of the clusters formed by material 2 becomes infinite. At this point the assumption of a mixed medium as defined previously seems to be valid. This fact could be an interesting field for further investigations, at this point the result provides an additional motivation to analyze the effective conductivity in terms of means.

The approximations by means fail at some distance of the percolation threshold and for high conductivity contrasts. The medium then is better described by the concepts of percolation theory. It is obvious that simple means cannot explain effects such as the percolation threshold – if one component is a perfect isolator the means would be zero independent of the frequency of the material. In this special example our model structure corresponds to the standard medium discussed in percolation theory, and the average conductivity above a typical threshold obeys a power law equation:

$$\overline{K} = k_2\left(\frac{p_2 - p_c}{1 - p_c}\right)^D \qquad p_2 > p_c \tag{15}$$

$$\overline{K} = 0 \qquad p_2 \le p_c \tag{16}$$

with $p_c \sim 0.317$ and $D \sim 1.72$ for our models. Percolation theory predicts a site percolation threshold of 0.3116 for the simple cubic lattices (Stauffer, 1985). The difference between the theoretical value and the one we determined may be explained by calculation errors and the fact that the size of the representative volume increases theoretically to infinity at the percolation threshold (*pc*). Formulae such as Equation (15) have been discussed repeatedly in a similar context (Staufer, 1985; Peitgen, Jürgens, and Saupe, 1992). Actually, the identical expression was determined for a percolation model derived from log normal distributions by hard clipping.

Percolation theory is related to the situation of a 0:1 conductivity contrast. It describes a singular situation similar to the iterative mean M [Eq. (14)]. In geological applications one may be faced with two or even more components with finite conductivities. If both components are conductive in our model, however, with high conductivity contrasts (>20:1, see previous discussion) then Equation (15) has to be modified. We determined that the following equation is a sufficient approximation to describe a generalized isotropic medium if the frequency of the material with higher conductivity (p_2, k_2) is above the percolation threshold:

$$\overline{K} = (k_2 - k_1)\left[(1-\overline{K}'(p_c))\left(\frac{p-p_c}{1-p_c}\right)^{D(k_1/k_2)} + \overline{K}'(p_c)\right] + k_1 \qquad ,p > p_c \qquad (17)$$

The indices are selected in a way so that the higher conductive material has the index 2. p_c is the percolation threshold for material 2. $\overline{K}'(p_c)$ denotes the normalized effective value at p_c:

$$\overline{K}'(p_c) = \frac{\overline{K}(p_c)}{k_1 + k_2} = \frac{M(k_1, k_2, p_2)}{k_1 + k_2} \qquad (18)$$

In this example, the exponent D is not a single constant, but it is a function of conductivity contrast. The terms (k_2-k_1) and k_1 scale the term in square brackets into the interval 0–1. If the conductivity contrast increases towards infinity, Equation (17) approaches asymptotically Equation (15), that is the percolation case. This illustrates that the classical percolation situation really should be taken as a singular situation which, however, can be accessed by taking the limit toward an infinite conductivity contrast. Although Equation (17) describes the effective conductivity for finite conductivity contrasts, it depends on the percolation threshold. The discussed limit has to be taken from frequencies above the percolation threshold. The formula is only valid for $p_2 > p_c$. We have not determined yet a similar simple transformation for the region below the percolation threshold. It seems that the limit from the area below the percolation threshold is of a different nature than the limit from above the percolation threshold, so p_c may mark a real discontinuity even in the generalized situation discussed here.

At last we consider the generalized anisotropic medium. It may be understood as an intermediate medium between the extreme situations of the statistically isotropic medium with the effective conductivity I and the perfectly layered medium with its effective conductivities A and H. The mean I can be calculated from Equation (17) as far as this equation is valid, or the value computed from the numerical flow model may be used. Thus, the effective conductivity should be limited by the corresponding limit values:

$$I \leq \overline{K}_{//} \leq A \qquad \text{for flow } \parallel \text{ to structure,} \qquad (19)$$

$$H \leq \overline{K}_{\perp} \leq I \qquad \text{for flow } \perp \text{ to structure,} \qquad (20)$$

These equations provide some analogy to Equation (12), and by an heuristic argument one may assume that the effective conductivity is once more a mean between the given bound, at least as long the conductivity contrasts are low enough. A first approach to describe $\overline{K}$ as a function of (H, I, A) is the weighted geometric mean which also provides a power law:

$$\overline{K}_{\mathsf{I}} = I^{\alpha} * A^{1-\alpha} \qquad \text{flow } \parallel \text{ to structure,} \qquad (21)$$

$$\overline{K}_{\perp} = I^{\beta} * H^{1-\beta} \qquad \text{flow } \perp \text{ to structure,} \qquad (22)$$

These equations, indeed, approximate the data as Figure 8 illustrates. The figure shows the data points and the associated power law correlation lines for conductivity contrasts from 1:1 to 20:1. The parameter given for every curve is the structural anisotropy P as defined in Ondrak, Bayer, and Kahle (1994):

$$P = \frac{L_{\max}}{L_{\min}} \tag{23}$$

It should be mentioned that the approximation fails for higher conductivity contrasts especially at high structural anisotropies, a situation encountered earlier during the discussion of the iterative mean for statistically isotropic media. At high anisotropies calculation errors may play an important role which, in addition, bias the approximation. These errors may be related to the fact that the representative volume increases rapidly with increasing anisotropy. The parameters α and β depend on the degree of anisotropy and the frequency of the materials. This dependency, again, is well described by a power law :

$$a = P^{x}, \tag{24}$$

where P denotes the structural anisotropy. The power x should be a complex function of material frequencies. Again, a simple functional relationship has not yet been determined and the situation close to the percolation threshold needs further research.

In summary, we determined that means and power laws provide good and fast approximations for effective conductivities of statistically isotropic and for anisotropic media within some limits. At least, the geometric mean can be understood as another power law so that a concise approach seems possible in order to describe media with a wide range of conductivity contrasts. A special and, perhaps, interesting result is that the iterative mean provides a perfect approximation for statistically isotropic media at the percolation threshold. This result introduces a 'fix point' which could provide a base for further studies.

CONCLUSIONS

The paper provides power law equations which relate the effective transport properties of a randomly generated two-component medium to the frequencies of the materials and to simple statistical measurements of anisotropy. This approach allows 'easy estimations' of the transport tensor for geological media by approximate functions and for different degrees of anisotropy of the medium.

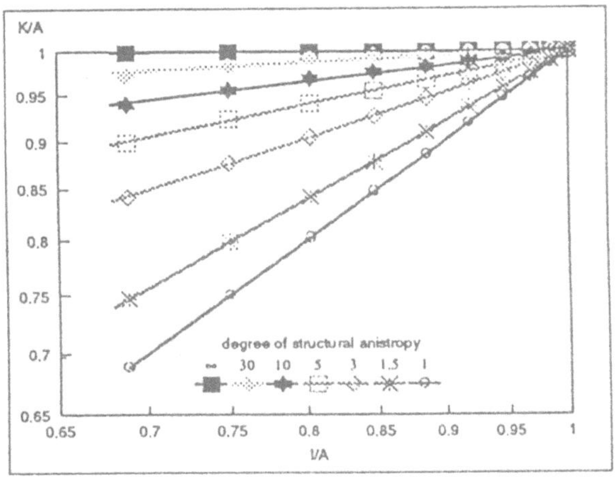

A

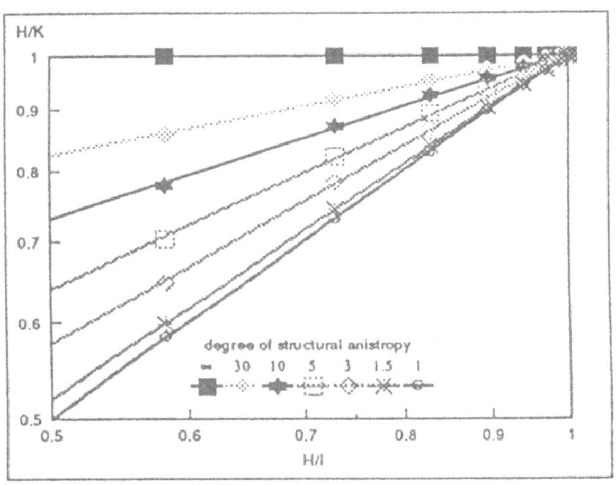

B

Figure 8. Effective conductivity ($\overline{K}$) for anisotropic media related to arithmetic mean (A), harmonic mean (H) and effective conductivity of statistically isotropic medium (I) by a power law. Points denote values calculated from flow model, lines show power law fits; (a) $\frac{\overline{K}}{A} = \left(\frac{I}{A}\right)^{\alpha}$ for flow || structure and (b) $\frac{H}{\overline{K}} = \left(\frac{H}{I}\right)^{\beta}$ for flow $\perp$ structure.

The random generation mechanism of the media includes the statistically isotropic medium. For such media with infinite conductivity contrasts, the classical descriptions used in percolation theory apply. In extension to percolation theory it was possible to determine an unbiased description for arbitrary conductivity contrasts at the percolation threshold, given by the limit of an iterative mean, combining the arithmetic and harmonic means. In addition, this iterative mean provides good approximations for any composition of the medium as long as the conductivity contrasts are low. In situations of high conductivity contrasts an extension of the concepts of percolation theory, including the iterative mean as a parameter, is best suited to describe the effective transport properties if the material with higher permeability has a frequency $p > p_c$. The rules of percolation theory, therefore, are not only applicable to black and white media, but can be extended to describe a continuous space of permeability contrasts.

The anisotropic medium was related to the isotropic and the perfectly layered medium by a simple power law respectively the weighted geometric mean. This approximation holds also for low conductivity contrasts. This result may provide a first step towards transformation rules which allow to map anisotropic media onto an equivalent problem involving statistically isotropic media. Such rules would allow to reduce the space of possible structures drastically.

A main result, however, is that the point of percolation threshold p_c also is meaningful in the context of media with finite conductivity contrasts. The coincidence with the perfect description by the iterative mean at the point of percolation can be considered an analogy to the geometric-arithmetic mean algorithm which forms a base for the theory of elliptical integrals (Beckenbach and Bellmann, 1965). Here could be an interesting connection between different theories such as percolation, fractals, effective medium theory, and classical calculus.

REFERENCES

Beckenbach, E.F., and Bellmann, R., 1965, Inequalities: Springer-Verlag, New York, 198 p.

Dikow, E., and Hornung, U., 1991, A random boundary value problem modelling spatial variability in porous media flow: Jour. Diff. Equations, v. 92, no. 2, p. 199-225.

Hilfer, R., 1992, Local-porosity theory for flow in porous media: Phys. Rev. B, v. 45, no. 12, p 7115-7121.

King, P.R., Muggeridge, A.H., and Price, W.G., 1993, Renormalization calculations of immiscible flow: Transport in Porous Media, v. 12, p. 237-260.

Kinzelbach, W., 1986, Ground water modeling: Elsevier, Amsterdam, 333 p.

Kyte, J.R., and Berry, D.W., 1975, New pseudo functions to control numerical dispersion: Soc. Pet. Eng. AIME Jour., v. 15, p. 269-276.

Morland, L.W., 1992, Flow of viscous fluids through a porous deformable matrix: Surveys in Geophysics, v. 13, p. 209-268.

Ondrak, R., Bayer, U., and Kahle, O., 1994, Characteristics and evolution of artificial anisotropic rocks, *in* Kruhl, J.H., ed., Fractals and dynamic systems in geoscience: Springer-Verlag, Berlin/Heidelberg, p. 355-367.

Peitgen, H.O., Jürgens, H., and Saupe, D., 1992, Chaos and fractals-new frontiers of science: Springer, New York/Berlin/Heidelberg, 220 p.

Prakouras, A.G., Vachon, R.I., Crane, R.A., and Khader, M.S., 1978, Thermal conductivity of heterogeneous mixtures: Intern. Jour. Heat Mass Transfer, v. 21, no. 8, p. 1157-1166.

Sahimi, M., 1993, Flow phenomena in rocks: from continuum models to fractals, percolation, cellular automata, and simulated annealing: Review of Modern Physics, v. 65, no. 4, p. 1393-1534.

Schoenberg, I.J., 1982, Mathematical time exposures: Math. Assoc. America, 270 p.

Schwartz, L.M., and Kimminau, St., 1987, Analysis of electrical con duction in the grain consolidation model: Geophysics, v. 52, no. 10, p.1402-1411.

Stauffer, D., 1985, Introduction to percolation theory: Taylor & Francis, London and Philadelphia, 89 p.

Stone, H.L., 1991, Rigorous black oil pseudo functions: Proc. 11th SPE Symp. on Reservoir Simulation, p. 57-68.

Wong, P., Koplik, J., and Tomanic, J.P., 1984, Conductivity and permeability of rocks: Phys. Rev., v. B30, p. 6606-6614.

THREE-DIMENSIONAL MODELING OF GEOLOGICAL FEATURES WITH EXAMPLES FROM THE CENOZOIC LOWER RHINE BASIN

Rainer Alms, Christian Klesper, and Agemar Siehl
Geologisches Institut der Universität Bonn, Bonn, Germany

ABSTRACT

Concepts and software systems for interactive free form modeling of geologically defined geometries are presented, which utilize the up-to-date availability of high performance techniques for data handling, 2D and 3D interpolation and computer graphics. Under control of a programmable interface, commercial programs (ONTOS, GMP) and research developments (GEOSTORE, GRAPE, GOCAD) have been implemented in a workstation network using Remote Procedure Call for data exchange. This system has been developed using examples from the Lower Rhine Basin.

INTRODUCTION

The present structure of geological surfaces, bodies, and property distributions is a product of interacting processes during geological evolution. The grade of complexity of the pattern depends on the number of structure-forming effects superimposed on one another. Complexity

Geologic Modeling and Mapping
Edited by A. Förster and D. F. Merriam, Plenum Press, New York, 1996

usually increases with the passing of time, but traces also may fade out and vanish. The recent assemblage of surfaces and bodies contains all information we have, and the geologist must recognize the dynamics of its evolution in the actual stage of development. The assemblage defines boundary conditions for all derived models, for example dynamic modeling of sedimentary and structural basin development.

The fabric of rocks intersected by many surfaces of different meaning thus is the key in revealing the nature and interaction of earlier processes, tracing them back through time, understanding how the structure was formed, and how it probably will develop in the future. This characteristic geoscientific process of inverse reasoning is outlined in Figure 1 (Siehl, 1993), but it is well known that inversions do not have unique solutions. This is the main reason why we make geometric models in geology and test them against various aspects of reality: to determine the most likely solution.

Geological data typically are sparse compared to the complex structures they are derived from and are supposed to describe, that is, the essential parts of the structure we want to model are not supported by data as they are hidden and cannot be verified. Therefore, the first step in modeling always is the problem of pattern recognition with few constraints. Usually the geologist must decide between contradictory assumptions. Thus, in addition to the usual inspection of geological maps and sections, drawings of alternative versions in perspective views, especially when structures are complicated, are needed. Modern computer-aided geometric design techniques combined with high-performance computer graphics can be powerful tools that assist the expert in the investigation of unknown domains. Various types of graphic representation, animation, and even stereoscopic views contribute decisively towards the rendering and understanding of modeled virtual geological objects in three and more dimensions.

When a geologist makes use of such modeling systems, he/she needs an interface and interactive tools to enter the conceptual geometric model together with the supporting data into the computer. He/she must bring in the background knowledge in order to formulate geometric hypotheses by which geological processes a surface might have been formed in regions where the data support is insufficient. This can be regarded as a type of intelligent interpolation and extrapolation, which up to now could not be achieved by an algorithm. Rendering of the designed structure permits the control of all geometrical consequences of the initial

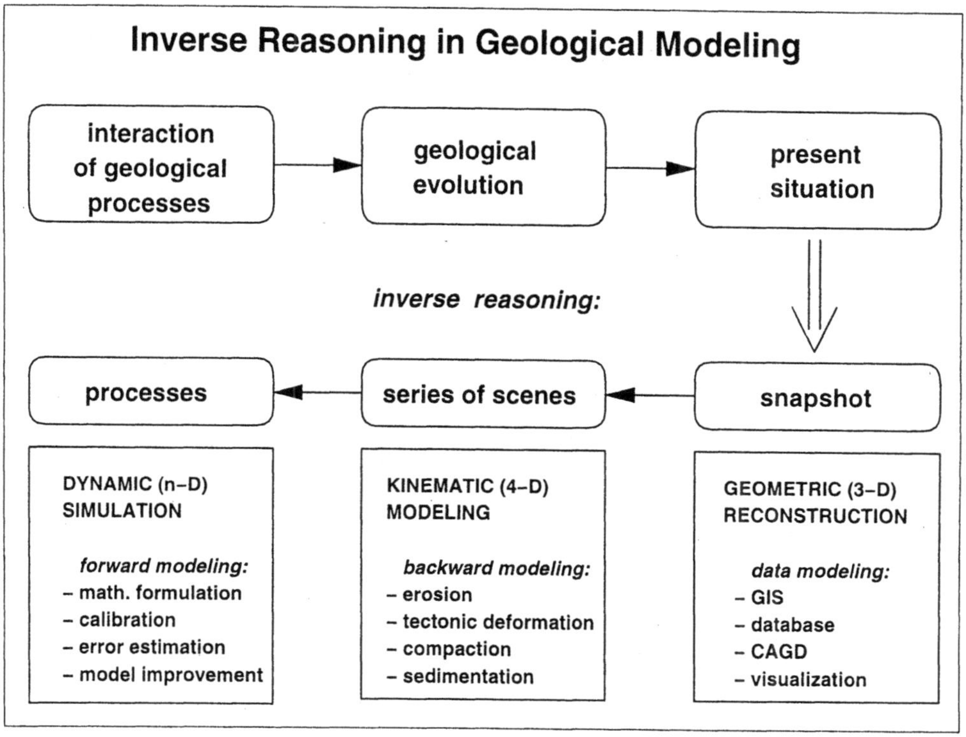

Figure 1. Inverse reasoning in geological modeling.

assumptions. Each of the designed elements and then the entire assembled model are tested subsequently in regard to geological plausibility and goodness of fit to the primary and derived data. This modeling process is one of interactive learning and increased insight into the spatial interrelationships of geological surfaces and bodies. It leads to better understanding of the history of formation which, in turn, is incorporated in the improvement of the model to increase structural resolution.

When, finally, this process has gone through several cycles, it may result in a consistent and up-to-date model of subsurface reality according to the state of knowledge, with a high-predictive power for practical purposes.

It is stored in a databank for application such as the execution of updated cartographic editions of maps and sections. One application of such models is the construction of geological maps by intersecting the assembly of geological surfaces with a terrain model and testing the resulting pattern against field and subsurface data (Siehl and others, 1992).

Such models consequently lead to the second step of inversion: the kinematic backward modeling to access paleogeological states, rearranging and smoothing out the deformed and disrupted pile of strata. Actual three-dimensional backward modeling of sedimentary basin evolution is now within reach, true to scale and evolution rates, and can provide details of sedimentation, burial, compaction, and subsequent structural development.

DESIGN OF THE SYSTEM

An interactive computer system for modeling geological geometries in 3D (Fig. 2) requires the following essential functionalities:

- storage and retrieval of large 3D data sets with their topological structure,
- generation of irregular triangular and tetrahedral meshes,
- interactive visualization and graphic edition in 3D,
- intersection algorithms for 3D objects,
- 3D interpolation algorithms,
- numerical evaluation of model properties.

Many of these features are realized in a variety of available programs. However, there exists no single 3D-GIS which completely meets the fundamental requirements for adequate geometric modeling of geologic structures in 3D. Procedures of database management, interactive visualization, methods for mesh generation, and 3D interpolation are rapidly developing. To be able to adapt the appropriate software product in each situation, an open system with a programmable user interface was created which

- establishes a problem-oriented user interface with graphics support of database management,
- ties different programs together,
- manages data exchange, and
- provides an interface to implement algorithms for new and specific requirements.

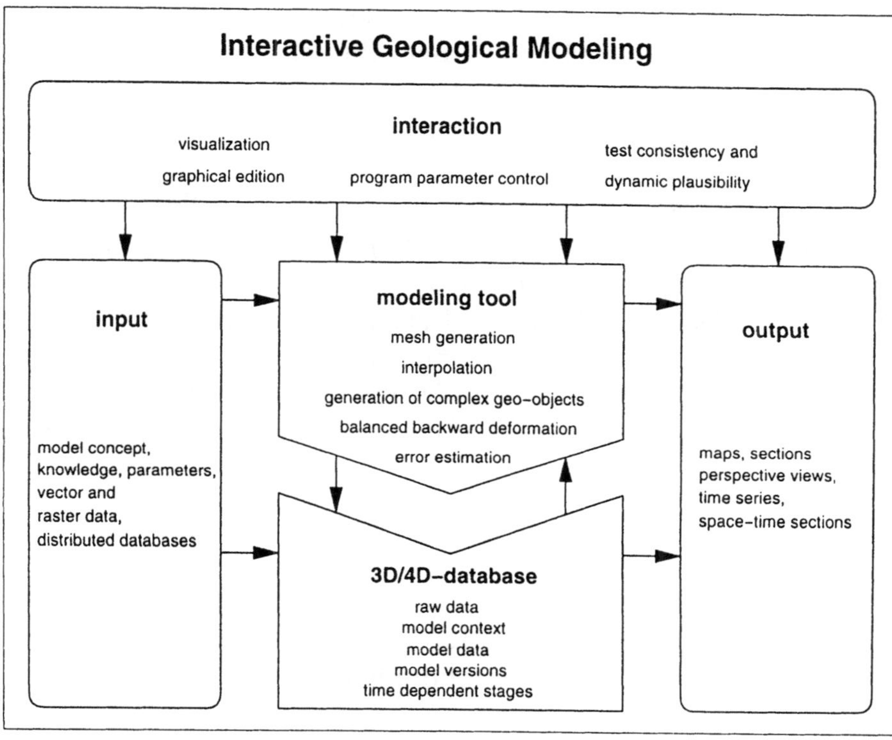

Figure 2. Interactive geological modeling.

The object-oriented 3D graphical programming system GRAPE (GRAPE, 1993) was used to build the framework (Fig. 3). The system kernel provides the possibility of integration of user defined projects with specific functionalities. Our GEOCON project (Alms, Klesper, Siehl, 1993) allows the utilization of other software systems such as GOCAD and GMP. In combination with powerful visualization facilities, newly developed modules such as a 3D cursor, interactive stereo mode, and volume discretization based on tetrahedrons increase its usability. Interfaces to all 3D modeling software tools used are implemented to hide data exchange behind the user interface.

As in our situation, most of the databanks are located on machines other than those with the modeling tools. Remote procedure calls (RPC) can be used to simplify and accelerate the online data exchange between local and remote programs. A RPC-connection is established by a calling sequence from the modeling tool to the remote database. The window of the database-management system will be displayed on the local machine where the data are selected by the user. The selected data are then sent to the local modeling tool for further processing. Such connections enable direct data exchange between the running programs without the creation and storage of data files.

For the database management of 3D objects, the ONTOS database-management system (ONTOS DB, 1992) is used. GEOSTORE (Bode, Breunig, and Cremers, 1994) is a tool kit that serves as an interface for 3D geological data to ONTOS and offers specific functions for geological data. A special GIS functionality is implemented to guide the formulation of spatial oriented retrieval queries by the visualization of the database content, for example the position of cross sections and fault lines in map

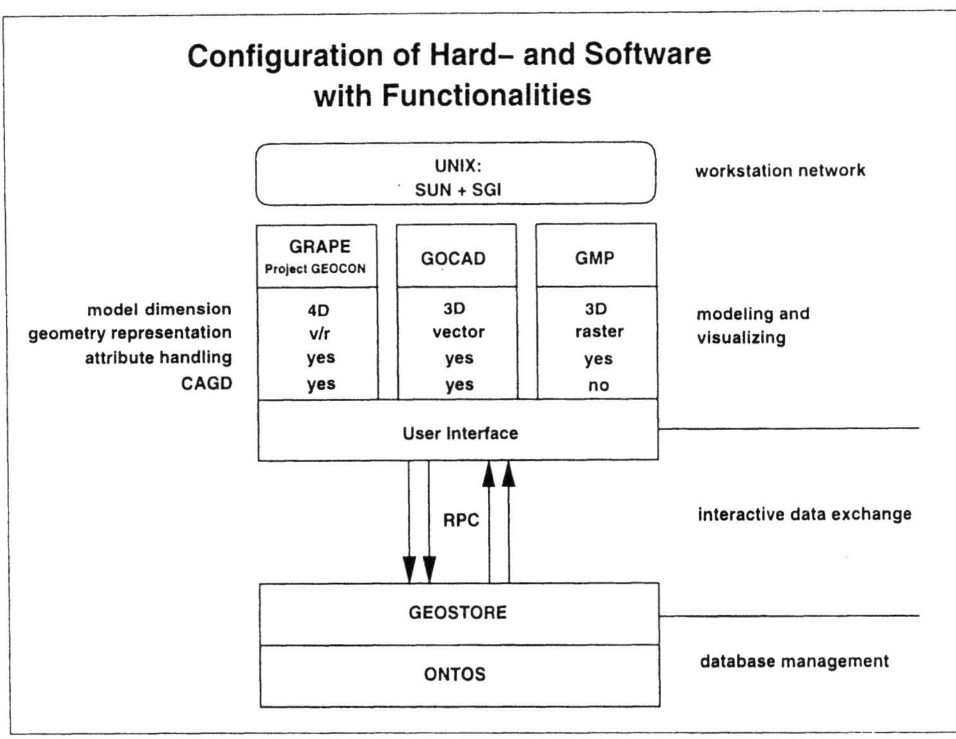

Figure 3. Configuration of hardware and software with functionalities.

view. It can provide cartographic orientation of the stored data and facilitates the selection of special objects. The user may obtain more information by selecting visualized objects on the screen and browsing deeper into the data structure. A specific geologic function is the possibility of defining any cross section through the three-dimensional data structure stored in the database and their visualization. Replacing objects in the database is controlled by a set of user-defined integrity constraints to maintain the consistency of the database.

PROCEDURES FOR SURFACE AND VOLUME MODELING

Models based exclusively on regular grids and on conventional CAD methods are not suitable for modeling the geometry of complex geological objects (Mallet, 1993). For mesh generation and interactive surface design, the program GOCAD (GOCAD report, 1993) is integrated in a manner as described. GOCAD is a program especially devoted to geological requirements using the Discrete Smooth Interpolation (DSI) method to fit surfaces interactively to points which represent complex structures. By this method surfaces are modeled as sets of triangular facets, minimizing a global roughness criterion. Local structures are modeled introducing constraints, for example fixing nodes, directing intersection lines, vectorial linking of surface nodes to other objects. This enables the user to include his/her knowledge and ideas immediately in the interpolation process and to visualize and analyze the results directly in 3D. Space-filling tetrahedrons correspond to the flexibility of irregular triangulation and provide the discretization for subsequent application of finite element methods for backward modeling of deformation.

The volume modeling program GMP (GEOCALC/GEODRAW, 1992) is used to calculate the distribution of attributes through space. The volume is discretized with a regular 3D grid using 2D grids to represent stratigraphical boundaries. Attribute values are interpolated with a "minimum tension" algorithm at the nodes of the grid. Defining the cutoff level for attribute ranges, the distribution can be displayed as isosurfaces. More complex structures have to be modeled separately by different volumes which must be merged after calculation. Three-dimensional grid spacing must be equal for all different volumes building one model. Description of shapes and internal structures of the volumes cannot be done by the 3D grid discretization itself. Additional surfaces for bounding and clipping are necessary to define parts of the volume.

GEOLOGICAL SETTING

We have developed our system in conjunction with the geological modeling of actual examples from the Lower Rhine Basin (LRB). The LRB is situated in NW Germany and the Netherlands and consists of a large-scale Cenozoic graben system filled with Oligocene to Quaternary sediments. In this location the LRB links the Upper Rhine Graben southwest of the Rhenish Massif and the North Sea Rift (Zagwijn, 1989). It can be regarded as the northwestern branch of an intraplate rift system with a triple junction located in the Mainz Basin (Ahorner, 1975; Illies, 1977) (Fig. 4).

The basement of the LRB is part of the Central European Variscan foldbelt and consists of the same Devonian and Carboniferous rocks as the surrounding Rhenish Massif. Permian and Mesozoic sediments of low thickness are known from drillholes and some outcrops at its southwestern boundary in the Eifel Mountains. Graben subsidence started in Middle Oligocene and continues today. The deepest parts of the LRB subsided more than 1200 m below present sealevel, while the adjacent Rhenish Massif was uplifted to several hundreds of meters (Fig. 5). Extension tectonics with a slight dextral wrench component created a characteristic block fault pattern striking in general NW-SE and dipping 55° to 65° to the west, whereas the strata are dipping eastwards.

Considerable deposits of lignite occur in the deltaic to fluvio-lacustrine series of Miocene age, overlain by the gravel and sand of braided river systems. Extensive exploration by drilling as well as by the mining of brown coal have provided an excellent three-dimensional database for the reconstruction of basin development.

EXAMPLE AND DATABASE

The database are lithological cross sections (horizontal scale 1:10 000, vertical scale 1:2 000) from the Rheinbraun AG mining company. The general orientation of the sections is SW-NE with a parallel distance between 1.5 to 4.5 km (Fig. 6).

We have used these interpreted cross sections and not the original well data because the primary data are extremely heterogeneous in resolution and quality. Rigorous interpretation by the Rheinbraun mine geologists, who also provided background knowledge not available in borehole logs, has been essential in providing a consistent database for 3D modeling.

Before digitizing, the sections must be inspected carefully and reinterpreted. This involves testing selected boundaries between strata in respect to their consistency as well as generalizations where horizons pinch out or where their assignment is not clear. A complete encoded description of the topological interrelations is attached to each data point before the set is stored in ONTOS via GEOSTORE.

The plausibility of the digitized geometry can be tested by visualizing the vectors as polygons and comparing them with the original cross sections. Additionally an automatic check of the consistency in regard to its explicit geometric accuracy relative to the internally defined data model is done by GEOSTORE using the topological information attached.

SURFACE REPRESENTATION TECHNIQUES

To represent surfaces, irregular triangulation meshes are used. As mentioned previously they provide a better fit to the irregular boundaries of geological objects than rectangular grids or voxel discretization.

Depending on the data distribution and the nature of the surfaces, two alternatives in designing the meshes were considered:

- a direct triangulation, in which each data point represents exactly a single node of the mesh, or
- an indirect meshing with the construction of the net on the basis of a convex hull around the data points; the triangulation then is fitted to the points.

With the given irregular data distribution in clusters along the section lines, direct meshing will evoke extremely adverse mesh geometry which will produce artifacts during further processing. For this reason indirect meshing was applied which yielded a more regular net geometry. The initial triangulation mesh is fitted closely to the points using the 3D interpolation algorithm DSI of GOCAD. Special constraints while modifying the triangulations have been applied to match the mesh exactly to the designer's ideas, for example:

- fixing of areas for local modification,
- minimizing distances between surface and data point, or
- directing intersection lines along faults.

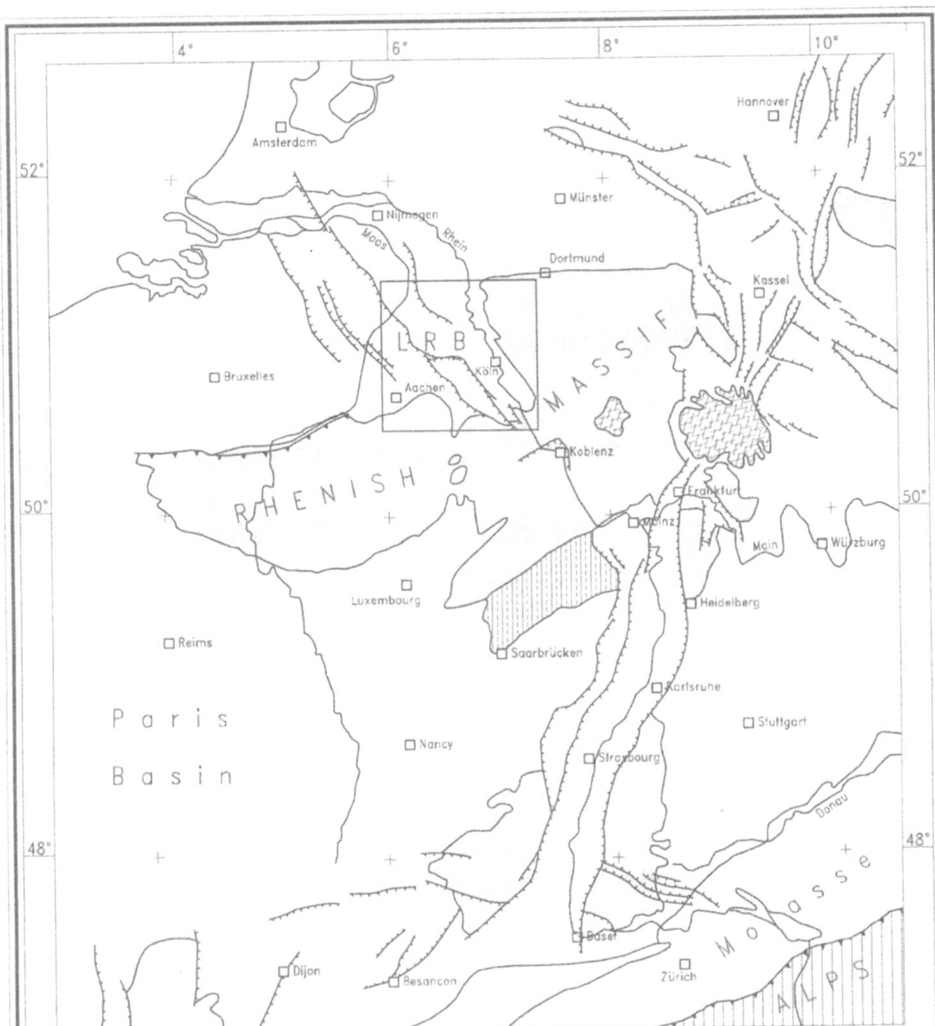

Figure 4. Location of Lower Rhine Basin (LRB) within generalized geological framework. Marked area is enlarged in Figure 6.

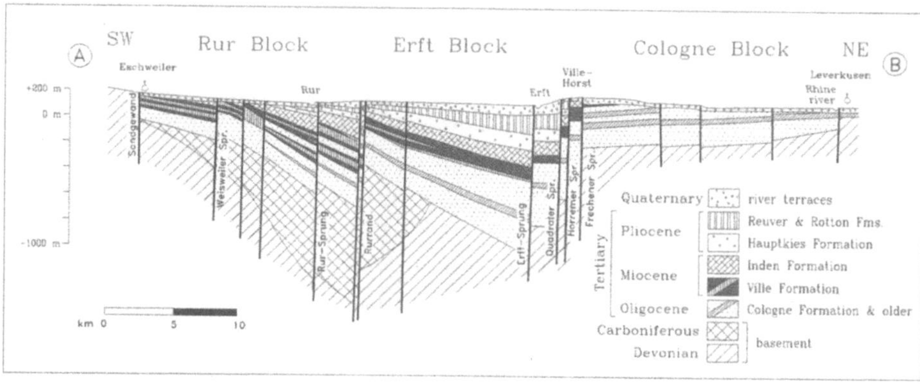

Figure 5. Generalized section trending SW-NE through center of Lower Rhine Basin, for location of A and B see Figure 6.

There generally are two approaches in modeling faults as discontinuities of bedding planes:

- creation of one coherent triangulation net which is deformed to a sharp edge along the intersection lines of faults and strata, or
- modeling fault and stratigraphic surfaces separately as independent objects which are related to each other by special operations (for example intersections) or constraints (for example movement vectors).

For the mere visualization of surfaces, the first method is adequate. However, for detailed structural analysis or for the intended 3D balanced backward modeling, the second method must be used. Here the geometrical elements can be assigned to specific geological events, and it is possible to experiment with alternative hypotheses of structural development, the most common way in which geological models are formed.

The construction of a set of faults and stratigraphic surfaces shown in Figure 7 will be demonstrated by some examples.

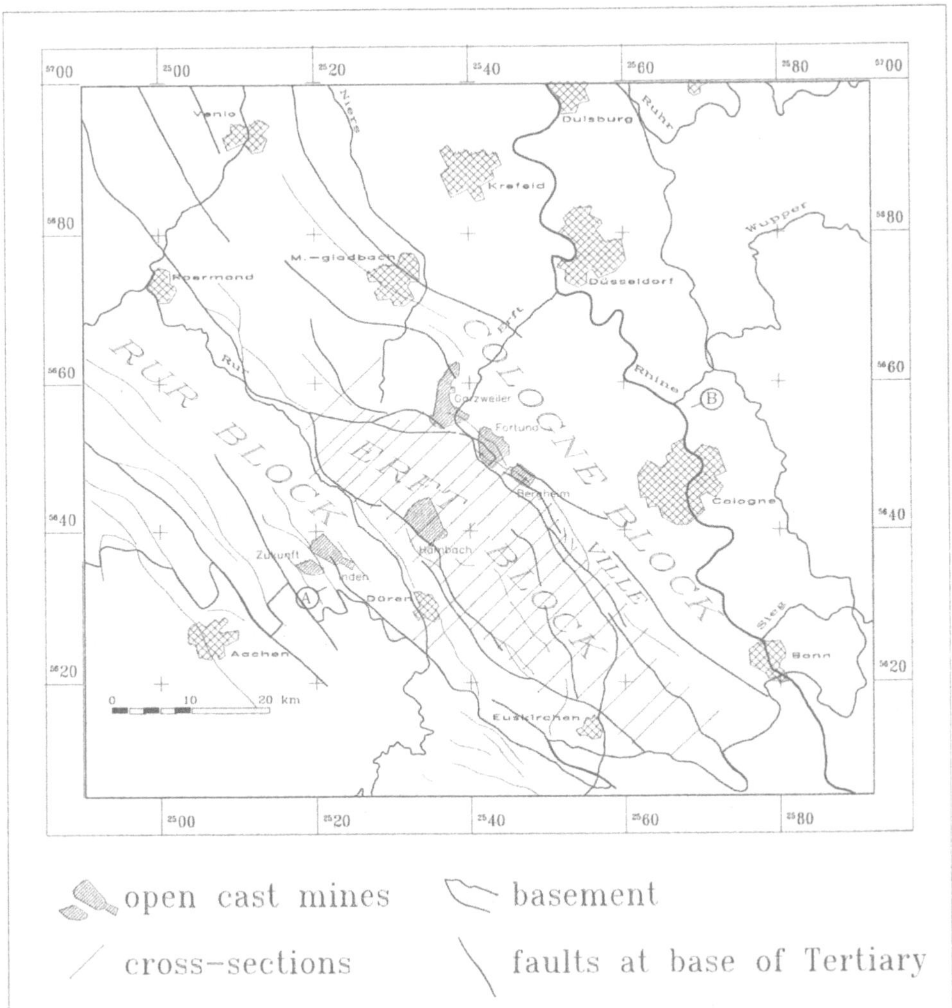

Figure 6. Area under investigation is located in southeastern part of Lower Rhine Basin. Thirty-eight cross sections (parallel lines) represent database of model. Section shown in Figure 5 is marked with A and B.

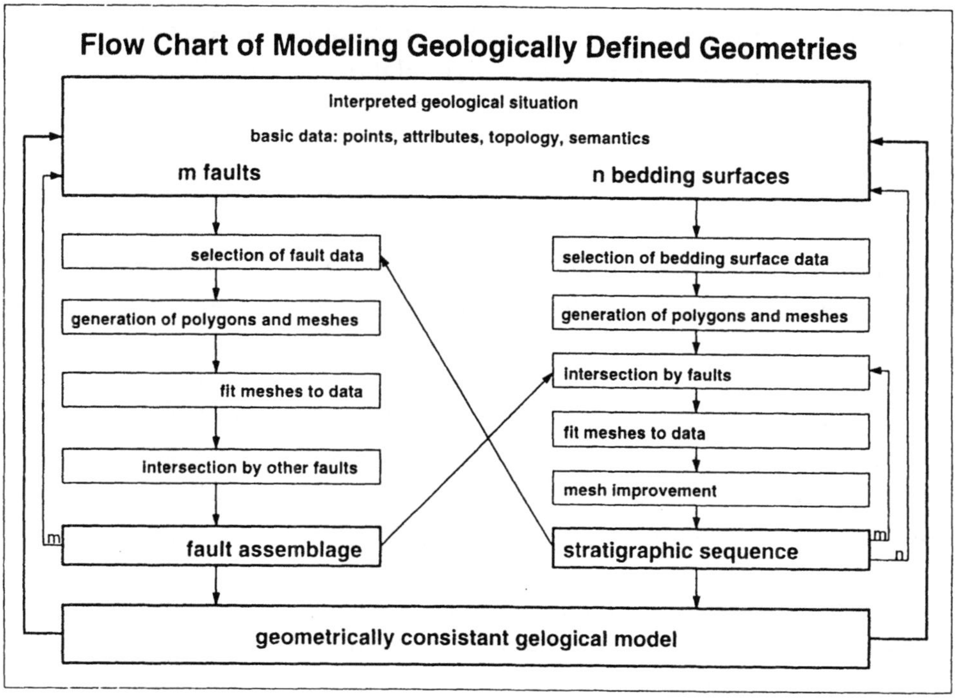

Figure 7. Construction of a complex fault and strata assemblage.

FAULT SURFACES

As all faults are younger than the bedding planes which they intersect and the general information of the fault position is given by the database, the construction of the surface assemblage starts with fault modeling. To construct a fault surface, the respective data points will be selected with GEOSTORE from the database, followed by automatic triangulation with GOCAD based on a convex hull around the selected points (Fig. 8).

Design of this convex hull depends on the data distribution and the 3D orientation of the surface to be constructed. A virtual plane can be defined by the user through a normal vector or by determining the surface with a minimum sum square distance to all points. The bounding polygon then is seen as the enclosure of the projection of the data points on the plane.

After construction of each fault, the intersection of faults with each other must be considered. With increasing complexity of the fault pattern, interactive control by the designer becomes more and more essential.

After numerical checks of the approximation quality and a visual comparison with known fault patterns, for example from isovalue maps, the fault surfaces can be stored as triangulation nets in ONTOS.

STRATIGRAPHIC SURFACES

The construction of stratigraphic surfaces is based on a convex hull around the data points. In this situation the initial plane defining the convex hull normally has subhorizontal orientation. The mesh is intersected by the previously constructed faults and is refined subsequently in several stages. Improvement of the approximation quality of the intersected stratum plane to the data points will be accomplished with DSI under consideration of several constraints. In particular the intersection lines of the layer must be fitted to the corresponding fault surfaces (Fig. 9).

These processes normally can be done automatically, but with increasing complexity of the pattern, it again is necessary to check and to correct the mesh interactively.

If the stratigraphic surfaces are more or less flat and nearly parallel, and the faults are subvertical and not too complex in their intersection pattern, a faster and simpler procedure for designing the stratigraphic layers can be applied. In this situation, a reference surface will be generated as a plane covering the maximum extension of the vertical projection of all other surfaces. This surface is intersected with the faults and fitted to the data points of one horizon in the center of the sequence. It then will be duplicated, vertically shifted to the levels of all other horizons, and fitted to their respective data points. The stratigraphic surface is finished interactively by adding new faults or clipping the surfaces according to the extension of the data supporting points.

ANALYZING THE MODEL

The practical investigation and rigorous testing of modeled 3D-surface structures normally is done, for example, by 2D visualiziation of isovalue maps with fault patterns (Fig. 10).

Two-dimensional isovalue maps provide direct comparison with conventional maps and sections. Improvement and updating of the

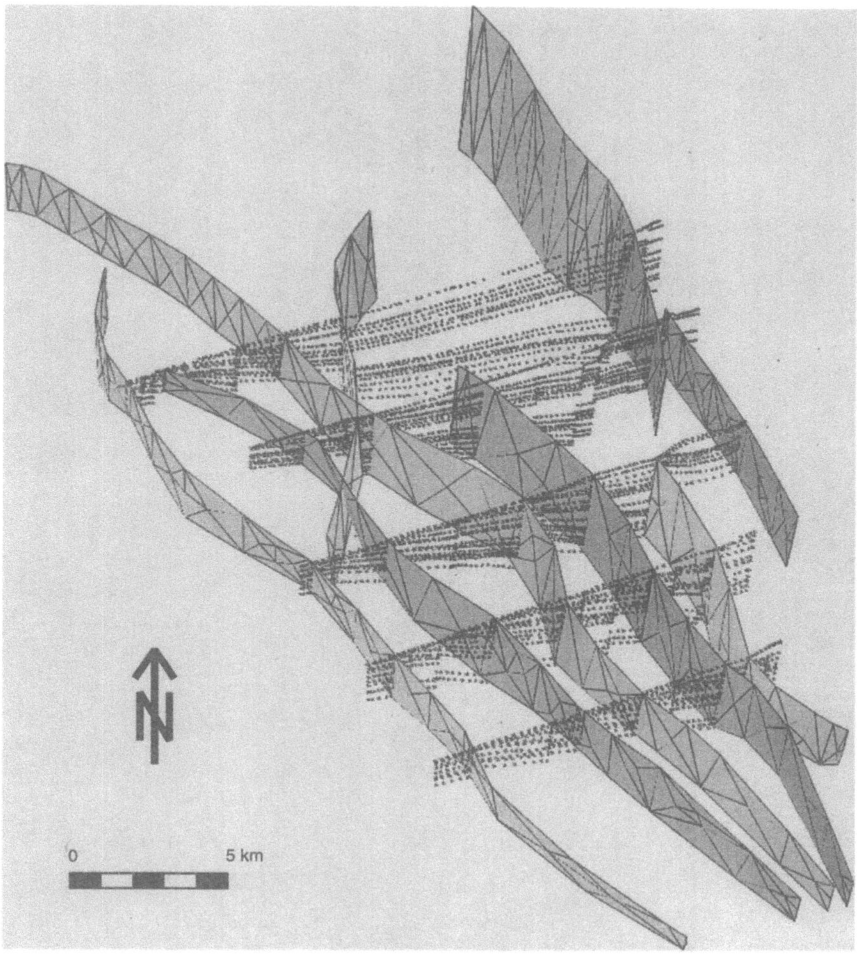

Figure 8. Digitized cross sections in southern part of Erft Block with triangulated fault surfaces.

computer model is continued until its consistency is satisfactory relative to the supporting data and the given assigment.

An important component of analyzing the model and model evaluation is interactive 3D visualization. The combination of different functionalities, such as translation, rotation, zooming, transparency, colors,

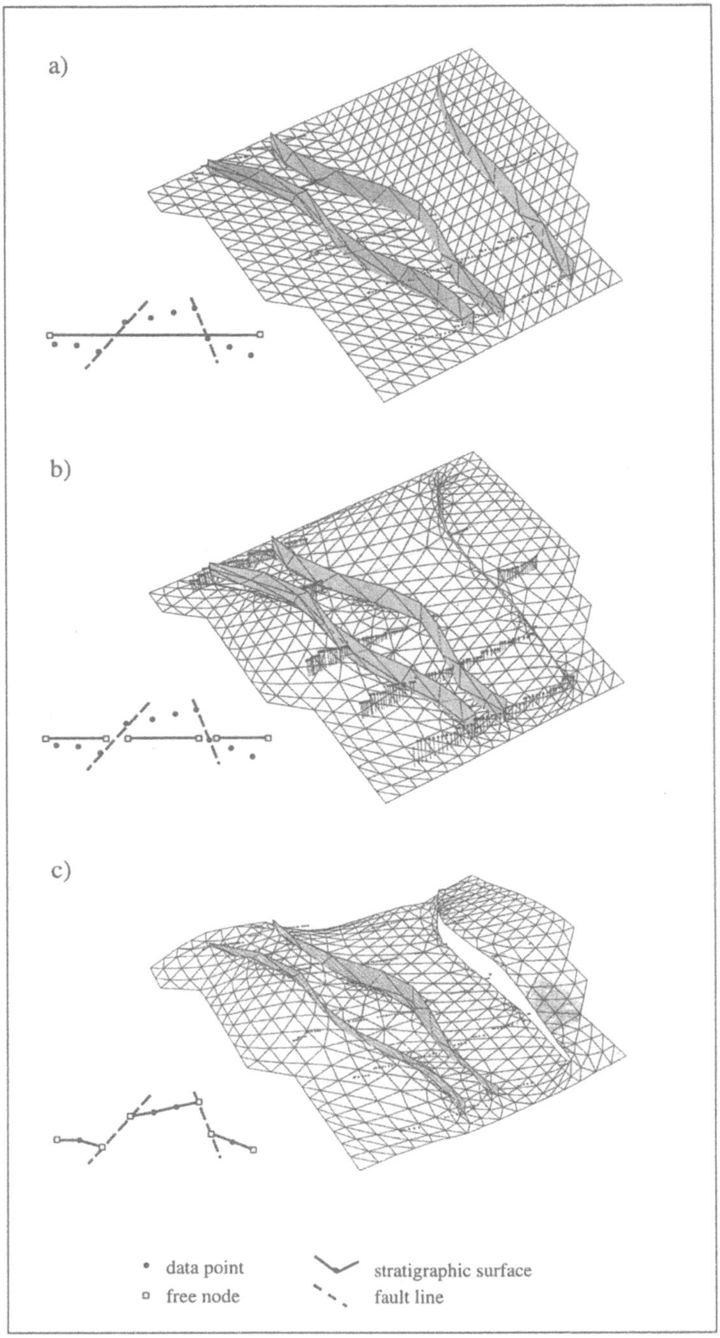

Figure 9. Successive steps in fitting stratigraphic surface dissected by three faults to data points in oblique view and in schematic sections on left.

and lighting, or animated stereo views, allows the inspection and understanding of complex 3D objects and derived models. Some examples of 3D visualizations are shown in Figures 11 and 12.

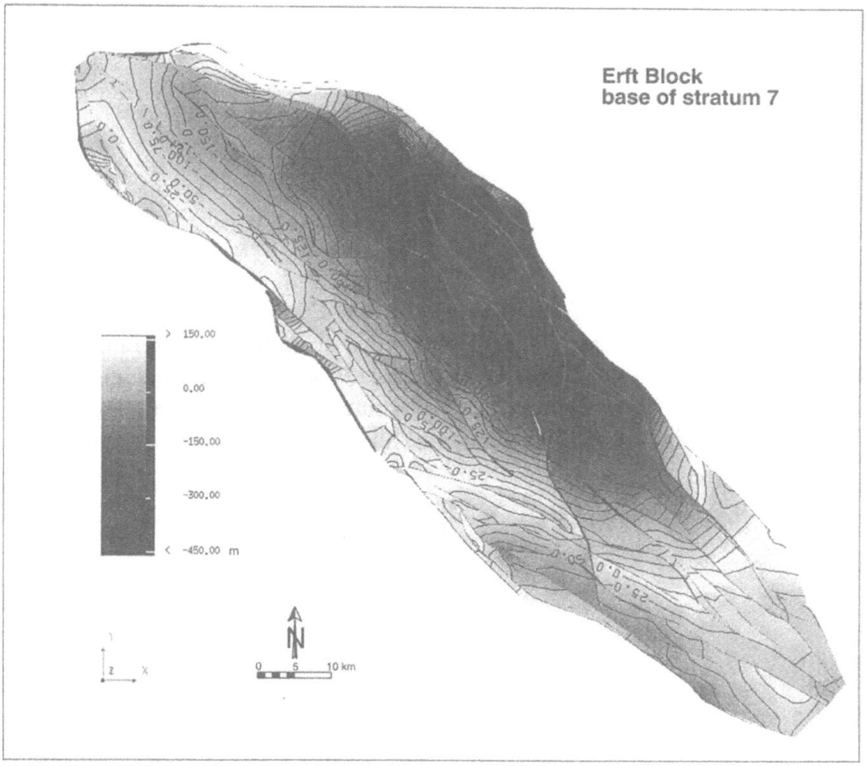

Figure 10. Depth line contour map of base of stratum 7.

DISCUSSION

Computer-aided geometric modeling in 3D is not yet usually employed in geology because of the following reasons:

- relevant data are inaccessible to electronic processing.
- data exchange between 3D modeling systems is not properly supported.
- much interactive input is needed to model complex geologic structures.

Figure 11. Oblique view from NW into Lower Rhine Basin. Lower layer represents base of Tertiary according to Hager (1988) with structural units "Krefeld Block," "Venlo Block," and "Rur Block." Upper layer represents digital terrain model with horizontal resolution of 50 x 50 m. Deepest parts of Rhine River Valley are about 30 m above sealevel, highest of Eifel mountains about 200 m.

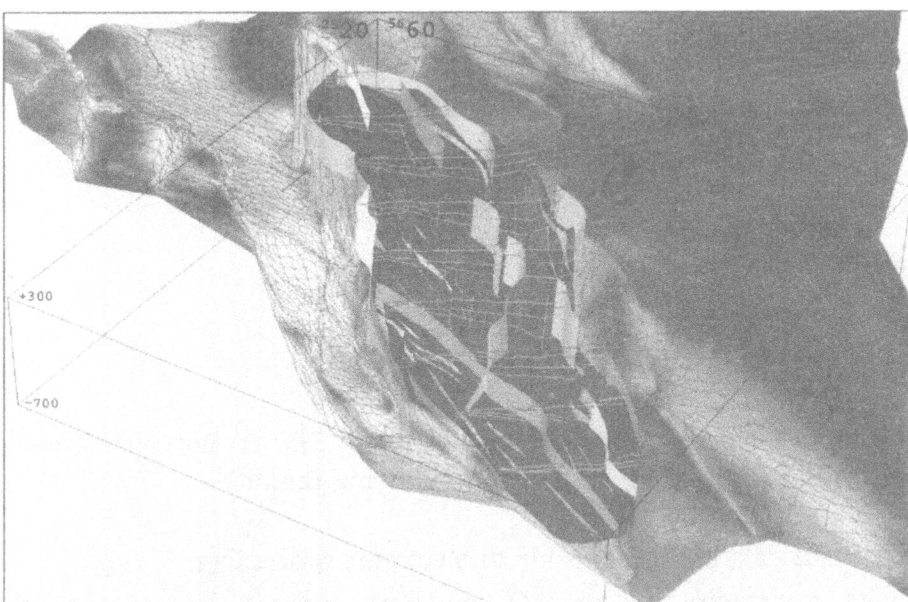

Figure 12. View from SW to model space. Gray base of Tertiary with triangular mesh. In area of "Erft Block" some of strata and faults with cross sections are modeled. Upper surface indicates Pliocene "Hauptkies-Formation" (Fig. 5).

- integration of expert knowledge still is done by direct manual interaction. This may introduce an unwelcome subjective component.
- geometric models are often too simple in relation to complex reality.
- methods of evaluation and error estimation of geometrical models are still rudimentary.

To model complex geological structures in 3D effectively, it is desirable to represent the geometrical and topological relationships as exactly and completely as possible in the data structure managed by the database system. Only then can local modeling and updating of parts of the structure be performed maintaining both the consistency of the entire model and the integrity of the database. This requires the design of an open, modular system with modeling tools interoperating closely with databases. Different functionalities implemented in distributed modeling systems can be tied together by an interface which gives online graphical control of the modeling process. Interactive data exchange for raw data and model data must also be managed if the systems and databases are on different machines.

Geometric modeling is normally done for a particular region, depending on the problem to be solved and the available data. In the course of further treatment, combination and merging of such models might be of great importance, but is seldom possible. It raises the problem of appropriate up and downscaling involving the subsequent refinement of structures and integration of additional elements, on one hand, as well as the generalization and omission of details on the other.

A critical point in 3D modeling is the recognition of objectives and the purpose of a particular model. Data models, as defined in Figure 1, demand high spatial resolution and close-to-nature complexity, because such models may be made to solve practical problems in predicting subsurface structures for exploration, mining, or geotechnical purposes. When bringing in kinematic or even dynamic components into the modeling, spatial resolution and complexity normally is reduced. In these situations we may be merely in the state to model simplified structures in order to understand basic principles of interacting processes and geological evolution.

The excellent quality of available graphical rendering sometimes may conceal the fact that in the individual case we know very little about accuracy and precision of our models.

ACKNOWLEDGMENTS

We would like to express our appreciation to the Rheinbraun AG mining company for providing the database of the Erft Block. Substantial financial support was received by the German Science Foundation (DFG) within the framework of the special research project "SFB 350: Wechselwirkungen kontinentaler Stoffsysteme und ihre Modellierung".

REFERENCES

Ahorner, L., 1975, Present-day stress field and seismotectonic block movements along major fault zones in Central Europe: Tectonophysics, v. 29, p. 233-249.

Alms, R., Klesper, C., and Siehl, A., 1994, Geometrische Modellierung und Datenbank-entwicklung für dreidimensionale Objekte: Nachrichten aus dem Karten- und Vermessungswesen, Reihe I, v. 111, p 9-16.

Bode, T., Breunig, M., and Cremers, A.B. 1994, First experiments with GEOSTORE, an information system for geologically defined geometries: submitted to Proc. Workshop on Advanced Research in Geographic Information Systems (Monte Verita, Ascona).

GEOCALC/GEODRAW, 1992, Geologic modeling program, DYNAMIC GRAPHICS, INC., rel 5.0.

GOCAD Report, 1993, Association Scientifique pour la Géologie et ses Applications, vers. 9.3, 1, p. 1-231; 2, p. 233-487.

GRAPE, 1993, GRAPE - A Graphical programming environment for mathematical Problems: Universität Bonn, SFB 256; Universität Freiburg, Inst. f. Angew. Mathematik, vers. 4.0.

Hager, H., 1988, Die Unterfläche der tertiärzeitlichen Schichten: Geologie am Niederrhein, GLA Krefeld, p. 37.

Illies, J.H., 1977, Ancient and recent rifting in the Rhinegraben: Geol. en Mijnbouw, v. 56, p. 329-350.

Mallet, J.L., 1992, GOCAD - a computer aided design program for geological applications: three-dimensional modeling with Geoscientific Information Systems: NATO ASI, C 354, p. 123-142.

ONTOS DB, 1992, User Guide (Release 2.2) ONTOS INC. (Feb. 1992).

Siehl, A., 1990: Geological maps by interactive spatial modeling: Three-dimensional computer graphics in modeling geologic structures and simulating geologic processes, Freiburger geowissenschaftliche Beiträge, v. 2, p. 95-96.

Siehl, A, 1993, Interaktive geometrische Modellierung geologischer Flächen und Körper: Die Geowissenschaften, v. 11, p. 342-346.

Siehl, A., Rüber, O., Valdivia-Manchego, M., and Klaff, J., 1992, Geological maps derived from interactive spatial modeling, *in* Digital map series in geosciences to geo-information systems: Geol. Jb., A122, p. 273-289.

Zagwijn, W.H., 1989, The Netherlands during the Tertiary and Quaternary: Proc. Symp. on Coastal lowlands; geology and geotechnology (Univ. Utrecht): Kluwer Acad. Publ., Dordrecht, p. 107-120.

VOLUMETRICS AND RENDERING OF GEOLOGIC BODIES BY THREE-DIMENSIONAL GEOMETRIC RECONSTRUCTION FROM CROSS SECTIONS OR CONTOUR LINES

Helmut Schaeben, Stephan Auerbach, and Esther U. Schütz
Universität Trier, Trier, Germany

ABSTRACT

In applied earth sciences a three-dimensional object may be given by a set of parallel cross sections. More specifically, the body is given by a set of contour lines, which may represent the intersection of its boundary surface(s) with the plane of the cross section thus distinguishing the body's interior and exterior, or a given value of some property of the body. Graphically displaying the body's geometry and topology, that is its internal structure in terms of contour lines discriminating the body's interior and exterior, or quantifying some of its features, for example arbitrary intersections or volumes corresponding to given contours, essentially requires to reconstruct the corresponding boundary surfaces of the three-dimensional body.

A computer-aided method of geometric reconstruction of closed three-dimensional bodies from their cross sections or contour lines, respectively, is presented, the essentials of which are implicit representation of surfaces and bodies, and set-valued interpolation. This approach suggests to approximate the interior of a three-dimensional body as an entity and thus allows the recovery of bodies of arbitrary shape and topology. In particular, a cross section may comprise more than one closed contour line.

Geologic Modeling and Mapping
Edited by A. Förster and D. F. Merriam, Plenum Press, New York, 1996

Set-valued interpolation is performed numerically by consecutively solving a sequence of univariate interpolation problems. These problems are solved by applying quadratic splines of class C^1 and employing their favorable properties of local support, smoothness, accuracy, and preservation of monotonicity.

The method is automatic and does not require any interaction by the user. Its capability to reconstruct complex geologic bodies from their contour lines for rendering and volumetrics is exemplified using synthetic data.

INTRODUCTION

This contribution is concerned with the problem of reconstructing a three-dimensional body (cf. Auerbach and Schaeben, 1990a, 1990b) in terms of its outer and inner boundary surfaces given by a collection of planar contours representing cross sections through the body. This problem has important applications beyond geosciences. The suggested method allows immediate applications to computer-aided graphical display or volumetrics once the body is reconstructed. It applies to bodies of arbitrary shape and topology and thus to complex geologic bodies. With respect to its range of applications it compares favorably with methods based on spatial triangulation (cf. Meyers, Skinner, and Sloan, 1992).

Rendering folded and faulted geological surfaces from series of cross sections is described in Klein, Pflug, and Ramshorn (1989). They apply interactive, that is user-controlled trial and error, spatial triangulation between digitized geological cross sections and corresponding interpolation algorithms to accomplish visual display of complex geological surfaces.

The review on computing volumes (Walsh and Brown, 1992) is summarized as follows:

* step method (contour areas × change in thickness): severely underestimates
* trapezoid rule (first-order Newton-Cotes numerical integration): slightly overestimates
* pyramid rule: slightly underestimates
* combination of trapezoid and pyramid rule

* Simpson's rule (second-order Newton-Cotes numerical integration, three terms): requires evenly spaced contours and an even number of contours

* 3/8 rule (second-order Newton-Cotes numerical integration, four terms): requires evenly spaced contours and the number of contours be a multiple of 3.

For details of numerical integration the reader is referred to (Press and others, 1990).

SET-VALUED INTERPOLATION

Let $B \subset IR^3$ be a closed and bounded body in the three-dimensional Euclidian space IR^3. Its intersections S_i, $i = 0,..., n + 1$, with $n + 2$ parallel planes, which are labeled by levels z_i according to a suitably selected coordinate system are defined as

$$B(z_i) = S_i = \{(x, y, z) \in B \mid z = z_i\} \subset B \subset I\!R^3, \quad i = 0, \ldots, n+1 \tag{1}$$

These cross sections are closed and bounded; however, it is not required that they are simply topologically connected. It is assumed that $S_0 = S_{n+1} = \emptyset$ and $S_i \neq \emptyset$, $i = 1,..., n$. The boundaries $\Gamma_i = \Gamma(S_i)$ of S_i, here termed contour lines, are assumed to be given as polygons, for example polygons which may be thought of as representing digitally recorded contouring lines of a map.

The problem is to reconstruct the body B from the contours $\Gamma(S_i)$, that is to determine a body $\hat{B} \subset IR^3$, which interpolates the cross sections for each level z_i, $\hat{B}(z_i) = S_i$, $i = 0,..., n + 1$. Because these interpolation conditions involve sets rather than numbers, the suggested method may be considered as "set-valued interpolation".

The following implicit representation of the body B is used. Let $d(\text{o,o})$ denote the usual Euclidian distance in the plane IR^2

$$d[(x, y, z), (x', y', z)] = \sqrt{(x - x')^2 + (y - y')^2} \tag{2}$$

Then, the distance of a point (x,y,z_i) to any nonempty set S_i corresponding to level z_i is defined by

$$d[(x,y,z_i),S_i] = \inf_{(x',y',z_i)\in S_i} d[(x,y,z_i),(x',y',z_i)] \tag{3}$$

For a given point $(x,y) \in IR^2$ and for every cross section level z_i the distance d_i of the point $(x,y,z_i) \in S_i^3$ to the contour $\Gamma_i = \Gamma(S_i)$ is given by

$$d_i(x,y) = \begin{cases} d((x,y,z_i),S_i^c) \text{ if } (x,y,z_i) \in S_i \\ -d((x,y,z_i),S_i) \text{ if } (x,y,z_i) \in S_i^c \end{cases} \quad i = 1,\ldots,n \tag{4}$$

$$= \begin{cases} d((x,y,z_i),\Gamma S_i) \text{ if } (x,y,z_i) \in S_i \\ -d((x,y,z_i),\Gamma S_i) \text{ if } (x,y,z_i) \in S_i^c \end{cases} \quad i = 1,\ldots,n \tag{5}$$

where S_i^c is the complement and $\Gamma(S_i)$ the boundary of S_i. $d_i(x,y)$ is well defined for $i = 1,\ldots, n$ and distinguishes the interior and exterior of S_i just by its sign. Because $S_0 = S_{n+1} = \emptyset$, $d_0(x,y)$ and $d_{n+1}(x,y)$ are not defined initially, they have to be set equal to a constant appropriatly chosen according to some heuristics, or determined according to some more sophisticated methods.

The sign of $d_i(x,y)$ indicates whether (x,y,z_i) is a point inside or outside the body B at intersection S_i, $d_i = 0$ if $(x,y,z_i) \in \Gamma(S_i)$.

The point of view of the three-dimensional reconstruction problem is changed now from the "horizontal" intersection planes to "vertical" lines in the following manner (see Fig. 1). For a given point $(x,y) \in IR^2$ the sequence $\{z_i, d_i(x,y)\}_{i=1,\ldots,n}$ may be thought of as data sampled along lines $(x,y,z_i) \in S_i$ through the body B orthogonal to its planes of intersection. Every sequence of data gives rise to a univariate interpolation problem. The univariate function $g(\circ;x,y)$ is a solution of an individual interpolation problem if

$$g(z_i;x,y) = d_i(x,y), \quad i = 1,\ldots,n \tag{6}$$

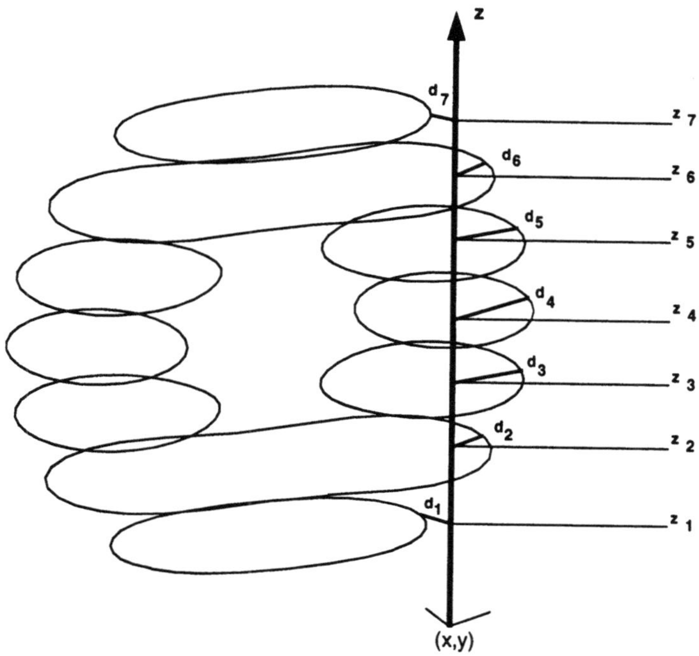

Figure 1. While two-dimensional view induced by given "horizontal" intersection planes at levels z_i is employed to determine $d_i(x,y)$, $(x,y) \in IR^2$, one-dimensional view along "vertical" lines (x_0,y_0,z), $z \in IR^1$ emphasizes to interpret $d_i(x,y)$, $i = 1,\ldots, n$, as given values read off univariate function $g(z;x_0,y_0)$: $g(z_i;x_0,y_0) = d_i(x_0,y_0)$ with parameters x_0,y_0.

Thus, there is an interpolant $g(\text{o};x,y)$ for any point $(x,y) \in IR^2$. These interpolants provide an implicit representation of the reconstructed body $\hat{B}$ by

$$(x,y,z) \in \hat{B} \text{ if and only if } g(z;x,y) \geq 0$$

$$(x,y,z) \notin \hat{B} \text{ if and only if } g(z;x,y) < 0 \qquad (7)$$

which can be employed easily for rendering. It is emphasized that the reconstructed body $\hat{B}$ depends only on the sign of the univariate interpolants $g(o,;x,y)$. The boundary $\Gamma(\hat{B})$ is represented by

$$\Gamma(\hat{B}) = \{(x,y,z) \in I\!R^3 \mid g(z;x,y) = 0\} \tag{8}$$

Thus, the initial three-dimensional problem has been reduced to a sequence of rather simple univariate problems.

By definition it is obvious that the reconstructed body $\hat{B}$ interpolates the cross sections S_i, that is $\hat{B}(z_i) = S_i$, $i = 1,\ldots, n$ (cf. Gmelig Meyling, 1988).

Moreover, the following theorem by Levin (1986) gives a convergence order for the distance between reconstructed and original body in terms of the cross-section distance h, dependent on the smoothness of the body's contours. It thus guarantees that $\hat{B}$ is a mathematically reasonable approximation of B. Let B be a closed body in the three-dimensional Euclidian space whose boundary $\Gamma(B)$ is composed of a finite number of surfaces, which are C^k -continuous and mutually disconnected. Let $z_i = z_0 + ih$, $0 \le i \le n + 1$, be the level of the cross sections and δ any positive constant. For each $h > 0$ consider any point (x^h, y^h, z^h) with $\min_{0\le i\le n} \mid z^h - z_i \mid \ge \delta$, which is a point of disagreement between B and the reconstructed $\hat{B}$

$$(x^h, y^h, z^h) \in B \cap \hat{B}^c \quad \text{or} \quad (x^h, y^h, z^h) \in B^c \cap \hat{B} \tag{9}$$

Then

$$d((x^h, y^h, z^h), \Gamma(B)) = O(h^{k+1}) \quad \text{as} \quad h \to 0 \tag{10}$$

For the theorem and its proof see Levin (1986).

VOLUMETRICS

An immediate application of set-valued interpolation is the estimation of the three-dimensional volume of B by the volume of the reconstructed body $\hat{B}$, which can be calculated according to

$$\begin{aligned}
\text{vol}_3(B) \sim \text{vol}_3(\hat{B}) &= \iiint_{\hat{B}} dz\,dx\,dy = \iiint 1_{\hat{B}}(x,y,z)\,dz\,dx\,dy \\
&= \iiint 1_{\{(x,y,z)\in \mathbf{R}^3 | g(z;x,y)\geq 0\}}\,dz\,dx\,dy \\
&= \iint \text{vol}_1(\{z \mid g(z;x,y) \geq 0\})\,dx\,dy \\
&= \iint l(x,y)\,dx\,dy \qquad (11)
\end{aligned}$$

Here $1_{\hat{B}}(x,y,z)$ denotes the indicator with respect to the set $\hat{B}$ and $l(x,y)$ denotes the total length of segments along the vertical line $\{(x,y,z) \mid z_0 \leq z \leq z_{n+1}\}$ for which $g(z;x,y) \geq 0$.

This estimate of the volume of the body B is supposed to be superior to the estimate

$$\text{vol}_3(\hat{B}) = \int \text{vol}_2(\{(x,y) \mid g(z;x,y) \geq 0\})\,dz \qquad (12)$$

provided by the step method (cf. Walsh and Brown, 1992).

THE INTERPOLANT g: UNIVARIATE HERMITE INTERPOLATION AND QUADRATIC C^1 SPLINES IN BEZIER FORM

As should be obvious, the reconstructed body $\hat{B}$ essentially depends on the univariate interpolants $g(\circ;x,y)$. More specifically, it sensitively depends on the roots of the interpolants $g(\circ;x,y)$, for which they change their sign. Therefore, the interpolants should not only satisfy the requirements in terms of local support, smoothness, and accuracy, but they also should be variation-diminishing and shape-preserving to yield good results.

Local support of the interpolant g implies that it only is affected locally by changes in the data. This feature provides serious numerical advantages when a cross section should be omitted or additionally included. The required smoothness of g should correspond to the assumed smoothness of the body to be reconstructed. For most practical applications it should be sufficient that g is continuously differentiable, that is $g \in C^1([z_0, z_{n+1}])$. According to Levin's theorem (1986) the scheme of set-valued interpolation is of second-order accuracy if the interpolant g is of second-order accuracy, that is if the interpolant exactly reproduces linear polynomials (including constants), then the set-valued interpolation exactly reproduces straight cylinders and cones with their axis of symmetry in $(x,y) = (0,0)$. The requirements of variation-diminishing and shape-preserving aim at an interpolant $g(z; o,o)$, which does not have more roots than the polygon is defined by $\{z_i, d_i(o,o)\}_{i=1,\ldots,n}$.

Furthermore, the interpolant should be monotone nondecreasing on intervals $[z_i,z_{i+1}]$ with $d_i \leq d_{i+1}$ and monotone nonincreasing on $[z_i,z_{i+1}]$ with $d_i \geq d_{i+1}$; this feature is referred to as comonotonicity with the data by Gmelig Meyling (1988). These requirements usually are met by additional constraints concerning the values of the derivative of the interpolant $g(z; o,o)$ at the data points $(z_i, d_i(o,o))$, $i = 1,\ldots, n$. An intuitively appealing and actually efficient choice for the slope estimates reflecting the shape of the original data is given by

$$d_i'(x,y) = \frac{d_{i+1}(x,y) - d_{i-1}(x,y)}{z_{i+1} - z_{i-1}}, \quad i = 1,\ldots,n \tag{13}$$

which in turn results in a Hermite interpolation problem

$$\begin{aligned} g(z_i; x,y) &= d_i(x,y), \quad i = 1,\ldots,n \\ g'(z_i; x,y) &= d_i'(x,y), \quad i = 1,\ldots,n \end{aligned} \tag{14}$$

Proper candidates for the interpolant g are splines, which may be thought of as piecewise polynomials and satisfy all requirements discussed previously. What springs to mind are quadratic splines because they provide a C^1 interpolant g and easy control of the sign of the first derivative, which is a piecewise linear function.

An early account of shape-preserving quadratic splines was given by Schumaker (1983). He suggested two-point Hermite interpolation on the interval $[z_i, z_{i+1}]$. However, a quadratic spline on $[z_i, z_{i+1}]$, which has only three parameters cannot generally solve the Hermite interpolation problem which requires to adjust four parameters. But a unique C^1 quadratic spline exists on $[z_i, z_{i+1}]$ with a simple knot at ζ_i, $z_i < \zeta_i < z_{i+1}$, which actually solves the Hermite interpolation problem (Gmelig Meyling, 1988). Usually, $\zeta_i = (z_i + z_{i+1})/2$ applies.

However, the splines provided by this constructive solution are not generally comonotone. Comonotonicity is satisfied only if the estimates of the first derivative $d_i'(o,o)$ have the proper sign and are sufficiently small in absolute value. Thus, the initial estimates d_i' provided by (13) usually need to be repeatedly improved (cf. Gmelig Meyling, 1988).

Here, a different construction of a quadratic spline satisfying all constraints, that is interpolating the given data and the estimated slopes, and being comonotone is suggested.

A quadratic spline may be represented by its Bezier control points (Farin, 1988). Thus, the basic idea is to construct the three Bezier control points for each interval $[z_i, z_{i+1}]$, $i = 0, \dots, n$. In the following, two cases have to be distinguished (see Fig. 2).

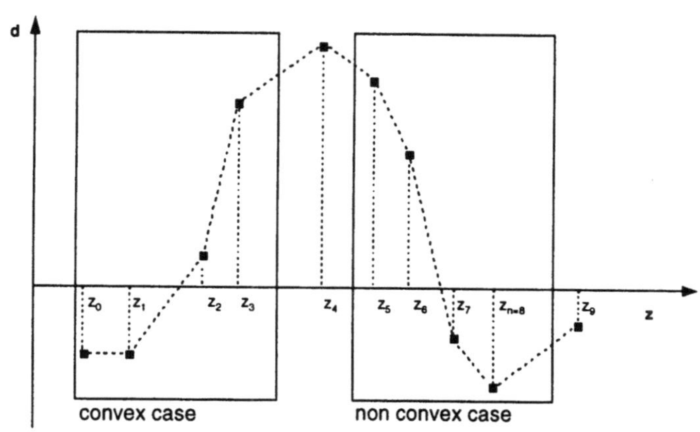

Figure 2. Univariate interpolation problem to determine function $g(z; x_0, y_0)$ interpolating given data $(z_i, d_i(x_0, y_0))$, $i = 0, \dots, n+1$, estimated slopes $d_i'(x_0, y_0)$, $i = 0, \dots, n+1$, and being comonotone; dashed line depicts polygon defined by $\{z_i, d_i(o,o)\}_{i=1,\dots,n}$.

Case (i) applies if the polygon $\{z_k, d_k(o,o)\}_{k=i-1,\dots,i+2}$ is convex, case (ii) applies otherwise.

Under the assumptions of case (i) the first and last Bezier control points $\mathbf{b}_0$, $\mathbf{b}_2$ are given by the two data points $\mathbf{b}_0 = (z_i, d_i(o,o))$ and $\mathbf{b}_2 = (z_{i+1}, d_{i+1}(o,o))$ themselves, the intermediate Bezier control point $\mathbf{b}_1$ is given as the point of intersection of the lines through $(z_i, d_i(o,o))$ with slope $d_i'(o,o)$ and $(z_{i+1}, d_{i+1}(o,o))$ with slope $d_{i+1}'(o,o)$ respectively (see Fig. 3).

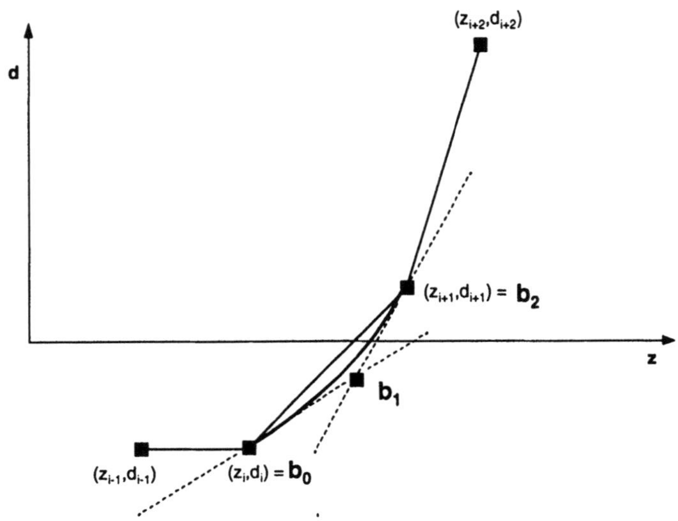

Figure 3. Geometrically constructive determination of intermediate Bezier control point in case (i) of convex polygon $\{z_k, d_k(o,o)\}_{k=i-1,\dots,i+1}$.

If ζ denotes the abscissa of this intermediate Bezier control point $\mathbf{b}_1$, then $z_i < \zeta < z_{i+1}$ holds.

Considering case (ii) this elementary geometric construction does not provide a useful intermediate Bezier control point because either the two lines do not intersect at all, or the point of intersection does not correspond to a spline function satisfying all required constraints. Therefore, in a first step of case (ii) an auxiliary point $\mathbf{b}_2 = \mathbf{b}_0'$ is defined by its abscissa $\zeta_i = (z_i + z_{i+1})/2$ and its ordinate

$\hat{d}(0,0) = (d_i(0,0)+d_{i+1}(0,0))/2$ provided by linear interpolation between $(z_i, d_i(0,0))$ and $(z_{i+1}, d_{i+1}(0,0))$. Further, the slope at this point has to be defined. Its value will control essentially the shape of the spline function. Any value larger (smaller) than the slope of the linear interpolant, that is the straight line through $(z_i,d_i(0,0))$ and $(z_{i+1}, d_{i+1}(0,0))$, if it is positive (negative), is a reasonable choice. Then, the intervals $[z_i, \zeta_i]$ and $[\zeta_i, z_{i+1}]$ both satisfy the assumptions of case (i). Therefore, in a second step, the procedure of case (i) is applied to each interval (see Fig. 4).

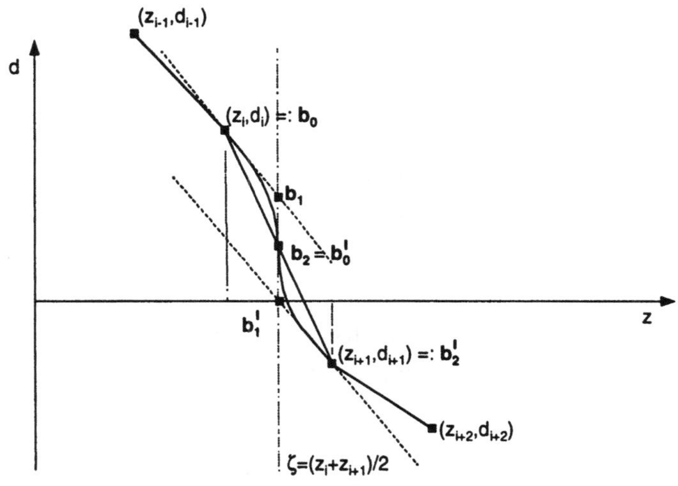

Figure 4. Geometrically constructive determination of intermediate Bezier control point in case (ii) of nonconvex polygon $\{z_k, d_k(0,0)\}_{k=i-1,\ldots,i+2}$ by applying case (i) to convex polygons $\{(z_{i-1}, d_{i-1}(0,0)), (z_i, d_i(0,0)), (\zeta_i, \hat{d}(0,0)), (z_{i+1}, d_{i+1}(0,0))\}$and $\{(z_i,d_i(0,0)), (\zeta_i,\hat{d}(0,0)), (z_{i+1}, d_{i+1}(0,0)), (z_{i+2}, d_{i+2}(0,0)\}$, respectively.

The quadratic spline thus defined on $[z_0, z_{n+1}]$ is C^1- continuous, co-monotone, and calculating its roots is straightforward.

As the boundary surface $\Gamma(\hat{B})$ is implicitly given by $g(z;x,y) = 0$, the roots of the interpolant $g(z;x,y)$ are required to render the body $\hat{B}$; more specifically, calculating the roots of the quadratic $g(z;x,y)$ on intervals $[z_i, z_{i+1}]$ with $g(z_i;x,y) \times g(z_{i+1};x,y) < 0$ is required.

NUMERICAL REALIZATION AND COMPUTATIONAL ASPECT

Gridding

Obviously, in any practical application the interpolant $g(z;x,y)$ can be evaluated only for some discrete points $(x_k,y_l) \in IR^2$. Therefore, a rectangle $[x_0, x_K] \times [y_0, y_L] = R \subset IR^2$ is defined such that $S_i \subset R$, $i = 1,\ldots,n$, and then gridpoints $(x_k,y_l) \in R$ are defined according to equidistant partitions of the intervals $[x_0, x_K]$ and $[y_0, y_L]$, respectively, by

$$x_k = x_0 + kh_1, \quad k = 0, \ldots, K \tag{15}$$

$$y_l = y_0 + lh_2, \quad l = 0, \ldots, L \tag{16}$$

The assumption of equidistant partitions has been made for the sake of notational simplicity only. All numerical operations are confined to grid points $(x_k,y_l,z_i) \in IR^3$.

A contour in the z_i plane is given in "digitized form" as an ordered sequence of a finite number of points to constitute a closed polygon, which in turn defines the cross section S_i. For each point (x_k,y_l,z_i) it is checked whether $(x_k,y_l,z_i) \in S_i$ or $(x_k,y_l,z_i) \in S_i^c$ by calculating its minimum distance from all contours of the cross section at level z_i according to Equations (4).

Then, the complete data $d_i(x_k,y_l)$, $d_i'(x_k,y_l)$, $i = 1,\ldots, n$, of each Hermite interpolation problem to be solved are available.

Roots of the interpolant $g(z;x_k,y_l)$

Because only the roots of the interpolant $g(z;x_k,y_l)$ are of major interest it is checked whether $d_i(x_k,y_l) \times d_{i+1}(x_k,y_l) < 0$. In the affirmative situation there is a root in $[z_i, z_{i+1}]$. Its actual numerical determination is then straightforward.

Definition of $d_0(x_k,y_l)$ and $d_{n+1}(x_k,y_l)$

The only problem left is to define $d_0(x_k,y_l)$ and $d_{n+1}(x_k,y_l)$ appropriately. To clarify the role of d_0, d_{n+1} let $(x_k,y_l) \in IR^2$ be fixed. If $d_1(x_k,y_l) > 0$, then $d_0(x_k,y_l)$ essentially determines the smallest of the roots

z^* of the corresponding interpolant $g(z;x_k,y_l)$; analogously, if $d_n(x_k,y_l) > 0$, then $d_{n+1}(x_k,y_l)$ essentially determines the largest of the roots z^* of the corresponding interpolant $g(z;x_k,y_l)$. Thus, it is reasonable to define $d_0(x_k,y_l)$, $d_{n+1}(x_k,y_l)$ such that $d_0(x_k,y_l) < 0$, $d_{n+1}(x_k,y_l) < 0$ as $S_0 = S_{n+1} = \emptyset$. If $d_1(x_k,y_l) < 0$, the actual value of $d_0(x_k,y_l)$ does not matter at all if $d_0(x_k,y_l) < 0$; analogously for $d_n(x_k,y_l) < 0$ and $d_{n+1}(x_k,y_l) < 0$.

The situation, when the numerical values of $d_0(x_k,y_l)$ and $d_{n+1}(x_k,y_l)$ actually matter, poses a problem of extrapolation. Levin (1986) suggested a global value independent of (x_k,y_l)

$$d_0(x_k, y_l) = d_{n+1}(x_k, y_l) = -d(S), \quad (x_k, y_l) \in R \tag{17}$$

where $d(S)$ denotes the diameter of the set $S = \cup_{i=1}^{n} S_i$,whereas Omwa (1987) and Gmelig Meyling (1988) favor local linear extrapolation

$$d_0(x_k, y_l) = \min(2d_1(x_k, y_l) - d_2(x_k, y_l), -\epsilon) \tag{18}$$

$$d_{n+1}(x_k, y_l) = \min(2d_n(x_k, y_l) - d_{n-1}(x_k, y_l), -\epsilon) \tag{19}$$

where ε denotes a small positive number.

Here it is emphasized that in special applications other values may be more useful; thus, it may be reasonable to set

$$d_0(x_k, y_l) = d_{n+1}(x_k, y_l) = -\infty \tag{20}$$

for example, when the intersections S_1, S_n coincide with the ``horizontal'' boundary surfaces of the body B.

Volumetrics

For the computation of the volume of $\hat{B}$, the integration formula simplified for the discrete situation applies

$$\begin{aligned} \mathrm{vol}_3(\hat{B}) &= \iint l(x,y)\,dx\,dy \\ &= h_1 h_2 \sum_{k=1}^{K} \sum_{l=1}^{L} \mathrm{vol}_1(\{z \mid g(z; x_k, y_l) \geq 0\}) \end{aligned} \tag{21}$$

where vol_1 $(\{z \mid g(z;x_k,y_l) \geq 0\})$ is the sum of the lengths $z^*_{j+1} - z^*_j$ of intervals with $g(z;x_k,y_l) \geq 0$ for $z \in \left[z^*_j, z^*_{j+1}\right]$ when z^*_j denote the roots of $g(z;x_k,y_l)$.

Rendering

The set of points $\{(x_k,y_l,z^*) \mid g(z^*;x_k,y_l) = 0\} \subset IR^3$ can be used to visualize the body, its outer and inner boundary surfaces by computer graphics involving voxels (Samet, 1990a, 1990b).

PRACTICAL EXAMPLES AND APPLICATIONS

In a first simple example, a sphere of unit radius is given by its contour lines in several cross sections. In this example, the plane of any intersection does not contain more than one (closed) contour line.

Numerical performance of the suggested scheme for approximating the volume of the unit sphere is summarized in Table 1.

Table 1. Numerical performance for approximation of volume of unit sphere: calculated value (relative error in percent: ranking of performance) [cpu time units]; true volume 4.188790.

# gridpoints	8 x 8	13 x 13	23 x 23
# cross sections (# points per counterline)			
5 (36)	4.149 (0.94%:5) [0.11]	4.396 (4.96%:14) [0.28]	4.417 (5.45%:18) [0.88]
5 (72)	4.138 (1.20%:7) [0.21]	4.375 (4.46%:13) [0.57]	4.409 (5.26%:17) [1.78]
9 (36)	4.043 (3.45%:11) [0.18]	4.253 (1.55%:8) [0.44]	4.270 (1.94%:10) [1.52]
9 (72)	4.028 (3.81%:12) [0.38]	4.237 (1.17%:6) [0.99]	4.263 (1.79%:9) [3.07]
19 (36)	3.979 (5.00%:15) [0.38]	4.201 (0.30%:2) [1.00]	4.208 (0.46%:4) [3.10]
19 (72)	3.977 (5.04%:16) [0.77]	4.178 (0.23%:1) [2.02]	4.207 (0.45%:3) [6.28]

For various choices of (i) the total number of gridpoints, (ii) the total number of cross sections, and (iii) the total number of points per counterline numerical performance is compared and ranked in terms of relative error in percent and cpu time units. It should be noted that the two-dimensional discretization of the planes of intersection and the distance of the sections should conform with one another to yield good results (cf. Levin's theorem). All results can be judged to be in good agreement with the "true" theoretical value of $4/3\ \pi = 4.188$.

In a second example, a torus with R = 500, r = 250 is considered to demonstrate the method's versatility. The planes of intersection have been selected parallel to a plane containing the torus' axis of largest inertia. Thus there are planes of intersections which contain more than one, that is two contourlines, the interior of each corresponds to the interior of the body to be reconstructed. Selecting two different discretizations (i) seven equidistant cross sections and a 10 × 10 square grid the approximate volume 6.287E+08 was calculated (see Fig. 5);

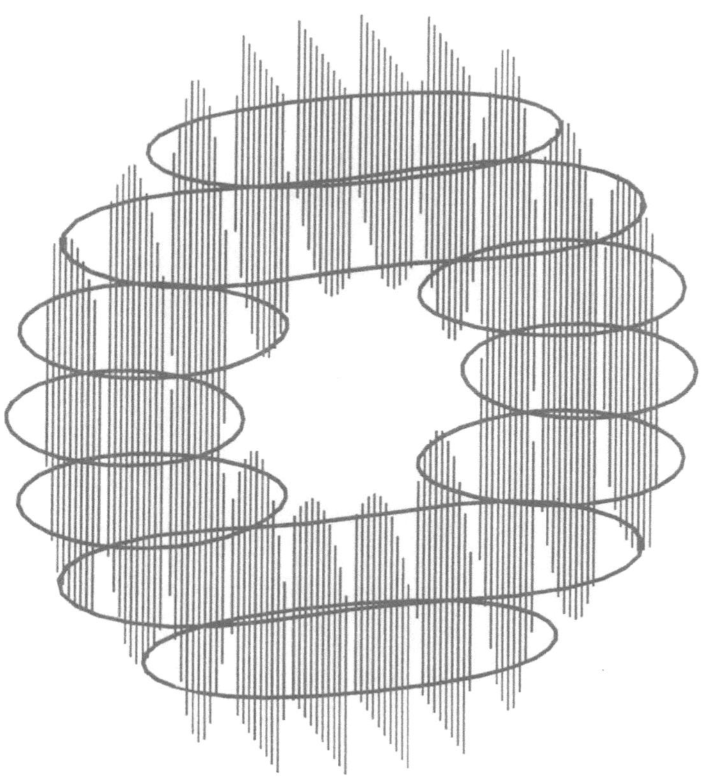

Figure 5. Approximately of volume of torus with R = 500, r = 250 from contour lines seven equidistant planes of intersection with grid of 10 x 10 points resulting in approximated volume of 6.287 E+08, true volume is $2\pi^2 Rr^2 = 6.168$ E + 08.

(ii) 17 equidistant cross sections and a 20 × 20 square grid the approximate volume 6.099E+08 was calculated (see Fig. 6), which are both in sufficiently good agreement with the "true" theoretical value $2\pi^2 Rr^2 = 6.168$ E+08 of the volume of a torus with $R = 500$, $r = 250$.

Had the planes of intersections been selected orthogonal to the torus' axis of largest inertia, then cross sections with more than one, that is two nested contour lines would have been generated with the interior of the inner contour lines corresponding to the body's exterior. In this situation the final volume of the torus is given by the difference of the volumes of the two bodies represented by the outer and inner contour lines, respectively, that is by applying the algorithm successively to each set of outer and inner contour lines, respectively. The numerical results do not differ significantly if the number of intersecting planes and the number of grid points are chosen appropriately.

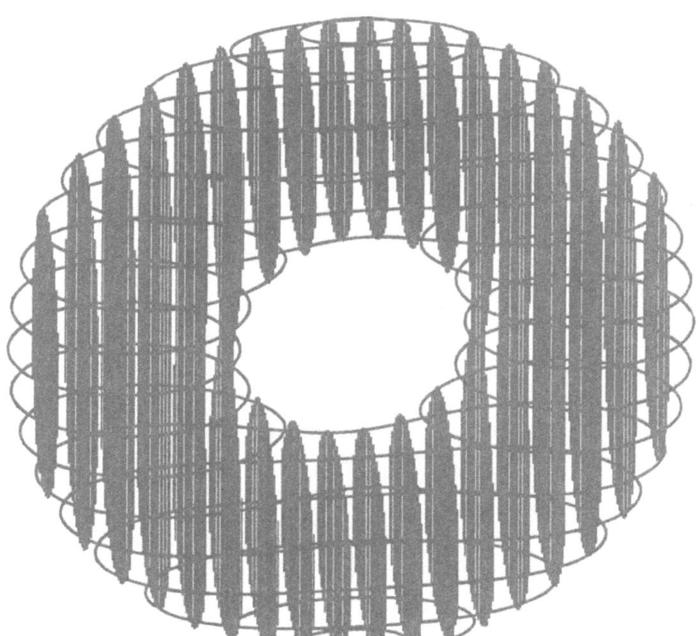

Figure 6. Approximation of volume of torus with $R = 500$, $r = 250$ from contour lines in 17 equidistant planes of intersection with grid of 20 × 20 points resulting in approximated volume of 6.099 E + 08, true volume is $2\pi^2 Rr^2 = 6.168$ E + 08.

CONCLUSIONS

The suggested method compares favorably with methods, which are more restrictive with respect to feasible shapes and topologies of the bodies to be reconstructed. Because the problem of reconstructing a complex body of nested partial bodies can be reduced to a sequence of three-dimensional problems, and because each three-dimensional problem can be reduced to a sequence of rather simple univariate problems, parallel processing provides a promising option.

REFERENCES

Auerbach, S., and Schaeben, H., 1990a, Surface representations reproducing given digitized contour lines: Math. Geology, v. 22, no. 6, p. 723-742.

Auerbach, S., and Schaeben, H., 1990b, Computer-aided geometric design of geologic surfaces and bodies: Math. Geology, v. 22, no. 8, p. 957-987.

Farin, G., 1988, Curves and surfaces for computer aided geometric design - a practical guide: Academic Press, Inc., New York, 334 p.

Gmelig Meyling, R.H.J., 1988, Three-dimensional reconstruction from serial cross sections by set-valued interpolation: Memorandum 754, Faculty of Applied Mathematics, Univ. of Twente, Enschede, The Netherlands, 30 p.

Klein, H., Pflug, R., and Ramshorn, C., 1989, Shaded perspective views by computer: a new tool for geologists: Geobyte, v. 4, no. 4, p. 16-24.

Levin, D., 1986, Multidimensional reconstruction by set-valued approximation: IMA Jour. Numer. Anal., v. 6, no. 2, p. 173-184.

Meyers, D., Skinner, S., and Sloan, K., 1992, Surfaces from contours: ACM Trans. Graph., v. 11, no. 3, p. 228-258.

Omwa, A.A., 1987, Reconstruction of a closed three-dimensional body by set-valued interpolation: Rept. Faculty of Applied Mathematics, Univ. of Twente, Enschede, The Netherlands, 42 p.

Press, W.H., Flannery, B.P., Teukolsky, S.A., and Vetterling, W.T., 1990, Numerical recipes: the art of scientific computing: Cambridge University Press, Cambridge, 702 p.

Samet, H., 1990a, The design and analysis of spatial data structures: Addison-Wesley, Reading, Massachusetts, 496 p.

Samet, H., 1990b, Applications of spatial data structures: computer graphics, image processing and GIS: Addison-Wesley, Reading, Massachusetts, 480 p.

Schumaker, L.L., 1983, On shape preserving quadratic spline interpolation: SIAM Jour. Numer. Anal., v. 20, no. 4, p. 854-864.

Walsh, J., and Brown, S., 1992, Computing areas and volumes from contour maps: Geobyte, v. 7, no. 3, p. 44-47.

THE EFFECT OF SEASONAL FACTORS ON GEOLOGICAL INTERPRETATION OF MSS DATA

D. Yuan
Desert Research Institute, Las Vegas, Nevada, USA

J. E. Robinson
Syracuse University, Syracuse, New York, USA

M. J. Duggin
SUNY College of Environmental Science and Forestry, Syracuse, New York, USA

ABSTRACT

Temporal variations are considered to be important factors in their effects on remotely sensed image data and discrimination analysis of the images as a basis for geologic interpretation. Some systematic temporal factors may be corrected by instrument adjustment but random and stochastic factors are not necessarily systematic. However, statistical methods can be used to determine the effect of the stochastic temporal factors. In this study, 29 MSS subscenes from northern Lincoln County, Nevada images were obtained and nine different lithologies were sampled. Two linear statistical models related to the stable background and seasonal factors were constructed and tested. The results suggest that (1) the band brightness values resulting from different lithologies are statistically different; (2) the seasonal effects do exist in a statistical sense; (3) the seasonal factors are such that the bands show higher values in summer and fall, but lower in winter and spring; (4) the seasonal effects are consistent within the four MSS bands for any given lithology area; and (5) the seasonal factors are consistent for different types of lithology within any given MSS band. These results provide statistical support for lithological classification using remotely sensed data.

TIME FACTORS

Duggin (1985) summarized the systematic and random factors that limit the discrimination and quantification of terrestrial features using remotely sensed radiance. If the band brightness from one ground target at one time is considered, the factors affecting the measured values result from the passage of radiant energy to the target, the reflection of radiance from the target, the passage of reflected target radiance to the detector, the effect of the detector, the effect of cloud and haze, and the effect of background radiance. The systematic variation across an image can be corrected but not the random variation. In a later study, Duggin, Sakhavat, and Lindsay (1987) examined systematic and random variations in recorded radiance for two LANDSAT Thematic Mapper bands of images obtained over a uniform agricultural area before and after harvest. They determined that the variations were large enough to be considered in the discrimination of the terrestrial cover types. The stated result also is true for lithological discrimination using satellite images, as long as the area has sufficient vegetation cover (Duggin and Robinove, 1990).

If the band brightness from a ground target in a fixed location is considered, the variation of the recorded radiance varies with respect to time. This variation primarily depends on the nature of the changing factors such as vegetation cover, cloud, snow, rainfall, and other random effects. If the detector, emitted energy, and satellite conditions are assumed constant when compared with that resulting from naturally changing factors, the variation across a series of images because of equipment factors is small and can be ignored. Therefore, factors affecting the band brightness from a constant ground target can be thought of as "stochastic".

The effect of vegetation varies on a seasonal basis, however, there are many uncertain factors, such as precipitation, temperature, and number of days of sunshine. Thus, the effect resulting from vegetation should be both "seasonal" and "stochastic". The snow effect is mainly "stochastic". Because there is more snow in winter than in other seasons. Snow also is a "seasonal" factor. Rainfall may be random, however, in the study areas, there is more rain in winter and spring than in summer and fall. Cloud and atmospheric turbulence usually are random factors, but some time dependent variation may need to be considered. If the band brightness from a ground target in a series of images is considered to be consistent, then the variation of the radiance will show both "seasonal" and "stochastic" characteristics. These characteristics should be detectable by statistical hypotheses testing methods. In order to simplify the discussion, only the seasonal factors are considered. However, the monthly, daily even hourly factors could be considered in the same way as long as there are enough images to perform statistical testing. Table 1 gives some of the time differing factors affecting the band brightness of MSS data. Note that the "systematic time factors" here are not the "systematic factors" that are dependent on the observation systems in the usual sense that have been discussed by authors such as Duggin (1985) and Duggin and Robinove (1990). Such factors also are stochastic in that they show some statistical trend or periodicity. Factors in this study seem to be systematic only when large numbers of images from a single area are examined.

It is necessary to estimate the effects of time factors, or more specifically, to estimate the seasonal factors, and determine their dependency on such factors as the frequency spectra of the indicating bands, the lithological type, and even the interaction of bands and lithologies. For this, it is necessary to construct statistical models.

In order to test the seasonal effects, a series of 29 MSS subscene images from northern Lincoln County, Nevada were collected. These images were recorded in the period from 13 September 1972 to16 June 1977. The study area is centered approximately at 114°30' West longitude, 30°20' North latitude, extending about 23.5 km on each side. The image size is 300 x 300 pixels with each pixel representing a 79 by 79 m^2 ground area after rectification. Figure 1 shows the coverage of these images and the selected subscene study area. Robinove and Chavez (1978) studied the temporal variation of the whole scene albedo of the same data set. Table 2 lists some basic information about this data set.

SURFICIAL GEOLOGY

The study area is typical southwest Valley and Range province topography with elongated north-south mountain ranges and intervening

Table 1. Summary of time factors affecting received lithological radiance.

Type of time factors	Subclasses	Characteristics	Examples
Systematic time factors	Trend time factors	Long term, linear, invisible	Greenhouse effects
		Long term, cyclic, invisible	Sunspot activity
	Cyclic time factors	Yearly (seasonal) Obvious to invisible	Vegetation, Precipitation, Snow, Wind, Cloud
		Monthly, maybe invisible	Lunar variation
		Daily	Sun angle, Brightness of the sky
Stochastic time Factors	Catastrophic stochastic time	Nondirectional, obvious visible	Floods, Snow Storm
	factors	Directional, obvious visible	Slide/Earth Quick, Human Activity
	Non-catastrophic stochastic time factors	Visible	Precipitation, Snow, Clouds
		Invisible	Atmospheric Disturbance

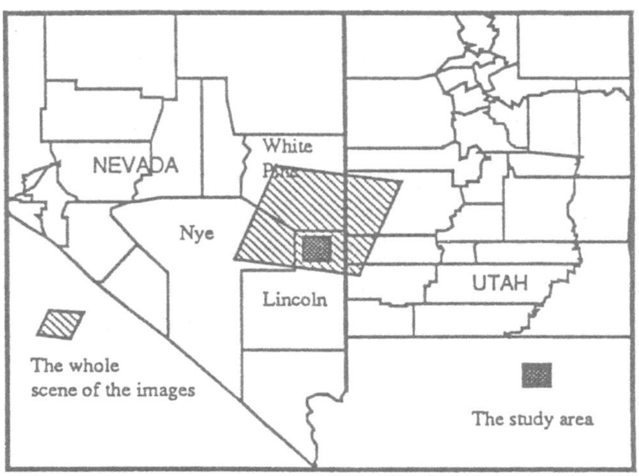

Figure 1. Approximate location of selected study area and area covered by satellite images used for study.

Table 2. Basic information about image data stack used for the study (data from Robinove and Chavez, 1978).

Date of exp.	Average And Minimum Brightness								Remarks
	B4 ave	B4 min	B5 ave	B5 min	B6 ave	B6 min	B7 ave	B7 min	
720913	47	13	50	6	50	4	23	0	
721124	54	9	50	1	47	0	19	0	snow widespread
721230	60	6	57	5	51	3	21	0	snow widespread
731014	36	8	39	1	39	0	17	0	
740623	55	16	61	8	61	7	28	0	
740816	48	11	53	7	53	3	24	0	
740903	47	12	51	6	50	5	21	0	
741202	25	1	24	0	24	0	10	0	snow on Mtns
750609	41	9	57	9	66	0	28	0	scattered clouds
750627	42	11	57	0	66	0	28	0	
750802	40	10	55	9	63	3	26	0	
751215	49	11	48	5	45	2	18	0	snow widespread
760120	23	9	22	3	27	2	9	0	snow on Mtns
760129	18	4	25	2	27	0	11	0	snow on Mtns
770410	39	8	53	8	58	6	24	0	snow on Mtns
760516	41	10	57	11	65	6	28	0	snow on Mtns
760621	42	10	59	9	67	5	29	0	
760709	41	11	58	10	64	6	26	0	
760805	43	12	46	6	47	3	20	0	
760823	38	11	40	5	42	3	17	0	
760901	35	6	49	4	54	2	22	0	
761007	25	4	35	2	40	0	16	0	
761016	29	9	29	3	30	2	13	0	clouds and wind trace
761025	25	5	34	3	37	0	15	0	scattered clouds
761130	15	1	21	1	24	0	10	0	
761218	14	1	19	0	21	0	8	0	
770405	35	9	47	6	52	4	21	0	snow on Mtns
770423	40	9	54	8	60	4	25	0	
770616	41	9	56	8	64	4	27	0	

dry valleys. The southern boundary of the subscene of approximately 11 miles north of Pioche. Climate and vegetation are related and altitude-dependent. Vegetation is scarce in the valleys, but is considerable on the mountains and uplands. The higher mountain areas with steppe vegetation have more rainfall, colder winters, and cooler summers. The lower areas are covered by sagebrush and the higher areas by single-leaf pinon and juniper forests (Tschanz and Pampeyan, 1970).

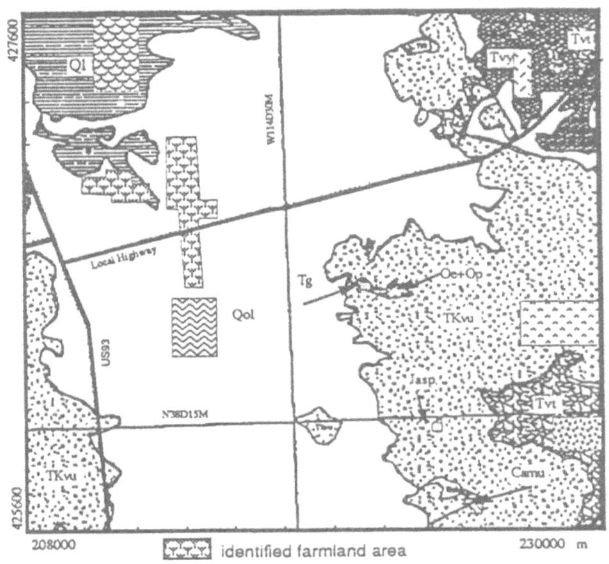

Figure 2. Brief lithological map and training area distributions. "Blocks" on map are training areas of correspondent lithologies. Piece of farmland in Lake Valley is identifiable from all images. Boundary differs with respect to seasons.

The eastern two-fifths of the subscene lies on the western side of the Wilson Creek Range, which is one of the highest and most inaccessible ranges in Lincoln County (Tschanz and Pampeyan, 1970). The part of this range in the subscene area is composed entirely of volcanic rocks. The western three-fifths of the subscene is Lake Valley. The majority of the valley surface consists of older alluvium deposits. The northwestern of the valley is recent lake sand deposits composed of material eroded from the nearby mountains. Alluvium fans and aprons occur continuously around the mountain front.

About 60 percent of the rocks exposed in the study area are Tertiary and Quaternary sedimentary deposits; the remaining 40 percent are volcanics. There are only a few patches of Cambrian and Ordovician sedimentary rocks. The northwestern portion of the subscene is the Lake Valley playa deposit. The eastern part of the subscene is Tertiary volcanic rocks that flank the southeastern side of Mount Wilson. There are 10

Table 3. Mean band brightness value by season for lithologies.

Lith.	Season	N	Band 1	Band 2	Band 3	Band 4
QL	All Season	29	51.28	61.98	62.13	25.79
	Spring	3	47.44	65.62	68.49	28.41
	Summer	7	58.74	78.21	82.87	35.31
	Fall	11	51.54	60.95	60.19	25.45
	Winter	8	45.81	47.84	44.28	16.96
QOL	All Season	29	36.42	44.75	47.29	19.86
	Spring	3	29.18	41.95	47.84	20.32
	Summer	7	36.03	50.20	58.10	25.26
	Fall	11	32.48	40.25	42.91	18.52
	Winter	8	44.89	47.20	43.65	16.82
TVT	All Season	29	28.56	30.90	40.79	18.85
	Spring	3	44.19	54.65	63.16	26.10
	Summer	7	24.93	30.65	48.21	23.59
	Fall	11	22.53	23.57	33.74	16.80
	Winter	8	34.19	32.31	35.60	14.79
TVY	All Season	29	25.61	29.24	35.80	15.56
	Spring	3	21.96	29.94	39.33	17.14
	Summer	7	28.97	38.29	48.72	21.34
	Fall	11	24.66	27.19	33.40	14.70
	Winter	8	25.34	23.86	26.49	11.06
TG	All Season	29	24.69	27.96	34.27	14.92
	Spring	3	21.99	28.84	37.69	16.25
	Summer	7	28.38	36.66	45.73	20.06
	Fall	11	25.02	27.19	33.12	14.52
	Winter	8	22.02	21.06	24.52	10.48

Table 3. (Cont.) Mean band brightness value by season for lithologies.

Lith.	Season	N	Band 1	Band 2	Band 3	Band 4
TKVU	All Season	29	21.09	21.85	29.99	14.12
	Spring	3	20.80	25.99	36.53	16.76
	Summer	7	23.52	28.14	41.45	19.92
	Fall	11	20.34	19.91	27.64	13.36
	Winter	8	20.09	17.48	20.73	9.10
OEOP	All Season	29	27.01	30.75	35.93	15.89
	Spring	3	25.35	32.53	40.61	17.89
	Summer	7	29.87	39.41	47.92	21.42
	Fall	11	25.56	28.06	32.04	14.32
	Winter	8	27.11	26.21	29.04	12.47
CAMU	All Season	29	27.02	31.33	38.16	16.87
	Spring	3	24.25	32.68	42.35	18.54
	Summer	7	31.05	40.57	51.05	22.73
	Fall	11	27.90	31.62	37.66	16.83
	Winter	8	23.31	22.35	26.00	11.17
JASP	All Season	29	25.65	29.05	36.58	16.17
	Spring	3	24.03	31.80	41.70	17.78
	Summer	7	27.95	36.10	47.92	21.35
	Fall	11	26.12	29.13	36.69	16.62
	Winter	8	23.58	21.74	24.60	10.40

mappable lithological units occurring on the surface in the study area. Classified lithologies are: Pliocene younger lake beds (Ql), Pliocene older alluvium deposits (Qol), Tertiary younger volcanic rocks, undivided, including intravolcanic sedimentary rocks and perlite (Tvy), Tertiary tuffs and tuffaceous sediments (Tvt), Tertiary granite stocks and dikes (Tg), Tertiary volcanic rocks, undifferentiated (TKuv), Middle Ordovician Eoreka Quartzite (Oe), Lower Ordovician Pogonip Group (Op), Upper Cambrian Limestone and dolomite (Ꞓu) and a small patch of Jasperoid (Tschanz and Pampeyan, 1970). Because two of them (Oe - Middle Ordovician Eoreka Quartzite and Op - Lower Ordovician Pogonip Group) were difficult to locate and distinguish from each other in the images, they were combined to form a "mappable unit" (Oe+Op or Ope in the figures) for this study. A simplified map of lithology and training area selection for the study area is shown in Figure 2. A detailed description of the lithological units is given in Tschanz and Pampeyan (1970). The mean seasonal band brightness values for the 9 lithologies are calculated and listed in Table 3. In order to actually view the temporal change over the selected temporal images, 14 of the 29 subscene images are converted into albedo images and displayed in Figure 3.

BAND-TIME INTERACTION TEST

The average brightness values for the 9 lithological units for the 4 bands from 29 MSS images in this study area have been computed and plotted against time of exposure in Figure 4. Although there are some irregularities, it is obvious that the time variant pattern for the 4 bands for any of the given lithologies are similar. Two time curves are considered to be "similar" or "in phase" if they reach their peaks and troughs almost simultaneously.

In order to analyze the effects of temporal factors on the band brightness for the lithologies, a specific testing procedure had to be designed. Observation of the band brightness values - time plots for different lithologies (Fig. 4) suggests that the effect resulting from time factors could be approximated by their linear portion. For example, considering the time effects to different bands of a given lithology, the suitable statistical model would be:

$$BVL_{lijk}=\mu_l+BAND_i+SEASON_j+BS_{ij}+E_{ijk} \qquad (1)$$

Where BVL_{lijk} is the k'th received band i radiance at time period j for lithology l, μ_l is the background value for lithology l, $BAND_i$ is the effect of band i, $SEASON_j$ is the effect of time period j, BS_{ij} is the interactive affect of band i and time period j, E_{ijk} is the random error for the individual image. The subscripts i = 1,2,3,4; j=Spring, Summer, Fall, Winter and k various depending on the number of images in correspondent period. We want to test the hypotheses:

$$H_{01bs}: \quad BS_{ij}=0, \qquad \textit{for any i and j.} \qquad (2)$$

If H_{01bs} could not be rejected, we will further test:

$$H_{01b}: \quad BAND_i=0, \qquad for\ any\ i=1,2,3,4. \tag{3}$$

and

$$H_{01s}: \quad SEASON_j=0, \qquad for\ any\ j=1,2,3,4. \tag{4}$$

The hypothesis H_{01bs}, H_{01b}, and H_{01s} can be tested by the SAS general linear model procedure GLM. Table 4 gives the tables for analysis of variance for the 9 lithological units. In GML procedure, the program first checks the validity of the model (i.e. compute the F-value for the model and compute the probability that the theoretic F-value is greater than the observed F-value). After the validity of the model is checked, the program then continuous to check the effects of the contributing factors (i.e. band passes and seasons). The former checks the effect of the result of "MODEL," whereas the later checks the effect of "FACTOR". For example, for lithology QL (Quaternary Lake Deposit), the F-value for checking the model is 10.88, the computed Prob ($F_{theoretic} > F_{observed}$) = 0.0001. This suggest that the model (is 1) valid for lithology QL. The same conclusion could be obtained for the rest 8 lithologies. All the computed the Prob ($F_{theoretic} > F_{observed}$) are smaller than 0.001 (8 of them are equal to 0.0001, and one of them (QOL) equals 0.0006). This suggests that all the 9 models are valid even at 99% significance level. The rest of the hypotheses can be tested in the same way.

For hypothesis (2) H_{01bs}: $BS_{ij} = 0$, the computed Prob ($F_{theoretic} > F_{observed}$) for the 9 lithologies range from 0.0770 (TKVU) to 0.9162 (QOL), that is 0.5758 (QL), 0.9162 (QOL), 0.1137 (TVT), 0.4877 (TVY), 0.2816 (TG), 0.0770 (TKVU), 0.7141 (OEOP), 0.2787 (CAMU), 0.4936 (JASP). 8 out of 9 lithologies with Prob ($F_{theoretic} > F_{observed}$) > 0.1 TKVU (0.0770) is the only exception. Actually, all are greater than 0.05. This suggests that at a given significance level, say 95%, the band-season interactions are insignificant statistically, regardless of the lithology. Also it can be inferred that the BAND-SEASON interaction is small compared with the effects directly the result of the BAND and SEASON factors in all situations. In other words, the seasonal effect is consistent for the different bands for any given lithology. This result can be observed from the band value v.s. time plots for each of the lithologies because the different bands have similar variant patterns with respect to time. For example, they have similar variation periods and attitudes, and reach the

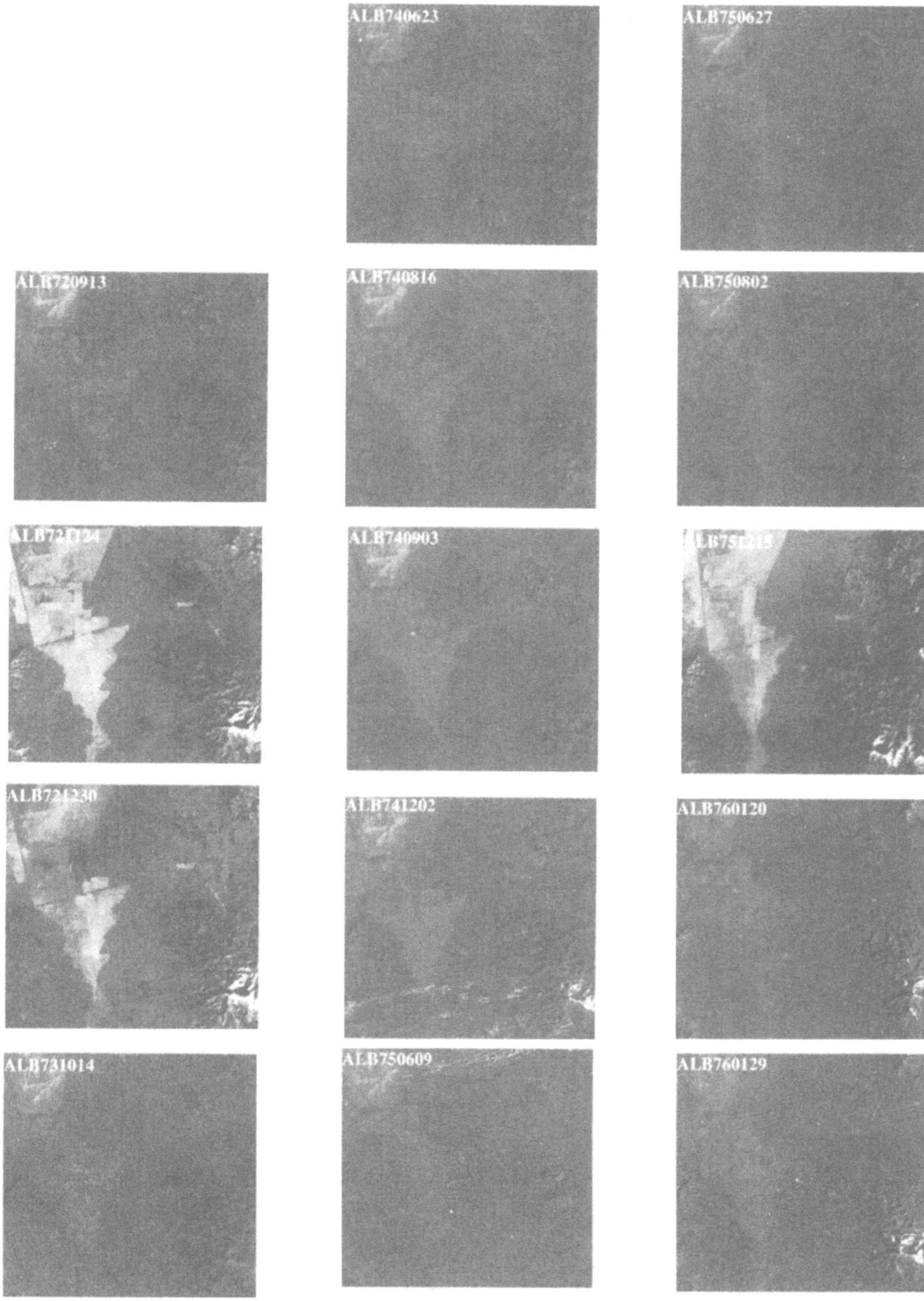

Figure 3. 14 albedo images converted from LANDSAT MSS images obtained from 13/9/1972 to 16/6/1977.

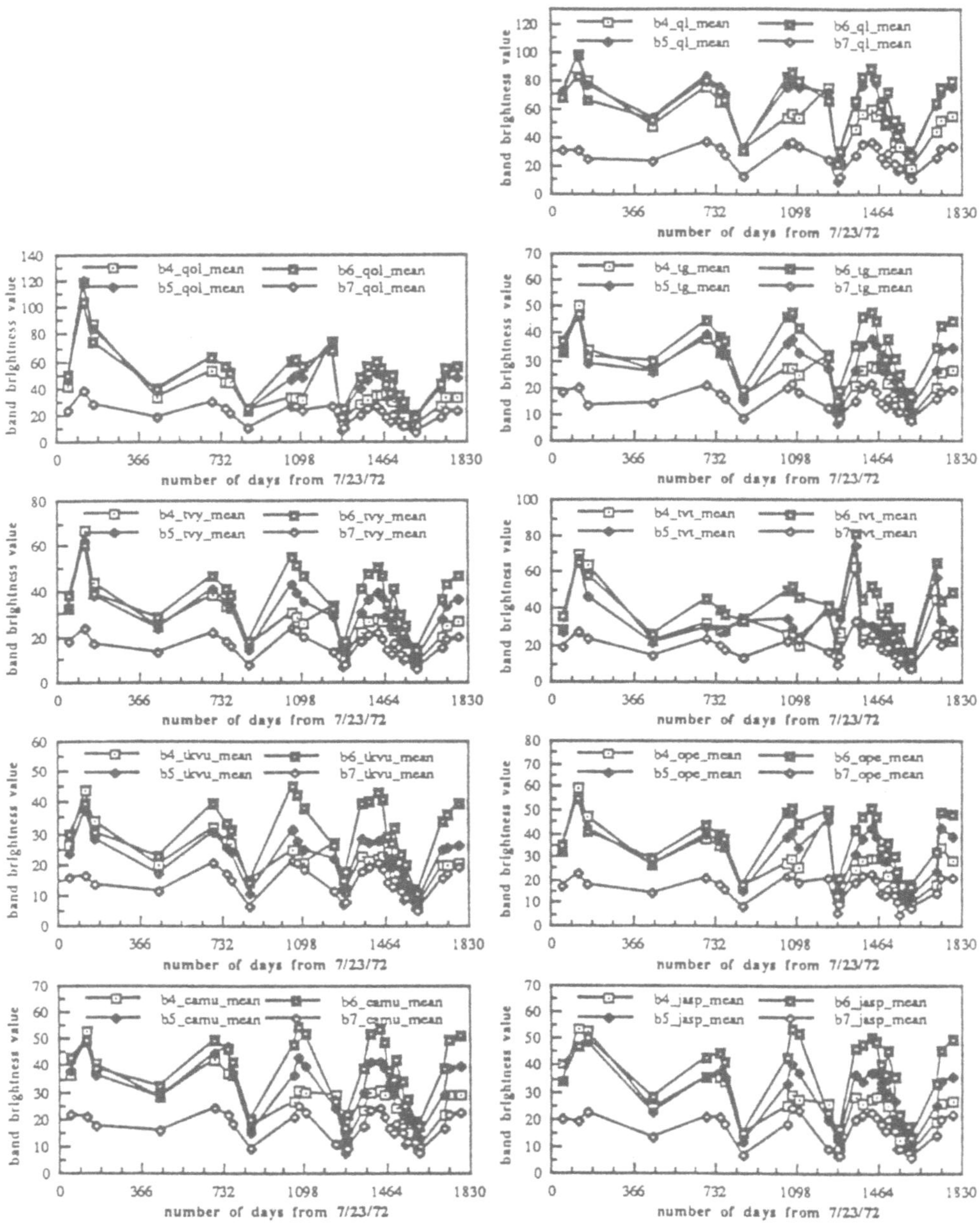

Figure 4. Band brightness value variant pattern for lithologies. Where b4_ql_mean stands for mean band 4 brightness value of lithology Ql, and so forth. "camu" stands for Cambrian carbonate rocks, "ope" stands for Ordovician rocks. Other lithologies see explanation in text.

Table 4. ANOVA for band - time interaction tests (for 9 lithologies).

A. ANOVA classification level information.

CLASS	LEVELS	VALUES
BAND	4	BAND 4, BAND 5, BAND 6, BAND 7
SEASON	4	SPRING, SUMMER, FALL, WINTER
NUMBER OF OBSERVATIONS IN DATA SET = 116		
DEPENDENT VARIABLE: BAND VALUE		

B. ANOVA table by lithology.

LITHO-LOGY	EFFECT	SOURCE	DF	SS	MS	F-VAL.	PR > F
QL	MODEL	MODEL	15	36636.98	2442.47	10.88	0.0001
		ERROR	100	22450.94	224.51		
		TOTAL	115	59087.92	513.81		
	FACTOR	BAND	3	25462.91	8487.64	37.81	0.0001
		SEASON	3	9464.63	3154.88	14.05	0.0001
		BAND*SEASON	9	1709.44	189.94	0.85	0.5758
QOL	MODEL	MODEL	15	16174.30	1078.29	2.98	0.0006
		ERROR	100	36175.41	361.75		
		TOTAL	115	52349.71	455.21		
	FACTOR	BAND	3	13334.96	4444.99	12.29	0.0001
		SEASON	3	1440.01	480.00	1.33	0.2699
		BAND*SEASON	9	1399.32	155.48	0.43	0.9162
TVT	MODEL	MODEL	15	13981.17	932.08	7.52	0.0001
		ERROR	100	12401.60	124.02		
		TOTAL	115	26382.77	229.42		
	FACTOR	BAND	3	7061.97	2353.99	18.98	0.0001
		SEASON	3	5087.24	1695.75	13.67	0.0001
		BAND*SEASON	9	1831.97	203.55	1.64	0.1137
TVY	MODEL	MODEL	15	9580.77	638.72	6.86	0.0001
		ERROR	100	9308.77	93.09		
		TOTAL	115	18889.54	164.26		
	FACTOR	BAND	3	6223.96	2074.65	22.29	0.0001
		SEASON	3	2562.69	854.23	9.18	0.0001
		BAND*SEASON	9	794.12	88.24	0.95	0.4877

Table 4 (Cont.)

B. (Cont.) ANOVA table by lithology.

LITHOLOGY	EFFECT	SOURCE	DF	SS	MS	F-VAL.	PR > F
TG	MODEL	MODEL	15	8841.06	589.40	11.76	0.0001
		ERROR	100	5010.93	50.11		
		TOTAL	115	13851.99	120.45		
	FACTOR	BAND	3	5667.87	1889.29	37.70	0.0001
		SEASON	3	2615.56	871.85	17.40	0.0001
		BAND*SEASON	9	557.63	61.96	1.24	0.2816
TKVU	MODEL	MODEL	15	6506.04	433.74	10.59	0.0001
		ERROR	100	4093.84	40.94		
		TOTAL	115	10599.88	92.17		
	FACTOR	BAND	3	3667.61	1222.54	29.86	0.0001
		SEASON	3	2174.40	724.80	17.70	0.0001
		BAND*SEASON	9	664.04	73.78	1.80	0.0770
OEOP	MODEL	MODEL	15	9114.25	607.62	5.97	0.0001
		ERROR	100	10172.01	101.72		
		TOTAL	115	19286.26	167.71		
	FACTOR	BAND	3	6281.13	2093.71	20.58	0.0001
		SEASON	3	2199.01	733.00	7.21	0.0002
		BAND*SEASON	9	634.11	70.46	0.69	0.7141
CAMU	MODEL	MODEL	15	11336.09	755.74	11.45	0.0001
		ERROR	100	6602.33	66.02		
		TOTAL	115	17938.42	155.99		
	FACTOR	BAND	3	6920.21	2306.74	34.94	0.0001
		SEASON	3	3678.20	1226.07	18.57	0.0001
		BAND*SEASON	9	737.68	81.96	1.24	0.2787
JASP	MODEL	MODEL	15	9711.32	647.42	7.05	0.0001
		ERROR	100	9189.71	91.90		
		TOTAL	115	18901.03	164.36		
	FACTOR	BAND	3	6240.34	2080.11	22.64	0.0001
		SEASON	3	2692.85	897.62	9.77	0.0001
		BAND*SEASON	9	778.13	86.46	0.94	0.4936

peaks and troughs almost at the same times. Their variations are harmonic.

For hypotheses H_{01b}: $BAND_i = 0$, the tests show that at a 95% significance level the hypothesis should be rejected (Prob ($F_{theoretic} > F_{observed}$) = 0.0001 for all the lithologies). These suggest that each lithology usually will have different values for different bands.

The test does produce anomalies for H_{01s}: $SEASON_j = 0$. For 8 out of 9 lithologies Prob ($F_{theoretic} > F_{observed}$) < 0.001, which suggests that the SEASON factors do have an effect at 99% level on the band brightness. However, the computed Prob ($F_{theoretic} > F_{observed}$) = 0.2699 for Qol, which suggests that the seasonal effect for lithology Qol is not as important as it is for other lithologies. However, there is some contribution. Generally, the seasonal factors do have some effect on band brightness for ground lithologies and in most situations, this effect is strong and should not be ignored.

LITHOLOGY-TIME INTERACTION TEST

The foregoing suggests that band brightness values for different lithologies differ statistically in MSS data and the seasonal effects should also have some effect on the variation for a given band and a given lithology. In Figure 5, the mean band brightness values for nine lithologies are plotted by seasons for the four bands. The actual mean seasonal band brightness values are listed in Table 3. For example, in the first diagram of Figure 5 (band 4), Quaternary Lake Deposits (Ql) has the highest band brightness value for all the seasons with average value 51.28. Tertiary Volcanics Undifferentiated (TKvu) has the lowest band brightness value for all the seasons with average value 21.09. The same regularity can be observed in the other diagrams of Figure 5. The general tendency for all the rocks is that they have a brighter value in summer and fall, whereas have darker value in winter and spring for all four bands. Also, the recent sedimentary rocks (Ql, Qol) have higher band brightness values than the older sedimentary rocks (Op+Oe, Ꞓu), whereas the older sedimentary rocks have greater band brightness values than volcanic rocks with possible exception of Tvt. The Tvt exception could be due to physical properties, mixed Tertiary tuffs and tuffaceous sediments, or due to sampling errors.

In order to test the temporal effect due to different lithologies for a given band, the time factors are grouped into four seasonal factors in the following lithology - time interaction model:

$$BVL_{gijk}=b_i+LITH_g+SEASON_j+LS_{gj}+E_{gjk} \tag{5}$$

Where BVL_{gijk} is the k'th received band i radiance at time period j for lithology g, b_i is the background value for band i, $LITH_g$ is the effect of lithology g, $SEASON_j$ is the effect of time period j, LS_{gj} is the interactive affect of lithology g and time period j, E_{gjk} is the random error for the individual image. The subscripts g = 1,2, ..., 9; j=Spring, Summer, Fall, Winter, and k various depending on the number of images in the corresponding period. It is necessary to test the hypotheses:

$$H_{02bs}\text{:}\quad LS_{ij}=0, \qquad \textit{for any i and j}; \tag{6}$$

$$H_{02b}\text{:}\quad LITH_i=0, \qquad \textit{for any } i=1,2,3,4; \tag{7}$$

$$H_{02s}\text{:}\quad SEASON_j=0, \qquad \textit{for any } j=1,2,3,4. \tag{8}$$

Model (5) and hypotheses (6), (7), (8) can be tested by applying the interactions between seasonal factors and lithologies. The SAS GML procedure is used for testing these hypotheses. Table 5 lists the results for analysis of variance for the four bands. The SAS program tests the validity of the models first and then checks the effects resulting from concerning factors. The following is a brief interpretation of these results.

All the models for the four bands based on (5) fit well within the Prob ($F_{theoretic} > F_{observed}$) = 0.0001. Thus, there is no problem in accepting model (5). The results for testing H_{02bs}: LS_{ij} = 0, for any i and j show that Prob ($F_{theoretic} > F_{observed}$) = 0.8521, 0.2546, 0.4551, 0.6666 for MSS band 4 to band 7 respectively. This suggests that, for a given significant level 95%, these lithology-time interactions are statistically insignificant. In other words, seasonal effects on band brightness do not depend on the actual lithology from which the radiance is obtained. Rather, seasonal factors are consistent in spite of different lithologies. This is an important result obtained from the statistical analysis as it provides the foundation for the possibility of extracting the "pure" radiance from the lithology out of a mixture of lithologies and other temporal factors. According to the statistical results, the temporal effects could be considered as time

Band 4 brightness value by seasons

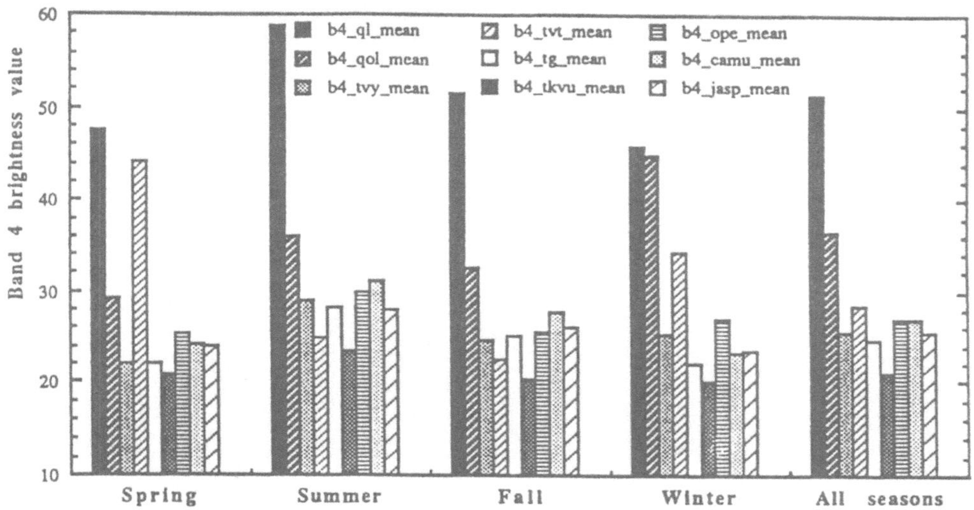

Band 5 brightness value by seasons

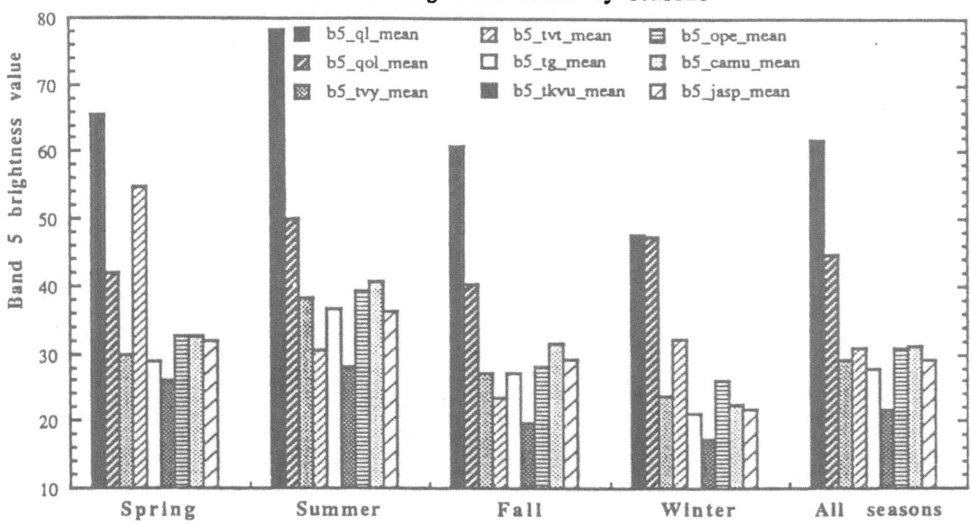

Figure 5. Mean seasonal band brightness value by lithologies (explanation see Fig. 3).

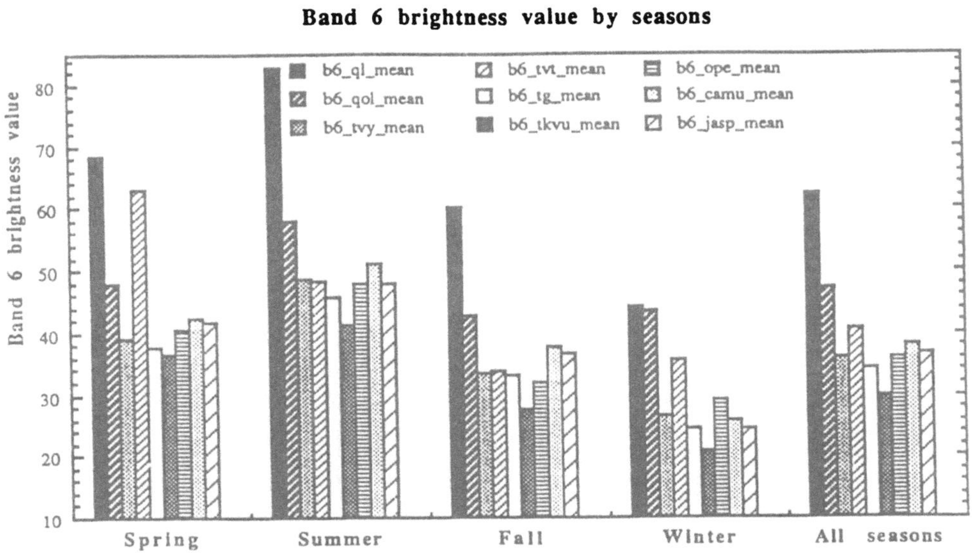

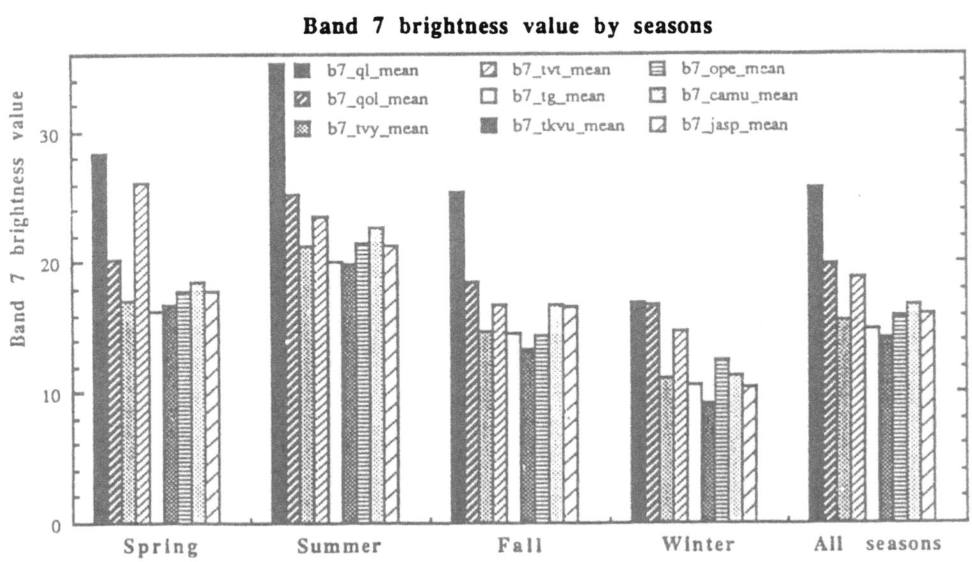

Figure 5. (Cont.) Mean seasonal band brightness value by lithologies (explanation see Fig. 3).

Table 5. ANOVA for lithology - time interaction tests (for 4 MSS bands).

A. ANOVA: CLASS LEVEL INFORMATION

CLASS	LEVELS	VALUES
SEASON	4	SPRING SUMMER FALL WINTER
LITH	9	CAMU JASP OEOP QL QOL TG TKVU TVT TVY
NUMBER OF OBSERVATIONS IN DATA SET = 261		

B. ANOVA table by band.

BAND	EFFECT	SOURCE	DF	SS	MS	F-VAL.	PR > F
4	MODEL	MODEL		22947.45	655.64	35	0.0001
		ERROR	225	44185.79	196.38		
		TOTAL	260	67133.24	258.20		
	FACTOR	SEASON	3	555.83	185.28	0.94	0.4204
		LITH	8	19105.27	2388.16	12.16	0.0001
		SEASON*LITH	24	3286.36	136.93	0.70	0.8521
5	MODEL	MODEL	35	44957.63	1284.50	7.94	0.0001
		ERROR	225	36383.18	161.70		
		TOTAL	260	81340.81	312.85		
	FACTOR	SEASON	3	6805.74	2268.58	14.03	0.0001
		LITH	8	33541.74	4192.72	25.93	0.0001
		SEASON*LITH	24	4610.15	192.09	1.19	0.2546
6	MODEL	MODEL	35	42198.54	1205.67	9.06	0.0001
		ERROR	225	29926.51	133.01		
		TOTAL	260	72125.05	277.40		
	FACTOR	SEASON	3	17923.66	5974.55	44.92	0.0001
		LITH	8	21052.60	2631.58	19.79	0.0001
		SEASON*LITH	24	3222.28	134.26	1.01	0.4551
7	MODEL	MODEL	35	7597.20	217.06	9.95	0.0001
		ERROR	225	4910.07	21.82		
		TOTAL	260	12507.27	48.10		
	FACTOR	SEASON	3	4170.51	1390.17	63.70	0.0001
		LITH	8	2980.18	372.52	17.07	0.0001
		SEASON*LITH	24	446.51	18.60	0.85	0.6666

dependent optical filters under which each lithology retains its own contribution.

For testing H_{02b}: $LITH_i = 0$ for any i, the computed Prob ($F_{theoretic} > F_{observed}$) =0.0001 for all the four models of the MSS bands. This suggests that the lithology effects are significant at 99% level. In other words, different lithologies do have different band values in a statistical sense. This is the foundation for lithological classification by using specific band brightness values for MSS data.

For testing H_{02s}: $SEASON_j = 0$, the computed Prob ($F_{theoretic} > F_{observed}$) = 0.0001 for band 5, 6, and 7. This suggests that the seasonal effects are significant for band 5, band 6, and band 7 at the 99% level. This suggests that the lithology effects and the seasonal effects can be considered significant at 90% level. However, the Prob ($F_{theoretic} > F_{observed}$) = 0.4204 for band 4 suggests that the seasonal factors do not have a significant effect on band 4 brightness in a statistical sense. In other words, seasonal factors have a weak effect on band four.

Summarizing the results, it can concluded that the time factors are consistent within different bands (except possibly band 4) for any given lithology, and are consistent for different lithologies for any given band. This provides a foundation for the possibility of extracting the lithological background from a series of actual images. However, determining a reasonable representation of the lithological trend from the temporal images remains a problem that must be considered.

CONCLUSIONS

Statistical results obtained in this study can be summarized as follows

(1) The different lithologies in the study area have different mean values of band brightness for each band. This provides the foundation for statistical classification based on the brightness values. Without the certainty of these differences any classification for lithologies based on band brightness values would not be significant.

(2) The received radiance or band brightness values show different mean values for each band for any given lithological formation. This suggests that band brightness from individual band represents different physical properties of the material.

(3) Seasonal factors usually cause important changes in band brightness. The rocks usually have higher band brightness values in summer and fall and lower values in winter and spring. Presumably these variations are the results of a clearer and drier atmosphere along with less vegetation in summer and fall, whereas there is greater rainfall and more vegetation in winter and spring.

(4) The seasonal factors do not have strong interaction with different bands. Thus, it can be concluded that the seasonal factors are consistent for individual bands and lithologies.

REFERENCES

Duggin, M.J., 1985, Factors limiting the discrimination and quantification of terrestrial features using remotely sensed radiance: Intern. Jour. Remote Sensing, v. 6, no. 1, p. 3-27.

Duggin, M.J., Sakhavat, H., and Lindsay, J., 1987, The systematic and random variation of recorded radiance in a LANDSAT thematic mapper image: Intern. Jour. Remote Sensing, v. 6, no. 7, p. 1257-1261.

Duggin, M.J., and Robinove, C.J., 1990, Assumptions implicit in remote sensing data acquisition and analysis: Intern. Jour. Remote Sensing, v.11, no.10, p.1669-1694.

Robinove, C.J., and Chavez, P.S. Jr. 1978, LANDSAT albedo monitoring method for an arid region: Presentation at the AAAS Intern. Symp. Arid Region Plant Resources, Lubbock, Texas.

Tschanz, C.M., and Pampeyan, E.H., 1970, Geology and mineral deposits of Lincoln County, Nevada: Nevada Bureau of Mines Bull. 73, 187p.

RECONSTRUCTION OF THE LEDUC AND WABAMUN ROCK SALTS, YOUNGSTOWN AREA, ALBERTA

Neil L. Anderson
University of Missouri–Rolla, Rolla, Missouri, USA

R. James Brown
University of Calgary, Calgary, Alberta, Canada

ABSTRACT

There are four relatively thick (>20 m) bedded Devonian rock salt units in the Youngstown area, south-central Alberta, Canada (T25-35, R5-20W4M), those of the Cold Lake Formation, Prairie Formation, Wabamun Group, and Leduc Formation. There is no substantive evidence that the former two Middle Devonian rock salts (Cold Lake and Prairie) have been leached post-depositionally in the study area. However, the latter two Upper Devonian rock salts (Leduc and Wabamun) have been leached extensively in places, and are preserved now as irregularly shaped bodies of variable areal extent, having maximum net thicknesses on the order of 45 m and 40 m, respectively.

The dissolution of the Wabamun and Leduc rock salts has caused the overlying strata to subside, more-or-less on a one-to-one basis. As a result, at any well location in the study area there is a direct correlation between the relative subsidence of a specific geologic horizon and the net thickness of all rock salt dissolved after the deposition of that horizon. This direct relationship between dissolution and subsidence is the key to reconstructing the paleodistribution of the Wabamun and Leduc rock salts and determining the timing and mechanisms of leaching.

Geologic Modeling and Mapping
Edited by A. Förster and D. F. Merriam, Plenum Press, New York, 1996

In this paper, a suite of present-day and original net-thickness maps for the Leduc and Wabamun rock salts in the Youngstown area of Alberta are presented. These maps were generated using well-log control. They are based mostly on the interval thicknesses of the residual rock salts and the encompassing Wabamun and Leduc intervals, and variations in structural relief at the Wabamun and Leduc levels and along the overlying post-Devonian horizons. Interpretation of the maps of net salt thickness supports the thesis that salt dissolution in the study area was initiated or enhanced by some or all of four principal processes: (1) the near-surface exposure of the rock salt, which resulted from the erosion of the overlying Paleozoic sediment during the pre-Cretaceous hiatus; (2) the partial dissolution of underlying rock salt; (3) regional faulting or fracturing during the mid-Late Cretaceous; and (4) glacial loading and unloading.

In support of the present-day net-salt-thickness maps, two seismic profiles also are presented. The first line images an isolated remnant of Wabamun rock salt; the second crosses a linear Wabamun salt-dissolution feature. Leaching at the second location was initiated during the mid-Late Cretaceous, probably as a result of faulting or fracturing.

INTRODUCTION

The residual rock salts of the Wabamun Group and Leduc Formation (Upper Devonian) in the Youngstown area, south-central Alberta (Figs. 1 to 4), are preserved as irregularly shaped bodies having maximum net thicknesses on the order of 45 m and 40 m, respectively. Both of these rock salts have been leached extensively and were deposited more widely than their present-day distribution might suggest (Anderson, 1991; Anderson and Brown, 1991a, 1991b, 1992; Anderson, Brown, and Hinds, 1988b; Anderson, White, and Hinds, 1989; Hopkins, 1987; Meijer Drees, 1986; Oliver and Cowper, 1983). Indeed, in a study of the adjacent Stettler area (T30-45, R10-25W4M), Anderson and Brown (1991b) conclude that about 40 m of net Wabamun rock salt probably was deposited throughout the entire Youngstown area (T25-35, R5-20W4M).

In our study of the Stettler area, we mapped the present-day distribution and reconstructed the paleodistribution of the Wabamun rock salt (Anderson and Brown, 1991b). On the basis of this suite of maps of

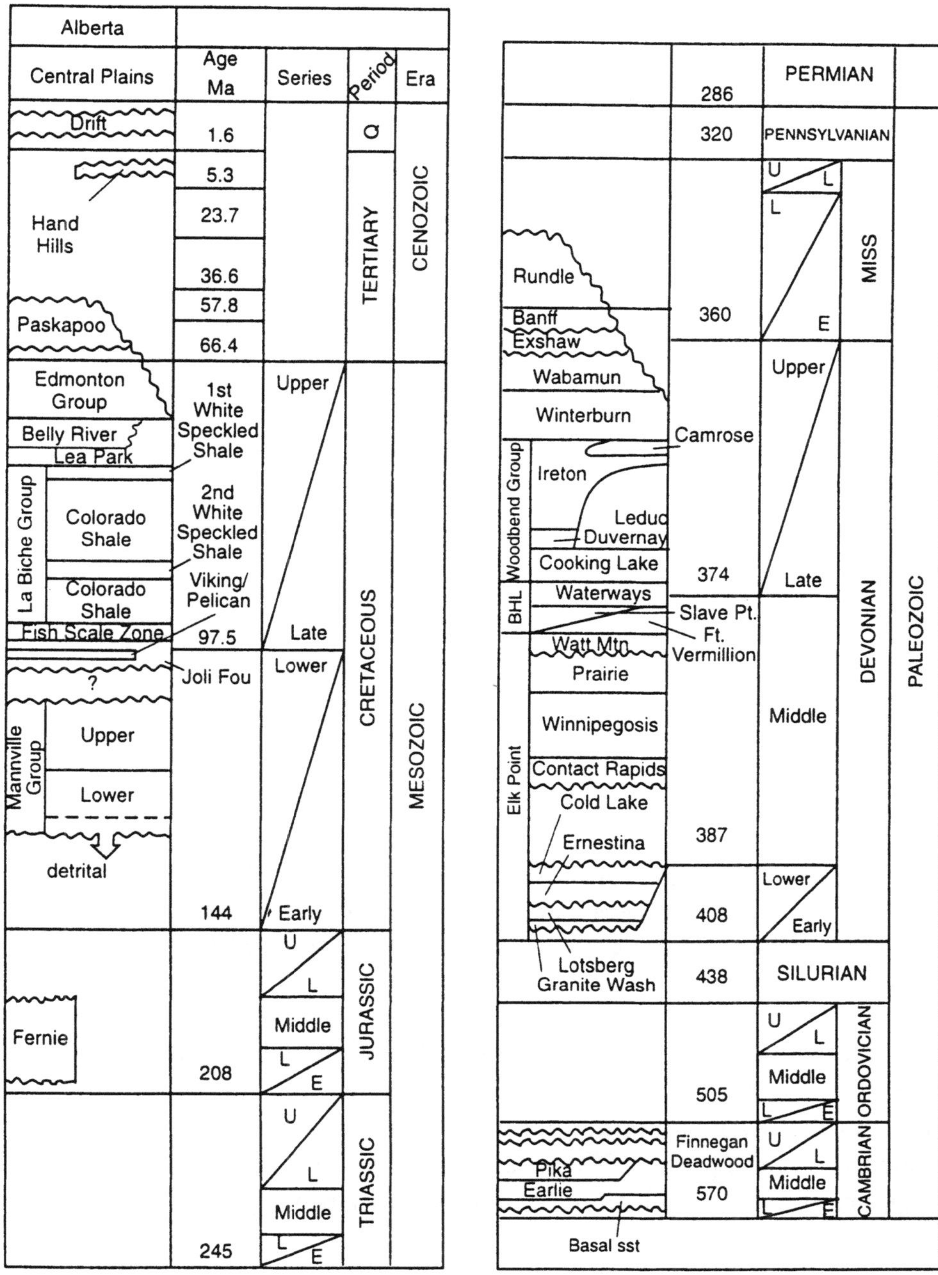

Figure 1. Stratigraphy of central Alberta (modified after AGAT Laboratories, 1988). In Youngstown study area (Fig. 2), relatively thick Devonian rock salt is preserved within several stratigraphic intervals: Cold Lake Formation, Prairie Formation, Leduc Formation, and Wabamun Group. In study area, Upper Leduc and Wabamun rock salts have been dissolved extensively; however, there is no conclusive evidence that underlying Middle Devonian Cold Lake or Prairie rock salts have been leached.

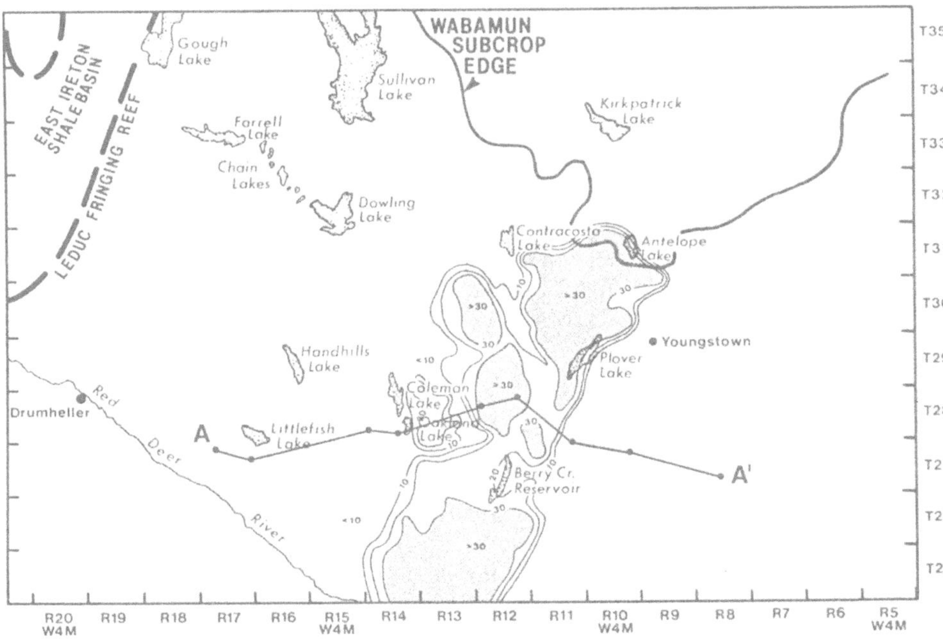

Figure 2. Isopachous map (in meters) depicting present-day net thickness of Leduc salt in Youngstown area. Net salt thicknesses are based on Leduc well control where available. Elsewhere thicknesses are based on estimates of salt-related subsidence at shallower levels. Areas of thickest salt (>30 m) are finely stippled. Only larger lakes are shown.

net salt thickness, we concluded that dissolution of the Wabamun salts in the Stettler area occurred episodically, although in some places (e.g. near the northeastern margin) more-or-less continuously from the late Paleozoic to the present. This is a result of several principal processes. In support of these conclusions Anderson and Brown (1991b) demonstrate that:

(A) On a local scale, the Wabamun interval ranges by up to 40 m (Fig. 3). These variations can be correlated directly to the thickness of the residual salt, which supports the interpretation that about 40 m of Wabamun rock salt was deposited in the Stettler study area and that this salt was extensively leached.

(B) Seismic and well-log data illustrate clearly that post-salt strata drape across residual Wabamun and Leduc salt. These observations indicate that dissolution and related subsidence have occurred.

(C) Generally, Wabamun salt is not encountered along the Wabamun subcrop edge (Fig. 3). This observation supports the interpretation that subsidence originated along the exposed Wabamun outcrop during the pre-Cretaceous hiatus.

(D) Basinward of the Wabamun subcrop edge, the Wabamun salts in the Stettler area have been leached preferentially along NNE-trending (and orthogonal) lineaments. This pattern is consistent with the theses that dissolution was initiated in places by regional faulting during the mid-Late Cretaceous and is self-sustaining. Leaching along these shear zones is envisioned as a cyclic and continuous process whereby the collapse of overlying strata enhances vertical permeability, thereby providing a conduit for unsaturated water and facilitating additional dissolution (Anderson and Brown, 1992; Anderson, Brown, and Hinds, 1988a).

(E) In places, present-day drainage patterns are correlatable to areas where one or both Upper Devonian salts are thin or absent indicating that dissolution probably has occurred in the post-Pleistocene, possibly as a result of glacial loading and unloading (Anderson and Brown, 1992; Anderson and Cederwall, 1993; Anderson and Knapp, 1993).

We have followed up these previous studies of salt dissolution in south-central Alberta and analyzed the leaching of the Wabamun Group and Leduc Formation salts in the Youngstown area (T25-35, R5-20W4M). Following the methodology of Anderson and Brown (1991b), we have reconstructed the distribution of these Upper Devonian salts at various times since their deposition (Figs. 2, 3, 5, 6, and 7). These paleoreconstructed net-salt-thickness maps are consistent with previous studies of salt dissolution in south-central Alberta and some of the large-scale mechanisms of salt dissolution reported by Anderson and Cederwall (1993) and Anderson and Knapp (1993) in their studies of other western Canadian Devonian rock salts.

ORIGINAL DISTRIBUTION OF THE LEDUC AND WABAMUN SALTS

Leduc Salts

The lithology of the Upper Devonian Leduc Formation varies significantly within the Youngstown area of Alberta. Near the edge of the East Ireton Shale Basin (Fig. 2), the Leduc is dolomitized fringing reef. Shelfward (to the east), the Leduc becomes increasingly evaporitic and consists predominantly of interlayered dolomites, anhydrites, and in places a basal halite (rock salt) unit (Belyea, 1964).

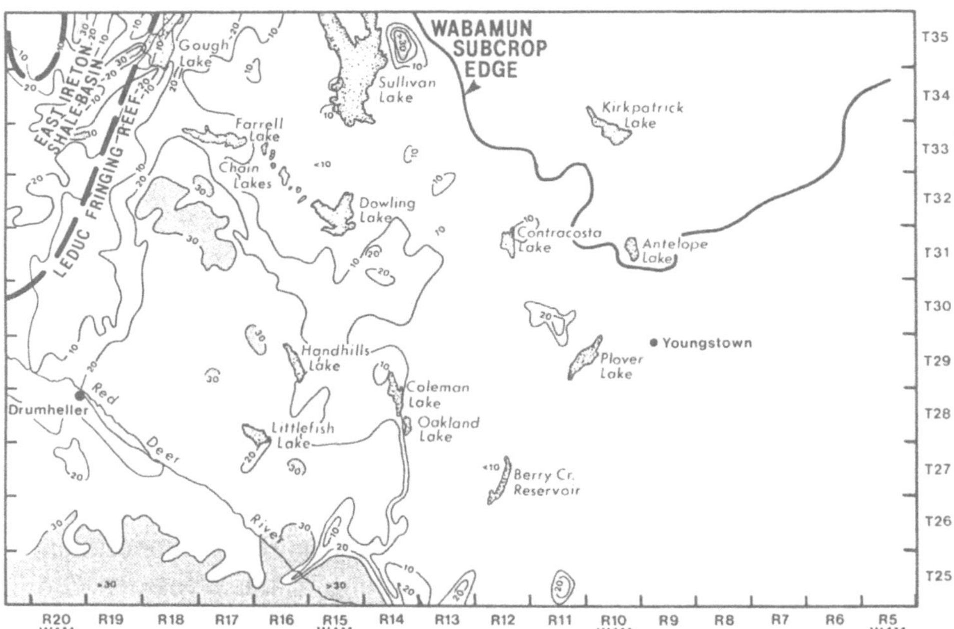

Figure 3. Isopachous map (in meters) depicting our interpretation of present-day distribution of Wabamun salts. Net salt thicknesses are based on Wabamun well control where available. Elsewhere, thicknesses are based on estimates of salt-related subsidence at shallower levels. Areas of thickest salt (>30 m) are finely stippled.

In order to estimate the original distribution of the Leduc salts (also referred to as the Cairn salts) in the Youngstown study area (Fig. 2) and following the methodology described by Anderson and Brown (1991b), we have constructed a suite of four maps: (1) net Leduc salt based on log control – sonic, density, and caliper only (Fig. 2 shows a modified version of this map); (2) Leduc isopach (not shown); (3) present-day Leduc structure (Fig. 8); and (4) restored Leduc structure (Fig. 9). This map suite indicates that there are direct correlations between relative structural relief at the top of the Leduc, thickness of the Leduc Formation, and net thickness of Leduc salt dissolved in post-Leduc time. For example, outside the interpreted Leduc salt basins (Fig. 5), the top of the Leduc (Fig. 8) is consistent with the restored Leduc structure map (Fig. 9), indicating that little, if any, Leduc salt was deposited in these areas. Within the confines of the salt basins, however, the present-day Leduc structure values are lower than the restored Leduc structure values, except at those well locations where thick residual salt is preserved (i.e. locations of little, if any, dissolution). Our interpretation is that, within the confines of the salt basins, the difference between the present-day and restored Leduc structure at any well location is a reasonable estimate of the thickness of Leduc salt that was dissolved at that well location in post-Leduc time. On the basis of this premise, the original distribution of the Leduc salt within the Youngstown study area (Fig. 5) was reconstructed.

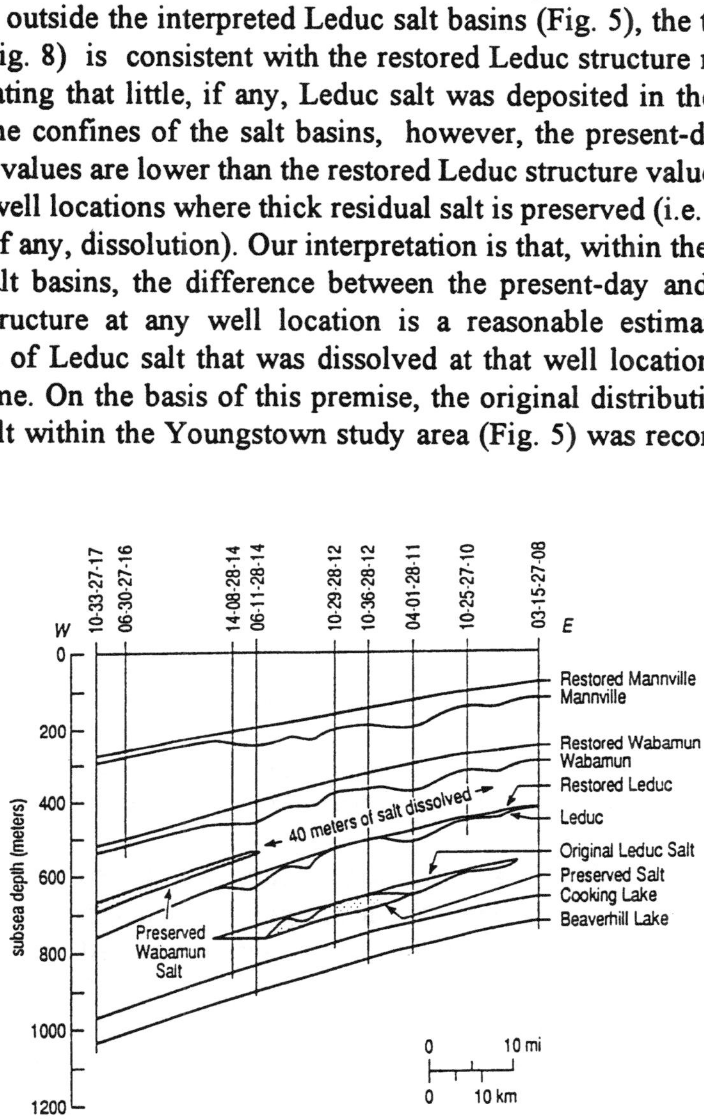

Figure 4. Geologic section from Youngstown study area illustrating discontinuous nature of Wabamun and Leduc salts. Both present-day and reconstructed profiles are displayed. Wells incorporated into geologic section are highlighted in Figure 2.

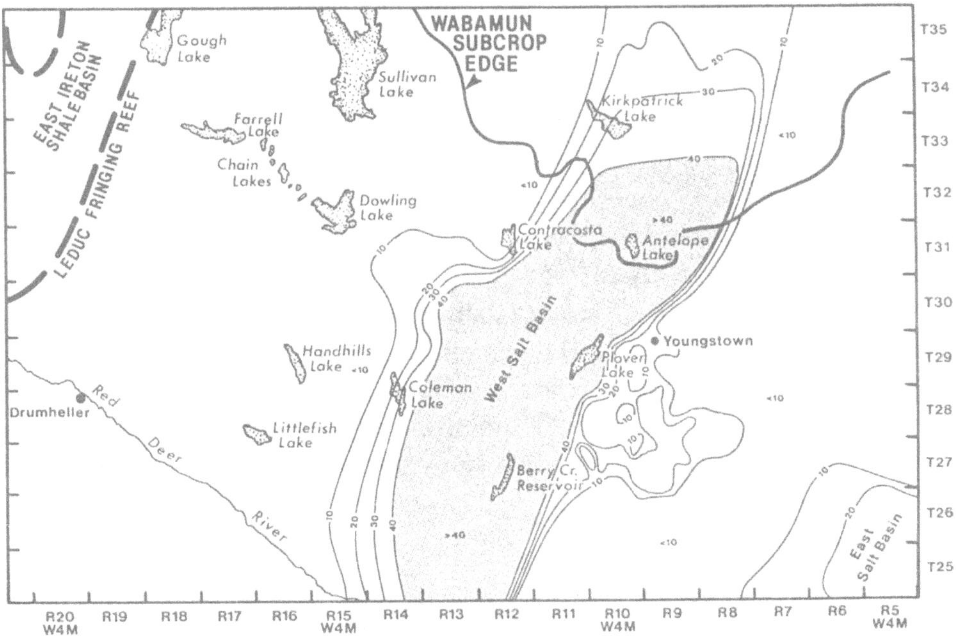

Figure 5. Isopachous map (in meters) depicting original distribution of Leduc salt in Youngstown study area. Net salt thicknesses are based on Leduc well control where available. Elsewhere, thicknesses are based on estimates of salt-related subsidence at shallower levels. Area of greatest salt thickness (>40 m) is finely stippled.

More specifically, at each well location where Leduc control was available, the difference between the present-day Leduc structure (Fig. 8) and the restored Leduc structure (Fig. 9) was calculated. At those well locations where Leduc salt is no longer present, this difference is interpreted to represent the original salt thickness. At those control points where salt remains, the original salt thickness was estimated by summing the thickness of the salt remnant and the calculated difference. At those shallow well locations where Leduc control is not available, the contouring of the present-day Leduc structure map was constrained by interpretive contouring trends and patterns of structural relief along post-Leduc horizons.

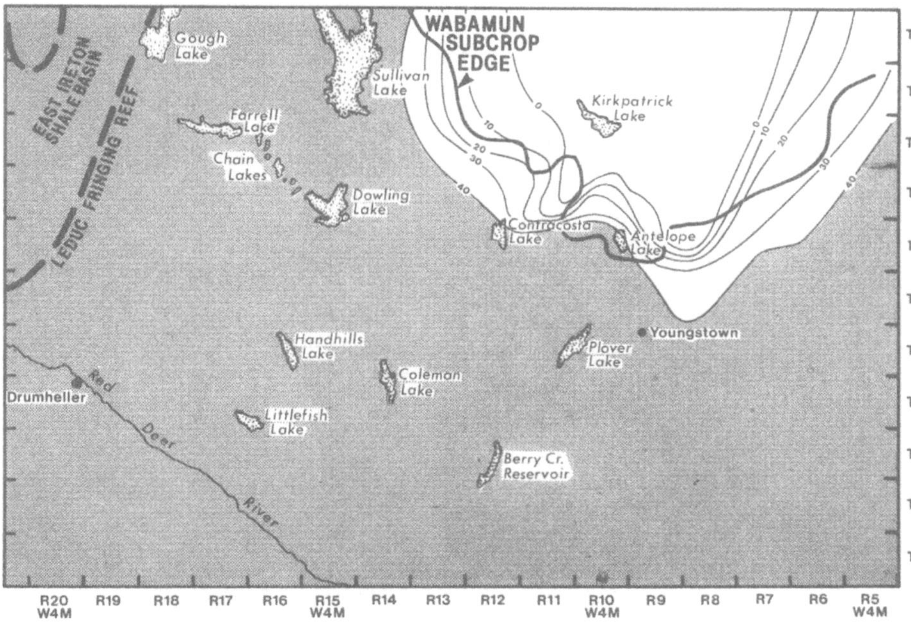

Figure 6. Isopachous map (in meters) showing our interpretation of distribution of Wabamun salts at end of Viking time. This map supports interpretations that: (i) about 40 m of Wabamun rock salt was deposited throughout Youngstown study area; (ii) earliest phase of dissolution initiated along Wabamun subcrop during pre-Cretaceous hiatus; and (iii) salt-dissolution front established along Wabamun subcrop migrated to southwest thereafter at variable rates and in self-sustaining manner. Because significant dissolution occurred only around Wabamun subcrop up to end of Viking time, this map also represents that original salt distribution where Wabamun is preserved. Area of greatest salt thickness (>40 m) is finely stippled.

Wabamun Salts

The Upper Devonian Wabamun Group in the Youngstown area is subdivided into the Stettler and Big Valley Formations (Fig. 1). The Stettler consists predominantly of interlayered dolomites, anhydrites, and residual rock salt; the Big Valley is composed of green shales and fossiliferous limestones (Andrichuk and Wonfor, 1953; Wonfor and Andrichuk, 1953).

The Wabamun rock salt (Stettler Formation) in the Stettler area (T30-45, R10-25W4M) has been mapped by Anderson, Brown, and Hinds (1988b) and Anderson and Brown (1991b), who conclude that about 40 m of rock salt was deposited throughout the Youngstown area, but that much of this salt was post-depositionally leached. In order to test this previous interpretation, we have constructed a suite of four maps: (1) net

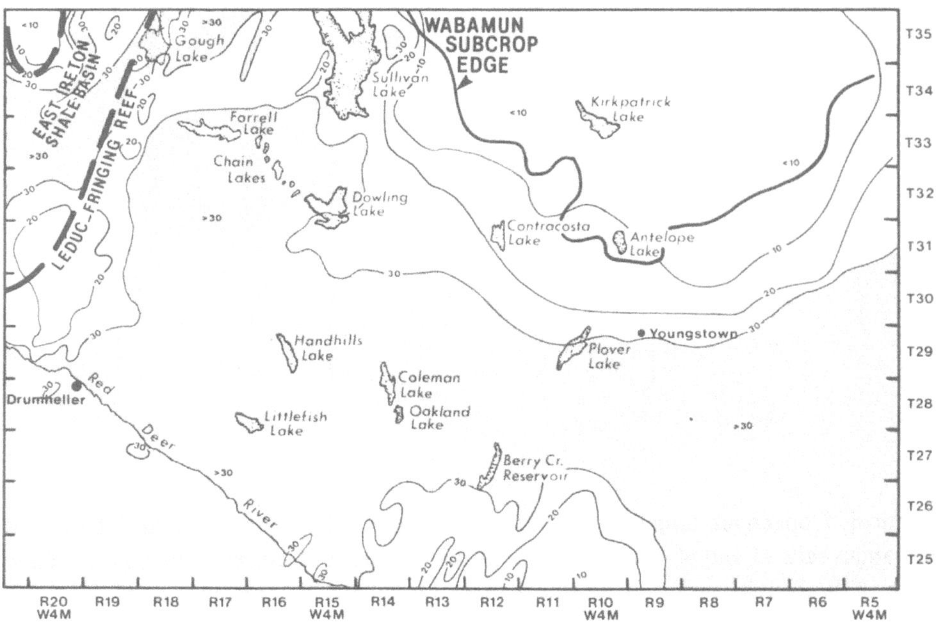

Figure 7. Isopachous map (in meters) showing our interpretation of distribution of Wabamun salt at end of Colorado time. Areas of thickest salt (>30 m) are finely stippled. Comparison of Figures 3, 6, and 7 supports thesis that extensive phase of salt dissolution was initiated by regional faulting during Colorado (mid-Late Cretaceous) time.

Wabamun salt based on log control – sonic, density, and caliper only (Fig. 3 shows a modified version of this map); (2) Wabamun isopach; (3) present-day Wabamun structure; and (4) restored Wabamun structure.

Outside the postulated Leduc salt basins (Fig. 5), there is a direct correlation between the structure at the top of the Wabamun, the thickness of the Wabamun interval, and the thickness of the residual Wabamun salt. More specifically, at well locations outside the interpreted

Leduc salt basins (Fig. 5), the top of the Wabamun is consistently up to 40 m lower than the respective restored Wabamun structure. At control well locations, the difference between these two structure contour maps when added to the thickness of the residual Wabamun salt is consistently about 40 m. At control locations within the Leduc salt basins, the difference between the two Wabamun structure contour maps, plus the thickness of any residual Wabamun salt, less our estimate of the thickness of Leduc salt that was dissolved in post-Wabamun time, is consistently about 40 m. These relationships support the thesis that about 40 m of

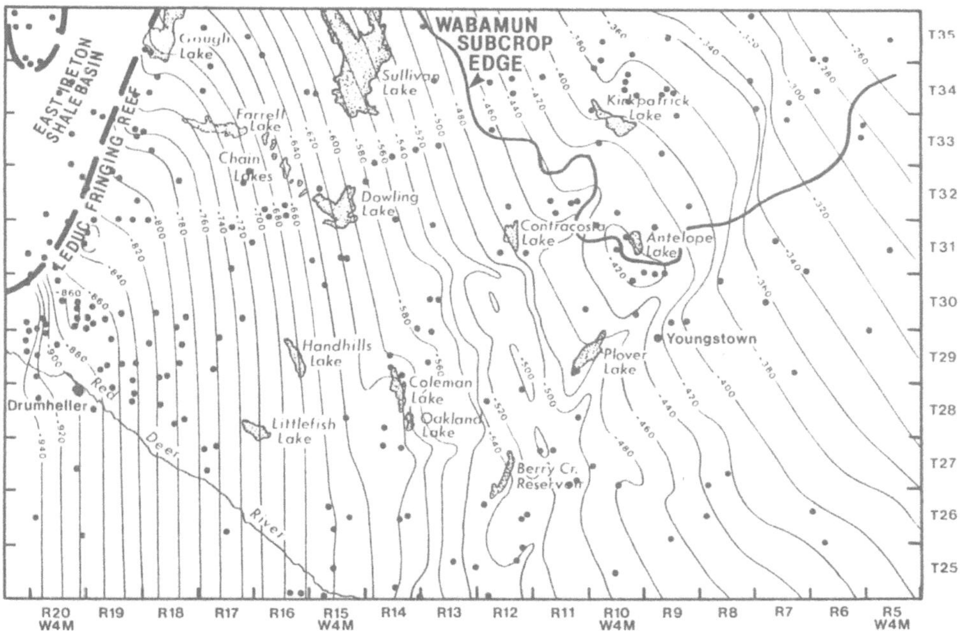

Figure 8. Structure map (in meters) of subsea depth to top of Leduc. Bold dots highlight those sections for which there is deep well control, that is, to top of Leduc. In those areas where deep well control is absent, contouring of present-day Leduc structure was constrained by apparent local trends and structural patterns along shallower horizons.

Wabamun salt was deposited uniformly throughout the Youngstown study area. At those shallow well locations where Wabamun control is not available, the contouring of the present-day Wabamun structure map was constrained by interpretive contouring trends and patterns of structural relief along post-Wabamun horizons.

PRESENT-DAY DISTRIBUTION OF THE LEDUC AND WABAMUN SALTS

Leduc Salts

The present-day distribution of the Leduc salt (Fig. 2), was estimated principally from three maps: (1) original Leduc salt thickness (Fig. 5); (2) present-day Leduc structure (Fig. 8); and (3) restored Leduc structure (Fig. 9). At each well control point, the present-day salt-

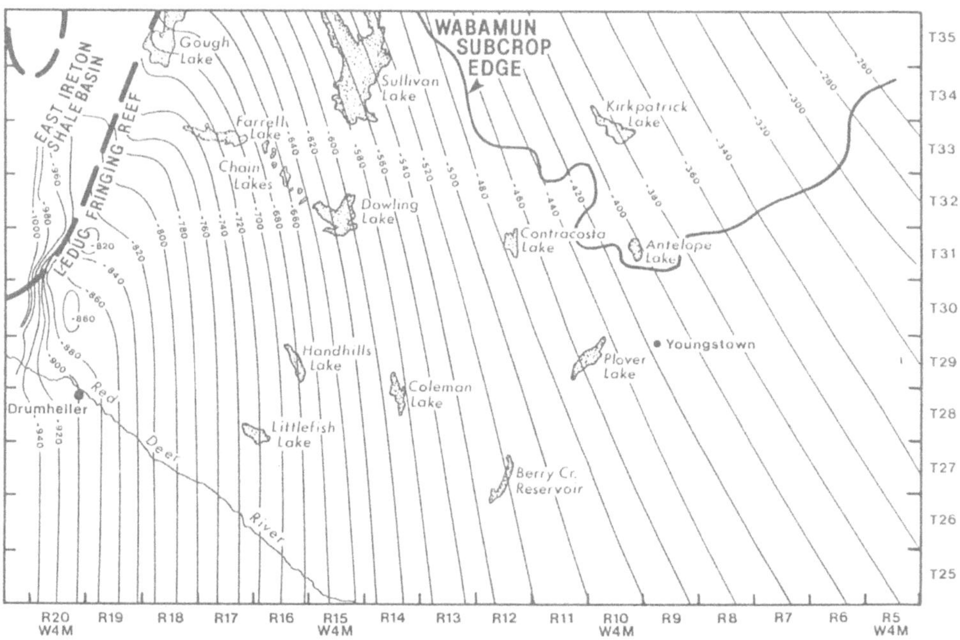

Figure 9. Restored Leduc structure map (in meters). Ideally, contours represent pattern of structural relief that would be observed if all of original Leduc salt (dissolved in post-Leduc time) were replaced.

thickness estimate (A_L) is equal to the original (end Leduc time) salt thickness (B_L) less the difference between the restored (C_L) and present-day (D_L) Leduc structure values. Following Anderson and Brown (1991b), we use the formula:

$$A_L = B_L - (C_L - D_L).$$

At those shallow well locations where Leduc control is not available, the contouring of the present-day Leduc structure map was constrained by local interpretive trends and structural relief along shallower horizons. At all available deep well control points, our present-day salt-thickness estimate (A_L) and the actual thickness of the residual salt (according to the well-log data) are effectively the same. The original and present-day thicknesses of the Wabamun salts were estimated concurrently with our analysis of the Leduc salts. This was necessary in order to ensure that our Leduc and Wabamun net-salt paleodistribution maps were compatible.

Wabamun Salts

The present-day distribution of the Wabamun salt (Fig. 3) was based largely on the analyses of three maps: (1) original salt thickness; (2) present-day Wabamun structure; and (3) restored Wabamun structure. Although a map of the original salt thickness is not shown here, Figure 6 is almost identical to it because up to the end of Viking time only slow dissolution occurred and it was localized around the Wabamun subcrop edge. At each well control point our estimate of present-day net salt thickness (A_W) is equal to the original salt thickness (40 m) plus our estimate of the thickness of Leduc salt dissolved in post-Wabamun time (B_W) less the difference between the restored (C_W) and present-day (D_W)Wabamun structure values. Following Anderson and Brown (1992b), we use the formula:

$$A_W = B_W - (C_W - D_W).$$

At all available deep well control points, our present-day salt-thickness estimate (A_W) and the actual thickness of the Wabamun salt remnant (according to the well-log data) were effectively the same. In those areas where Wabamun control is sparse, the contouring of the present-day Wabamun structure map was constrained by apparent local trends and structural patterns along shallower horizons.

PALEODISTRIBUTION OF THE LEDUC AND WABAMUN SALTS

Methodology

The paleodistributions of the Leduc and Wabamun salts were estimated following the three steps outlined here:

Step 1. The subsea depths to the eight horizons listed here (Fig. 1) were determined at about 3000 well locations within the study area; (increasing depth and age, top to bottom):

(A) top Lea Park;
(B) top Colorado;
(C) top Second White Speckled Shale (Second Specks);
(D) top Viking;
(E) top Mannville;
(F) base Cretaceous (top Paleozoic unconformity);
(G) top Wabamun;
(H) top Leduc.

Step 2. Both present-day and restored structure maps were drafted for each of these eight structural data sets. (Figs. 8 and 9 are the present-day and restored maps for the Leduc.) Ideally, the restored map for a particular horizon represents the pattern of structural relief that would exist if we replaced all of the Leduc and Wabamun salts that were dissolved after deposition of that particular horizon. Differences in structural relief between corresponding present-day and restored structure maps are therefore estimates of the thickness of all of the salt (both Leduc and Wabamun) removed after the deposition of the relevant strata.

Step 3. The paleodistribution of the Leduc and Wabamun salts were determined at the following times following the described methodology:

(A) end Leduc (Fig. 5);
(B) end Wabamun (Fig. 6);
(C) end Mannville;
(D) end Viking (Fig. 6);
(E) end Second Specks;
(F) end Colorado (Fig. 7);
(G) end Lea Park;
(H) present-day (Figs. 2 and 3).

The net salt thicknesses at these times were estimated on the basis of: (1) the original salt-distribution maps (Figs. 5 and 6); (2) the present-day structure maps (Fig. 8); and (3) the restored structure maps (Fig. 9). For example, the total thickness of Leduc salt at the end of Viking time (E_L) was calculated to be equal to the present-day thickness of the Leduc salt (F_L) plus the difference between the restored (G_L) and present-day (H_L) Viking structure values less the thickness of Wabamun salt leached in post-Viking time (I_L). We use the equation:

$$E_L = F_L + (G_L - H_L) - I_L.$$

In an analogous manner, we concurrently estimated the thicknesses of the Leduc and Wabamun salts at the other selected times during the Cretaceous. At most of our control well locations, this process was relatively straightforward (or at least constrained) for one of several reasons: (1) Leduc salts were not deposited (Fig. 5) and hence only Wabamun salts were leached; (2) thick (on the order of 45 m) Leduc salts were preserved (Fig. 2) and probably only Wabamun salts were leached; (3) the thicknesses of the Wabamun and Leduc salt remnants were known from well control; and (4) both salts had been leached totally from the section. The agreement between our present-day net-salt-thickness estimates and the actual thicknesses of the salt remnants (as determined from well control), and the compatibility of the suite of reconstructed salt-distribution maps indicates that the paleoreconstructed salt-thickness estimates probably are reliable to within 10 m or less. Our estimates of present-day net salt thickness are probably reliable to within 5 m.

Paleoreconstruction

Selected maps showing the past and present-day distribution of the Leduc and Wabamun salts are depicted in Figures 2, 3, 5, 6, and 7. On the basis of the full suite of such reconstructed net-salt-thickness maps we have concluded:

(1) The Leduc salt in the study area was deposited in two, possibly interconnected, fault-controlled, restricted basins on the shelfward side of the developing Leduc fringing-reef complex (Fig. 5). The maximum thickness of the Leduc salt in the study area was on the order of 45 m. There are several reasons for concluding that the

Leduc salt basins were fault-controlled: (i) prominent diffraction patterns and vertical offsets along near-basement events (pre-Leduc) are imaged on seismic data from the Leduc salt-basin areas; (ii) the interpreted Leduc salt basins are oriented more-or-less parallel to the edge of the fringing-reef complex (Fig. 5), indicating that both features could have been influenced by movement along preexisting orthogonal planes of weakness; (iii) the western edges of the Leduc salt basins are interpreted to be abrupt rather than gradational, a feature more characteristic of a fault-controlled basin than of a gradual change in the depositional environment; (iv) the dissolution of both the Wabamun and Leduc salts during the mid-Late Cretaceous initiated along an orthogonal set of lineaments; and (v) with respect to the western salt basin, dissolution of the Leduc salts has been most extensive near the basin margins, perhaps as a result of the reactivation of the faults that originally controlled the development of the basin.

(2) About 40 m of Wabamun salt was deposited uniformly throughout the Youngstown study area (Fig. 6). This conclusion is based on our observation that there is a direct correlation between structure at the top of the Wabamun, thickness of the Wabamun interval, and net thickness of Wabamun salt dissolved in post-Wabamun time. All three values vary, in a relative sense, by 40 m or less.

(3) The dissolution of the Leduc and Wabamun salts was initiated and enhanced by some or all of four principal processes: (i) near-surface exposure, as a result of erosion during the pre-Cretaceous hiatus; (ii) regional faulting/fracturing during the mid-Late Cretaceous; (iii) glacial loading and unloading (Fig. 10); (iv) partial dissolution of the underlying salts.

(4) With respect to erosion during the pre-Cretaceous, the earliest phases of Wabamun salt dissolution initiated along the Wabamun outcrop edge during the pre-Cretaceous hiatus (Fig. 6). As indicated on the suite of reconstructed maps, the dissolution front that was established along the subcrop edge migrated towards the southwest in a self-sustaining manner. An isolated remnant of Wabamun rock salt is seen in the seismic section of Figure 11. The observations that the Mississippian (top of the Paleozoic

subcrop) and earliest Cretaceous (Mannville) horizons drape across the residual salt support the interpretation that dissolution along and immediately to the southwest of the Wabamun subcrop occurred during Early Cretaceous time (Fig. 7).

(5) The orthogonal patterns displayed on the suite of Wabamun and Leduc salt paleodistribution maps (Figs. 2, 3, 5, 6, and 7) support the interpretation that dissolution fronts developed along a suite of orthogonally oriented (NNE-WNW) regional fault planes during the mid-Late Cretaceous. The apparent expansion of these areas of dissolution through time indicates that the dissolution fronts migrated laterally. It is noteworthy that the edges of the fringing reef complex, as well as edges of the Leduc salt basins, are consistent with the strikes of the hypothesized faults (Fig. 5), indicating that possibly these lineaments were reactivated planes of weakness. A seismic profile across an interpreted NNE-trending fault is presented as Figure 11. The observation that most of the dissolution occurred after the deposition of the Second Specks and prior to the end of Colorado time supports the interpretation that faulting was mid-Late Cretaceous in age (Fig. 1).

(6) Several lakes to the west of the postulated Leduc salt basins (Gough, Sullivan, and Dowling; Fig. 3) are situated in areas where the Wabamun salts are thin or absent; and all of the larger lakes in the west Leduc salt basin area (Plover, Antelope, Contracosta, Coleman, Oakland, and the Berry Reservoir; Fig. 2) are situated in areas where the Leduc salts have been extensively leached. These relationships indicate that significant salt leaching has occurred in Recent time, possibly in response to glacial unloading (Anderson and Brown, 1991b, 1992; Anderson and Knapp, 1993).

(7) Cold Lake and Prairie Evaporite rock salts of the Middle Devonian Elk Point Group are preserved within the study area. Although we have seen no evidence that these salts have been leached here, it is possible that the dissolution of these underlying rock salts could have triggered the leaching of the Wabamun or Leduc salt. Within the confines of the postulated Leduc salt basins, the dissolution of the Wabamun salt could have been triggered by the leaching of the underlying Leduc salt. Such

leaching could have occurred at any time after the deposition of the Wabamun salt.

(8) The dissolution of the Leduc and Wabamun salts has occurred at various times during the geologic past and, at least in places, has been more-or-less continuous since deposition. Several trigger mechanisms have been identified and it has been concluded that leaching is self-sustaining. With respect to self-perpetuation, we note that the established dissolution fronts do not advance at a uniform rate. These observations indicate that a number of secondary factors influence salt dissolution. Consideration should be given to effects of features or processes such as regional tectonism, periods of emergence, underlying reefs, the differential compaction of pre-salt sediment, uneven loading and unloading, gypsum to anhydrite conversion (and vice-versa), facies changes within both the salt and encompassing strata, the local hydrological and geochemical environment and changes therein, and the effects of oil and gas activity.

(9) The timing of salt dissolution is of significance to the explorationist for several reasons: (i) stratigraphic traps can form where reservoir facies were either preferentially deposited or preserved in salt-dissolution lows; (ii) reservoir facies can develop in high-energy environments such as topographic highs that are controlled by salt edges or remnants; (iii) structural traps can form where reservoir facies are draped across salt remnants or collapse features; and (iv) salt remnants can be misinterpreted as reefs, faults, or other structural features (Anderson, 1991; Anderson and Brown, 1987, 1991a; Anderson, Brown, and Hinds, 1988a, 1988b; Anderson, White, and Hinds, 1989).

(10) The methodology developed by Anderson and Brown (1991b), and employed in this study, is based on the premise that the dissolution of subsurface salt causes the contemporaneous collapse of the overlying strata. The thickness of leached salt and the magnitude of subsidence are assumed to be equal. Although the methodology is relatively robust, as evidenced by the suite of compatible, consistent, restored structure maps, in places it might not adequately account for processes such as: (i) nonuniform

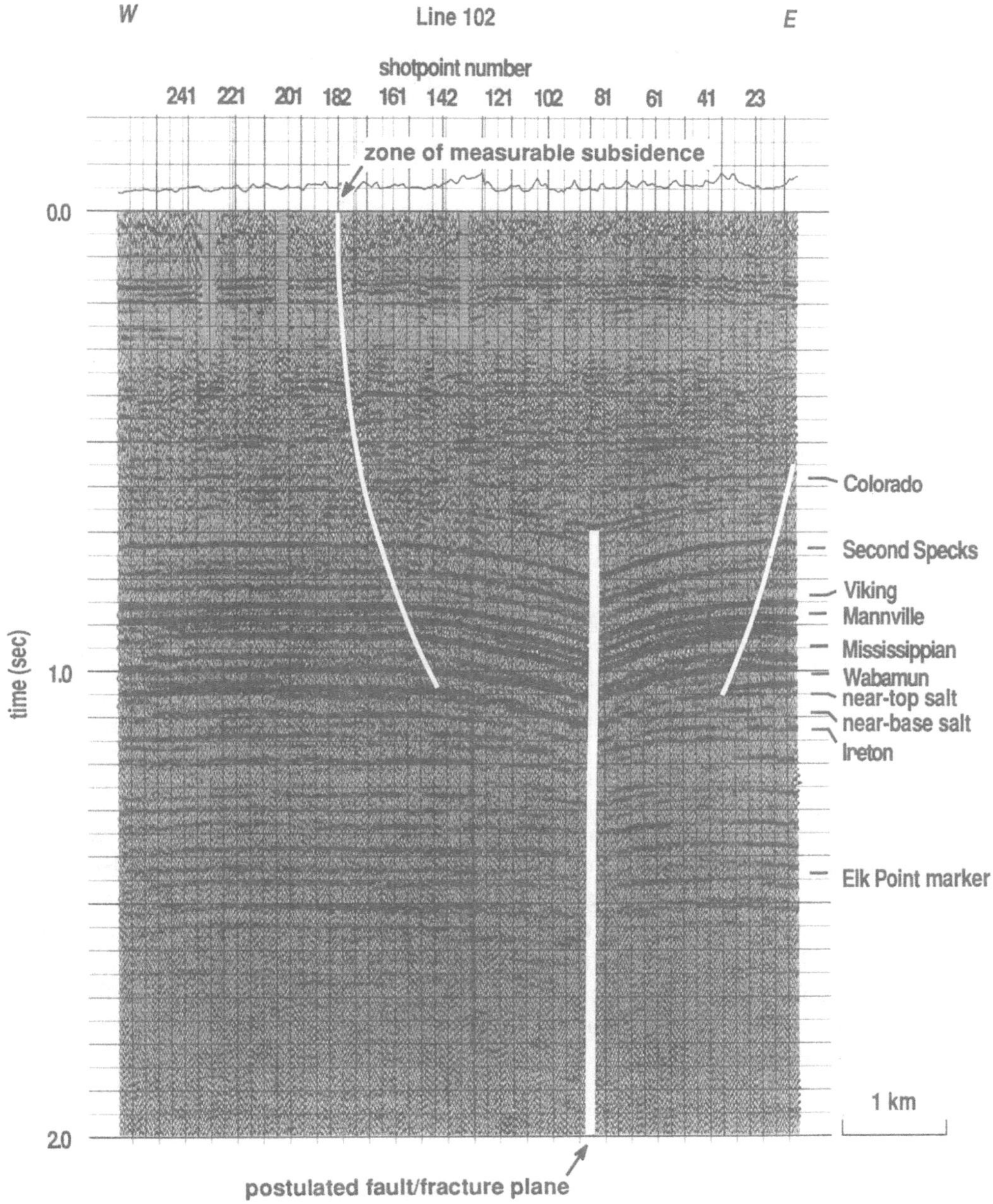

Figure 10. West-east seismic profile across NNE-trending salt-dissolution feature (T33, R19W4M; Fig. 3) in Youngstown area of Alberta (after Anderson and Brown, 1992). Dissolution is thought to have initiated along reactivated shear zone during mid-Late Cretaceous. Subtle time-structural relief along shallowest events may be indicative that leaching and subsidence has occurred in Recent time in response to glacial loading and unloading.

primary deposition; (ii) erosion; (iii) differential compaction; (iv) horizontal strain associated with collapse; (v) differential compaction of infill (compensation) sediments; (vi) salt flow; (vii) dissolution of underlying salts; and (viii) faulting.

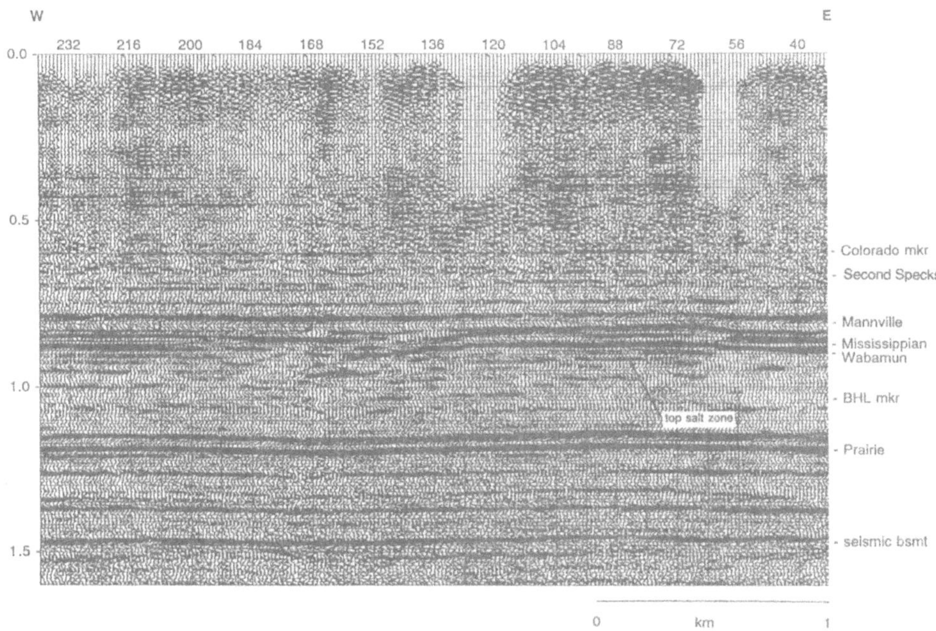

Figure 11. West-east seismic profile across isolated remnant of Wabamun rock salt (T34-35, R11W4M; Fig. 3) from Youngstown area of Alberta (after Anderson, Brown, and Cederwall, 1994). Salt-bearing interval is imaged as moderately high-amplitude peak-trough sequence. Relative to those areas where rock salt has been leached, salt-bearing interval is characterized by an anomalously thick Wabamun interval, up to 8 ms of velocity pullup along pre-salt events, and up to 25 ms of relative relief at Mississippian level. These observations indicate that up to 40 m of residual salt is preserved locally.

SUMMARY

About 40 m of Wabamun Group salt were deposited throughout the Youngstown area; the Leduc salts (up to 45 m thick), in contrast, are thought to have been deposited only within two principal, fault-controlled, restricted salt basins on the shelfward side of the developing Leduc fringing-reef complex. Both salts have been dissolved extensively and more-or-less continuously since deposition and by a variety of processes.

Our investigations indicate that dissolution of the Wabamun and Leduc salts was initiated and enhanced by some or all of four principal processes: (1) the near-surface exposure of these salts, as a result of the erosion of the overlying Paleozoic sediment during the pre-Cretaceous hiatus; (2) the partial dissolution of underlying salts; (3) regional faulting/fracturing during the late Cretaceous; and (4) glacial loading and unloading.

Here, we have presented a suite of reconstructed salt-distribution maps in order to substantiate the theses of continuous leaching and diverse mechanisms. We have discussed our reconstruction methodology and concluded that it is relatively robust. We are hopeful that this technique can be applied to other areas where the effects of leaching are not masked by processes such as erosion, faulting, compaction, or salt flow.

ACKNOWLEDGMENTS

We wish to thank Renaissance Energy, Gulf Canada Resources, Crestar Energy and Poco Petroleums, all of Calgary, for their generous support of this and related studies, and to acknowledge the skilled draftsmanship of Bill Matheson (Matheson Graphics Services, Calgary).

REFERENCES

AGAT Laboratories, 1988, Table of formations of Alberta: AGAT Laboratories, Calgary.

Anderson, N.L., 1991, Dissolution of the Wabamun Group salt: exploration implications, *in* Cavanaugh, T.D., ed., Integrated exploration case histories, North America: The Geophysical Society of Tulsa, Spec. Publ., p. 179-210.

Anderson, N.L., and Brown, R.J., 1987, The seismic signatures of some western Canadian Devonian reefs: Jour. Can. Soc. Exploration Geophysicists, v. 23, no. 1, p. 7-26.

Anderson, N.L., and Brown, R.J., 1991a, A seismic analysis of Black Creek and Wabamun salt collapse features, western Canadian sedimentary basin: Geophysics, v. 56, no. 5, p. 618-627.

Anderson, N.L., and Brown, R.J., 1991b, Reconstruction of the Wabamun Group salts, southern Alberta, Canada, *in* Cavanaugh, T.D., ed., Integrated exploration case histories, North America: The Geophysical Society of Tulsa, Spec. Publ., p. 145-178.

Anderson, N.L., and Brown, R.J., 1992, Dissolution and deformation of rock salt, Stettler area, southeastern Alberta: Can. Jour. Exploration Geophysics, v. 28, no. 2, p. 128-136.

Anderson, N.L., and Cederwall, D.A., 1993, Westhazel General Petroleums Pool: Case history of a salt-dissolution trap in west-central Saskatchewan, Canada: Geophysics, v. 58, no. 6, p. 889-897.

Anderson, N.L., and Knapp, R.W., 1993, An overview of some large-scale mechanisms of salt dissolution: Geophysics, v. 58, no. 9, p. 1375-1387.

Anderson, N.L., Brown, R.J., and Hinds, R.C., 1988a, A seismic perspective on the Panny and Trout fields of north-central Alberta: Can. Jour. Exploration Geophysics, v. 24, no. 2, p. 154-165.

Anderson, N.L., Brown, R.J., and Hinds, R.C., 1988b, Geophysical aspects of Wabamun salt distribution in southern Alberta: Can. Jour. Exploration Geophysics, v. 24, no. 2, p. 166-178.

Anderson, N.L., White, D.G., and Hinds, R.C., 1989, Woodbend Group reservoirs, *in* Anderson, N.L., Hills, L.V., and Cederwall, D.A., eds., Geophysical atlas of western Canadian hydrocarbon pools: Can. Soc. Exploration Geophysicists and Can. Soc. Petroleum Geologists, p. 101-132.

Anderson, N.L., Brown, R.J., and Cederwall, D.A., 1994, A seismic analysis of erosion, salt dissolution and structural relief at the Paleozoic subcrop in the Sullivan Lake area, south-central Alberta, Canada: Can. Jour. Exploration Geophysics, v. 30, no. 1, p. 51-59.

Andrichuk, J.M., and Wonfor, J.S., 1953, Late Devonian geologic history in Stettler area, Alberta, Canada: Alberta Soc. Petroleum Geologists, News Bull. 1, no. 12, p. 3-5.

Belyea, H.R., 1964, Woodbend, Winterburn and Wabamun groups, *in* McCrossan, R.G. and Glaister, R.P., eds., Geological history of western Canada: Alberta Soc. Petroleum Geologists, p. 66-88.

Hopkins, J.C., 1987, Contemporaneous subsidence and fluvial channel sedimentation: Upper Mannville C pool, Berry Field, Lower Cretaceous of Alberta: Am. Assoc. Petroleum Geologists Bull., v. 71, no. 3, p. 334-345.

Meijer Drees, N.C., 1986, Evaporitic deposits of western Canada: Geol. Survey Canada Paper 85-20, 118 p.

Oliver, T.A., and Cowper, N.W., 1983, Wabamun salt removal and shale compaction effects, Rumsey area, Alberta: Can. Petroleum Geology Bull., v. 31, no. 3, p. 161-168.

Wonfor, J.S., and Andrichuk, J.M., 1953, Upper Devonian in the Stettler area, Alberta, Canada: Alberta Soc. Petroleum Geologists, News Bull. 1, no. 9, p. 3-6.

USE OF THE COMPUTER FOR THE STRUCTURAL ANALYSIS OF THE ORDOVICIAN SEDIMENTARY BASIN IN ESTONIA

Alla Shogenova
Estonian Academy of Sciences, Tallinn, Estonia

ABSTRACT

Structural analysis of the Ordovician sedimentary basin in Estonia and estimation of fracture zones with small amplitude was made with a computer with an example from the Rakvere oil shale and phosphorite deposit. Trend analysis was the main applied method. A database containing the spatial coordinates of 500 boreholes and depth of 12 stages and formations, interpreted from gamma-ray and resistivity logs, was created using a Lotus spreadsheet. A BASIC program was written to read data from the database, calculate the depths and thicknesses of the appropriate formations, and calculate the trend coefficients. The first-degree trend coefficients were computed for modeling regional structure maps (Estonian Homocline) and regional isopachous maps of the sedimentary units. Structure and isopachous maps of 12 stages and formations in the Ordovician carbonate bedrock were made using kriging. The result of the structural and thickness trend analysis showed four new zones with rather small amplitudes established for the Rakvere deposit: three fracture zones with subsidence of beds and thinning, and one fold zone with the thinning over on anticlinal fold.

Geologic Modeling and Mapping
Edited by A. Förster and D. F. Merriam, Plenum Press, New York, 1996

INTRODUCTION

The geological structure of Estonia is essentially the Estonian Homocline - a regional structure on the south slope of the Baltic Shield, situated in the northwestern part of the Russian Platform (Suveizdis, 1979). A large amount of data is available in Estonia (20,000 boreholes) showing the structure of the relatively thin sediments (less than 1 km) in the early Paleozoic basin. Structural analysis of the area was completed traditionally by hand. At this time the main structural features of this region were displayed, and zones of disturbances were outlined (Puura, Vaher, and Tuuling, 1986, 1987). The Estonian Homocline slopes gently to the south (2.9 - 4.4 m/km). The slope is complicated by major and minor linear disturbances and fracture zones, caused by fault movements in the crystalline basement. Most disturbed areas are complicated by karsting, and carbonate rocks are dolomitized. As is known from the experience of exploitation of the Baltic oil shale and phosphorite basin (Fig. 1), fracture zones - nonamplitude disturbances, or those with rather small amplitudes -are the most dangerous for mining (Gazizov, 1971; Heinsalu and Andra, 1975). They usually are not located during detailed prospecting of deposits. For locating zones of disturbances, resistivity method with dipole configuration of electrodes was used (Vaher, 1983, 1986), but it did not provide exact and detailed information about location and structure of these zones.

Computerization of the structural analysis of the Ordovician carbonate bedrock, overlying the commercial seam of phosphorites (Pakerort Stage), including the commercial seam of the oil shale kukersites (lower part of Kukruse Stage), was completed by the example of the Rakvere deposit. Trend analysis of the structural surfaces and thicknesses of carbonate rocks revealed the zones of disturbances, characterized the structure of detected zones, compared their structure at different stratigraphical levels, and allowed conclusions to be made about geological processes which led to the formation of these zones.

INPUT DATA

For the structural analysis of the Rakvere deposit, the data from gamma-ray logs in 500 boreholes, resistivity logs in 290 boreholes measured by Geological Survey of Estonia were used. The logging data

Figure 1. Location of study area (Rakvere oil shale and phosphorite deposit).

available in the analog form were interpreted by a technique established by Shogenova (1989a), which permits determination of the depths of 12 stages and formations (Table 1) in the Ordovician succession and with accuracy high enough to the reveal the low-amplitude zones.

The analysis of fracturing of carbonate rocks by using these logging data has been described in earlier publications (Shogenova, 1989b, 1991a, 1991b).

For compiling the database, the spreadsheet from Lotus software, Autocad program, and BASIC programs were used. It is easy to use the spreadsheet for borehole data. The coordinates of 500 boreholes were digitized using a digitizer and Autocad program. They were imported to the database using the BASIC routine. The data from the wells and depth of stages and formations in the boreholes were input from the keyboard or digitized directly from the logs (Table 1).

After completion of the database it was essential to correct all errors. The sources of errors in this situation may be the result of the following: (1) incomplete depths of some formations in the database because of the absence of resistivity logs in some boreholes; (2) thinning of upper formations in some boreholes or their total absence as a result of erosion (for example in the north and northeast of the described region); (3) mistakes in the elevations of the boreholes; (4) inaccurate location of boreholes on the map.

If some of the stages or formations are absent in the upper part of the section, those depth are marked by zero and in the middle or lower parts of the section by the value of the previous depth. This permits the BASIC routine, used for exporting data from database, to eliminate those boreholes with an absence of the data.

Table 1. Stratigraphic scheme of study area according to Hints and others (1993). * - stages and formations included in database.

ERA	SYSTEM	SUBSYSTEM	STAGES	REGIONAL STAGES	INDECES	FORMATIONS	INDECES
Quaternary deposits *					Q		
EARLY PALEOZOIC	ORDOVICIAN	Upper	ASHGILL	Nabala	$O_{2-3}nb$	Saunja	O_2sn
						Paekna	O_2pk
		MIDDLE	CARADOC	Rakvere	O_2rk	Rägavere	O_2rg
				Oandu	O_2on	Hirmuse	O_2hr
				Keila *	O_2kl	Keila	O_2kl
				Jõhvi *	O_2jh	Jõhvi	O_2jh
				Idavere *	O_2id	Vasavere	O_2vs
						Tatruse	O_2tt
			LLANDEILO	Kukruse *	O_2kk	Viivikonna	O_2vv
				Uhaku	O_2uh	Kõrgekallas *	O_2kr
			LLANVIRN	Lasnamägi	O_2ls	Väo *	$O_2vä$
				Aseri *	O_2as	Aseri	O_2as
				Kunda *	$O_{1-2}kn$	Napa	O_2np
						Loobu	O_2lb
		LOWER	ARENIG			Sillaoru	O_1sl
				Volhov *	O_1vl	Toila	O_1tl
				Latorp *	O_1lt	Leetsu	O_1lt
			TREMADOC	Pakerort	O_1pk	Türisalu	O_1tr

In this manner every borehole is described in the database by the following parameters: N - number of borehole, X, Y - spatial coordinates of boreholes, elev. - elevation of the borehole, and depth of 12 stages or formations of the Ordovician carbonate section (Table 1), and Quaternary sediments.

To export data from the database, it was translated from the Lotus format (.WK1 file) into the Data Interchange Format (.DIF file).

TECHNIQUE OF THE SPATIAL ANALYSIS

The problem of the spatial data analysis has been described in publications by Davis (1986), Nikitin (1986), Phillip and Watson (1987), Merriam and Jewett (1988), Bohnam-Carter (1989), and Harff, Lange, Olea (1992). The possibilities of using the computer mapping procedures were tested with different gridding algorithms applied in commercial packages and using BASIC programs. The best results and those closest to the hand-contoured interpolation were obtained using a kriging algorithm with selection of the search radius corresponding to the object under study. Therefore, the kriging algorithm was used for mapping different parameters.

The main mathematical method which was used for the spatial and isopachous analysis was trend analysis as described by Davis (1986). The regional structure of the studied area is the Estonian Homocline and the structural and thickness data were modeled by the first-degree polynomial trend surfaces.

For the spatial analysis in the sedimentary basin, the following techniques were applied: (1) modeling of the structure maps, their three-dimensional views using the kriging algorithm; (2) computing of the first-degree polynomial trend coefficients with gridding on the map of the same density as structure maps and plotting of the first-degree trend surfaces; (3) plotting of the residual maps constructed by subtracting of first-degree trend surfaces from structure maps; (4) plotting of the isopachous maps for every formation and for sums of formations using a kriging algorithm; (5) computing of the first-degree trend coefficients with gridding of the map of the same density as the isopachous maps and plotting of the first-degree trend thicknesses; and (6) plotting of the residual isopachous maps constructed by subtracting first-degree trend thicknesses from the isopachous maps.

To export data from the database and calculate the trend coefficients, a BASIC program consisting of subroutines and working in an interactive mode with the following menu was used:

(1) Read data from .DIF file
(2) Calculate Z and print X,Y,Z on display
(3) Save X,Y,Z in .DAT file
(4) Read X,Y,Z from any .DAT file
(5) Calculate trend coefficients and residuals
(6) Calculate of statistical characteristics
(7) Exit.

Subroutine number 2 of this menu has three subroutines:

(1) Calculation of depth
(2) Calculation of thickness
(3) Calculation of sum of depths.

Numbers 5 and 6 of this menu are subroutines as described by Davis (1986).

After applying this program all the files and trend coefficients are computed for the gridding and plotting of maps and three-dimensional views by the commercial package according to the described sequence of procedures.

SPATIAL ANALYSIS OF THE ORDOVICIAN CARBONATE ROCKS

Spatial Structural Analysis

Structure of the 12 stages and formations (Table 1) in the Ordovician carbonate basin was studied by the structure contour, trend, and residual maps and three-dimensional views obtained using the described technique (Figs. 2 and 3). The residual maps, constructed by subtracting first-degree trend surfaces from the structure maps, show zones with subsidence (Figs. 2C, 3C). From the maps of the first-degree trend surfaces, the regional strike and dip of formations were defined. The regional NE-strike azimuth of formations changes from 83° (O_1lt) (Fig. 2B) to 86.5° (O_2*rg*)(Fig. 3B). The regional SE dip of formations is changing from 2.96 m/km (O_2*rg*) (Fig. 3B) to 3.6 m/km (O_2kk).

The negative anomalies on the residual structure maps define zones with subsidence or synclinal folds of structures. The positive

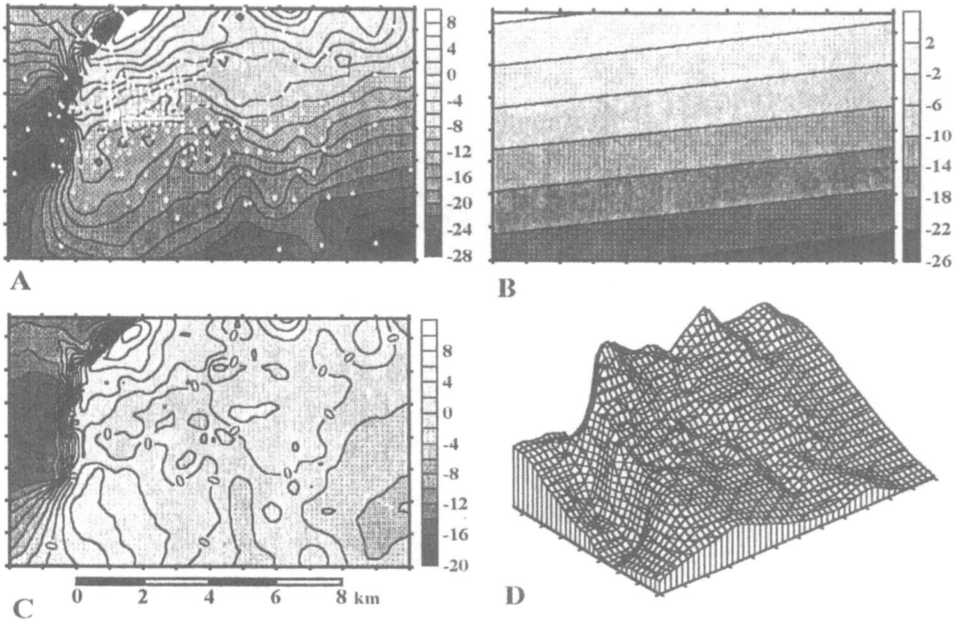

Figure 2. A, Structure contour map of Rakvere deposit, drawn on bottom of Latorp Stage (lower reference surface of section). Borehole control shown by white points. Contours in meters from sealevel; B, first-degree trend surface of Latorp Stage; C, residual map, constructed by subtracting first-degree trend surface from structure contour map of Latorp Stage (A). Contour interval is 2 m. (D) Three-dimensional view drawn on bottom of Latorp Stage.

anomalies on these maps reflect anticlinal folds. The well-known Aseri zone of disturbances (west part of the study area) is observed clearly by the more dense location of contours on structure maps and on residual maps as well. But zones with small amplitudes are determined on the structure maps with difficulty - usually by structural noses, seldom by isometric anomalies (Figs. 2A, 2C, 3A, 3C). On the residual maps, constructed by subtracting the first-degree trend surface from structure maps, zones with subsidence up to 5 meters, determined by contours with negative values, are revealed (Figs. 2C, 3C).

Zones with small amplitudes revealed on the residual maps change with depth and are not in similar locations in different zones; some decrease upwards and others are emphasized.

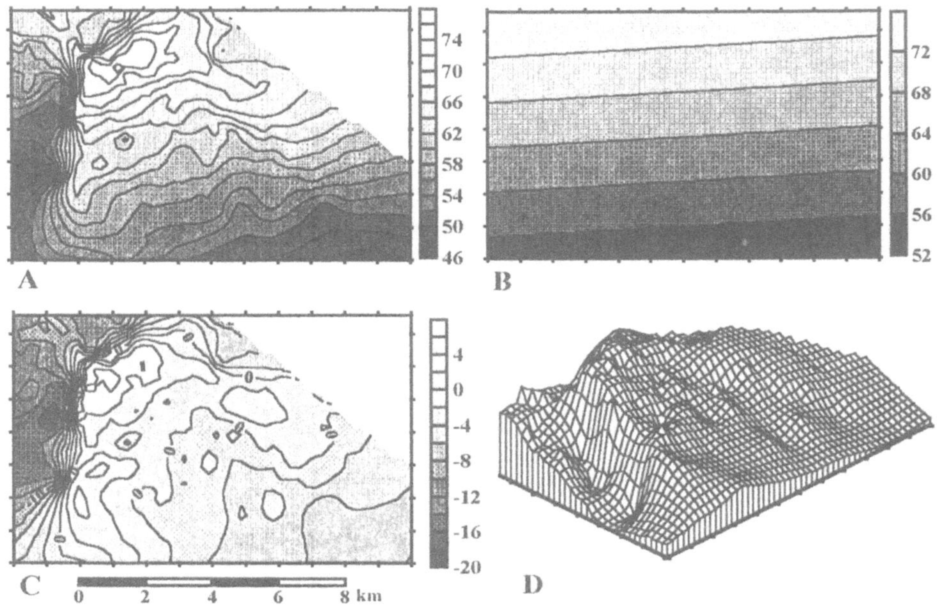

Figure 3. A, Structure contour map of Rakvere deposit, drawn on bottom of Rägavere Formation (upper reference surface of section). Contours in meters above sealevel. Original borehole control shown in Figure 2A. On blanked area Rägavere Formation is eroded. B, First-degree trend surface of Rägavere Formation; C, residual map, constructed by subtracting first-degree trend surface from structure contour map of Rägevere Formation (A). Contour interval is 2 m. D, Three-dimensional view drawn on bottom of Rägavere Formation.

Spatial Thickness Analysis

Using the described techniques, isopachous maps of stages and formations of the Ordovician carbonate basin (Figs. 4A, 5A), of different parts (Fig. 6C), and the widespread ones (Fig. 6A) were compiled, maps of the first-degree trend thicknesses (4B, 5B) and residual thicknesses were produced (Figs. 4C, 5C, 6B, 6D). As it observed on the maps (Figs. 4B, 5B), the regional trends of thickness and the amplitudes of these features are distinguished for different formations. The data conform generally to the described directions of thickness changes and their amplitudes (Puura, 1987). The change of these parameters is connected with different paleogeographical conditions of sedimentation.

It should be noted that thicknesses of the lower stages O_1lt (Fig. 4A-4C), O_1vl, $O_{1-2}kn$, and O_2as (Table 1) in the study area have undergone relatively slight changes and are distinguished weakly from the regional background. The amplitude of the alteration of the background thickness is there not more than 1 meter. Directions of thickness alterations coincide only in the Latorp and Volhov Stages and are distinguished in all other formations. The amplitude of the regional changes of thickness rise upwards and reach in $O_2v\ddot{a}$, O_2kr, O_2kk to some meters. As they compensate each other, the amplitude of regional alteration of their total thickness reaches to three meters. Also the high negative residuals of thicknesses (thickness loss) are observed in these formations, equal to 1.5 meters in $O_2v\ddot{a}$, up to 3.0 meters in O_2kr and O_2kk. The rocks of the upper formation O_2id and O_2jh are characterized by the common features of the regional background - nearly coinciding directions of thickness increase from southwest to northeast and also a small change of regional thickness, being about 0.8 meters in the area under study in O_2id and 0.7 m in O_2jh. In some places in these

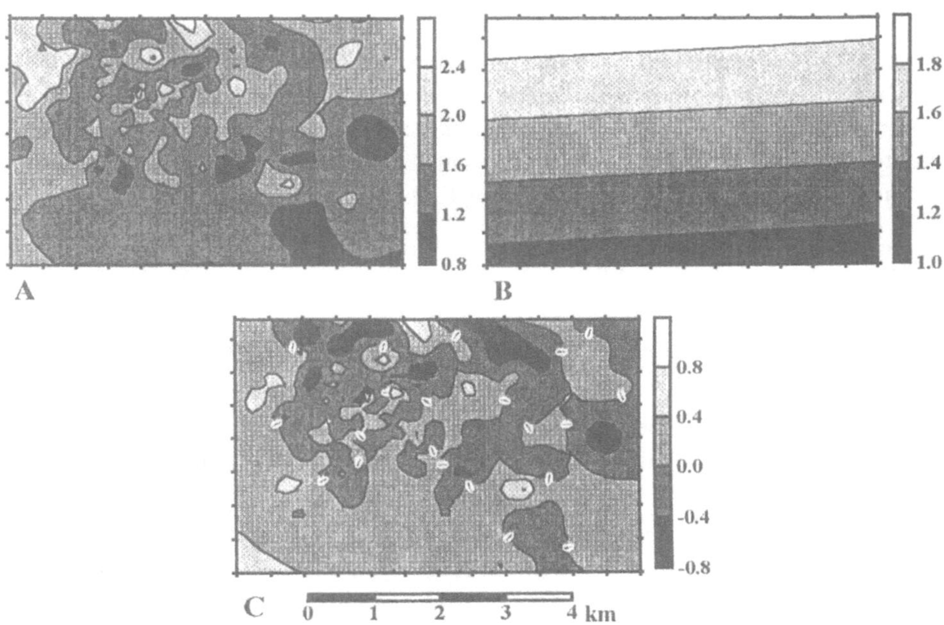

Figure 4. A, Isopachous map of rocks of Latorp Stage. Contour interval is 0.4 m. Original borehole control shown in Figure 2A. B, First-degree trend thickness of rocks of Latorp Stage. Contour interval is 0.2 m. C, Residual map constructed by subtracting first-degree trend surface from isopachous map of rocks of Latorp Stage. Contour interval is 0.4 m.

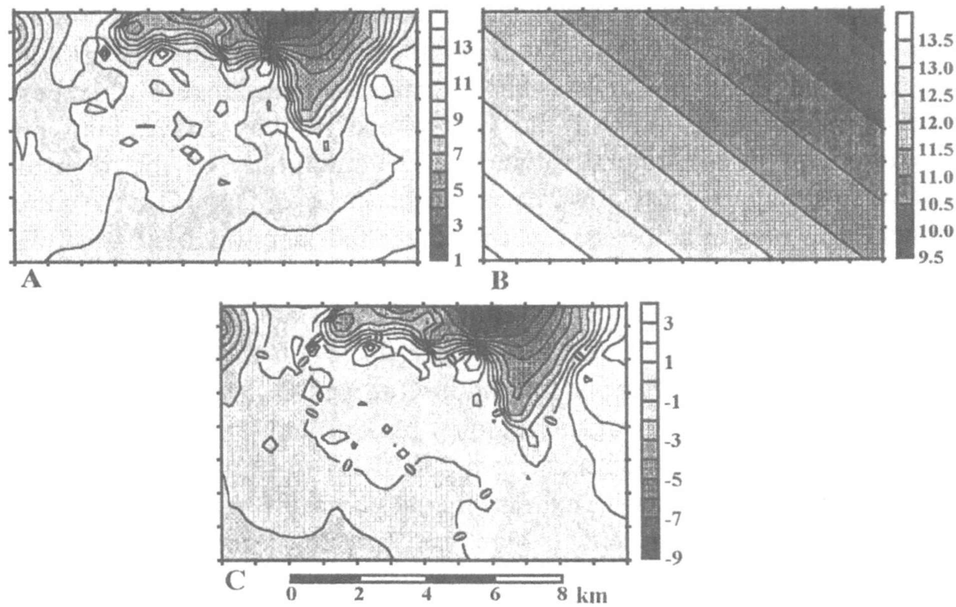

Figure 5. A, Isopachous map of rocks of Keila Stage (in north and northeast, in valley of Kunda River, rocks are partly eroded). Contour interval is 1 m. Original borehole control shown in Figure 2A. B, First-degree trend thickness of rocks of Keila Stage. Contour interval is 0.5 m. C, Residual map constructed by subtracting first-degree trend surface from isopachous map of rocks of Keila Stage. Contour interval is 1 m.

formations, decreasing thickness of 1.5 meters in comparison with the background thickness is observed. The same value of thickness loss is noted in O_2kl, increasing up to 6 meters in the north and northeast (in the valley of the Kunda River) because of erosion processes (Figs. 5A-5C). The rocks of the upper formations are eroded completely in the northeastern part of the study area (Fig. 3A). The erosion of the rocks of the Keila Formation has affected the calculation of its first-degree trend thickness (Fig. 5B). As a result of this erosion in the southwestern corner of the residual map thickness of O_2kl, we see negative values (Fig. 5C).

The isopachous map of the O_1lt-O_2*hr* (Fig. 6A) shows difference between two structural maps, drawn on bottom of the Latorp Stage, covering the commercial seam of the phosphorites (Fig. 2A), and bottom of the Rägavere Formation, the first reference surface in the upper part of the section (Fig. 3A). On the residual isopachous map, constructed by subtracting the first-degree trend thickness from the isopachous map of O_1lt-O_2*hr*, zones of thinness are revealed (Figs. 6B ,6C). Comparing them with the Aseri dislocation, which is indicated on this map by thinning of up to 8.5% of the total thickness of the carbonate beds, we can conclude that the other anomalies of decreasing thickness (thickness losses up to 4.4%) reflect also zones of disturbances. Comparison of the isopachous map of the beds O_1lt-O_2*hr* (Fig. 6A) with the total isopachous map of the O_2*vä*+O_2*kr*+O_2kk (Fig. 6C) has shown that the main part of the thinness in the basin is the result of karst which takes place in these formations, including the commercial seam of the oil shale kukersites (O_2kk).

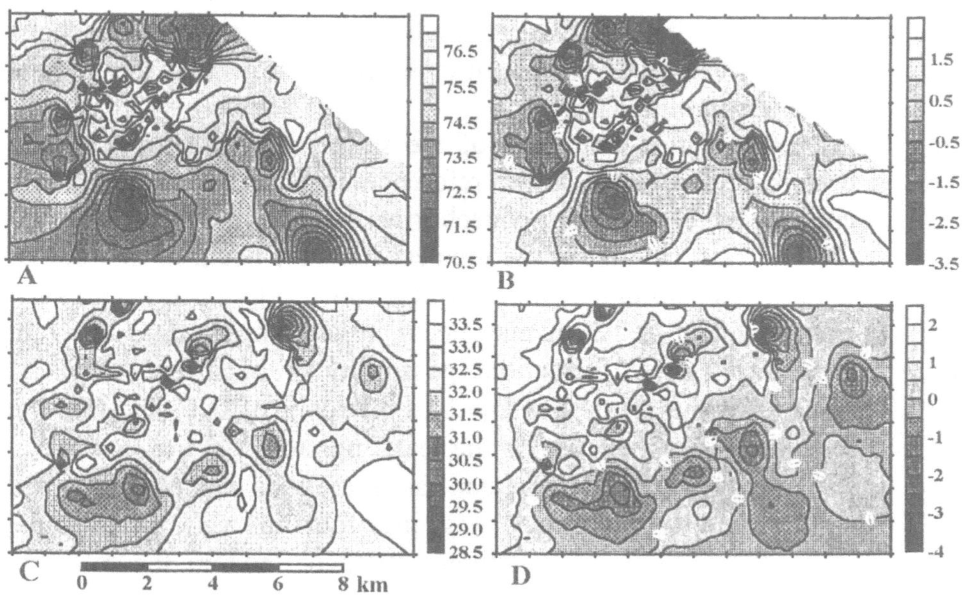

Figure 6. A, Isopachous map of beds O_1lt-O_2*hr*. Original well control shown in Figure 2A. Contour interval is 0.5 m. On blanked area rocks of some upper formations are eroded. B, Residual map constructed by subtracting first-degree trend surface from isopachous map of beds O_1lt-O_2*hr*.Contour interval is 0.5 m. C, Isopachous map of beds of O_2*vä*-O_2kk (O_2*vä*+O_2*kr*+O_2kk). Original borehole control shown in Figure 2A. Contour interval is 0.5 m. D, Residual map constructed by subtracting first-degree trend surface from isopachous map of beds of O_2*vä*-O_2kk. Contour interval is 0.5 m.

Results of the Structural Analysis

After collating data onto one map, the axes of the subsidence, axes of thinness, axes of the low-resistivity anomaly zone (Fig. 7), and the analysis of the structure and isopachous maps of all formations, we can observe the following:

(1) In the study area, except for the known Aseri dislocation, four new zones with small amplitudes were established: three fracture zones with the subsidence of the beds 2-5 meters and thickness loss up to 4.4% (1-3) and one folding zone (4) with thickness loss of up to 7.3% on its anticlinal fold.

(2) These zones partly coincide with the axes of the low-resistivity anomaly zones.

(3) Thickness losses in the zones of disturbances, which are reflected by the negative anomalies on the residual isopachous maps, are caused by karsting intensified by movement along faults in the crystalline basement.

(4) The greatest amplitude of the subsidence is defined by the bottom of O_1lt - the lower reference surface, which covers the commercial seam of the phosphorites.

(5) The karst cavities of the greatest size may occur in the rock of O_2*rg*-$O_{2\text{-}3}$nb (up to 4,0 m), in O_2kk, O_2*kr* (up to 3.0 m). The karst cavities of up to 1.5 m may occur in $O_{1\text{-}2}$kn, O_2*vä*, O_2id, O_2jh, O_2kl. From these formations possibility of occurrence of karst cavities is the greatest in O_2*vä*. In the rocks of O_1vl and O_1lt the location of karst cavities is the least possible.

(6) The investigated Ordovician carbonate beds may be subdivided by the rigidity of the rock mass into four intervals, whereas the most difficult mining conditions were defined in the oil-shale containing interval O_2kk-O_2*kr*.

CONCLUSIONS

Computerization of a structural analysis of log data by applying a kriging algorithm for the gridding procedure and trend analysis for subtracting the regional background have permitted:

(1) completion of a comparative analysis of all structural surfaces and of stages and formations in the Ordovician carbonate basin;

(2) characterization of regional trends of structural and thickness alterations;

(3) zones of disturbances to be revealed especially those with small amplitudes and erosion zones;

(4) description in detail of structures of these zones and determining places of karst location;

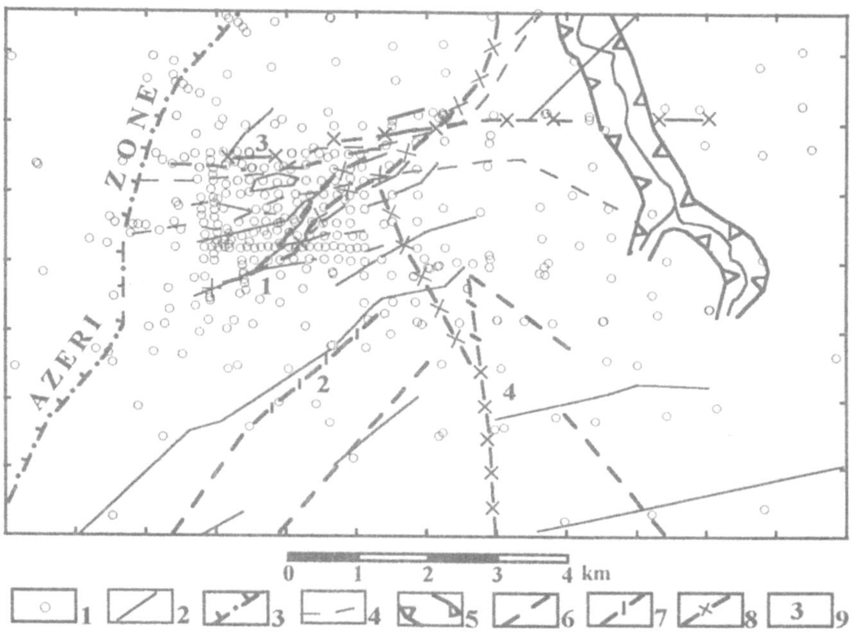

Figure 7. Scheme of axes of located zones of disturbances and low-resistivity anomaly zones.

Explanation: 1 - boreholes; 2 - axes of low-resistivity anomaly zones by Vaher (1983); 3 - axes of homocline fold of Aseri zone of disturbances by Vaher (1983); 4 - uncertain borders of Kantküla zone of disturbances (Puura, Vaher, and Tuuling, 1987); 5 - erosional cut of buried valley of Kunda River; axes of zones, revealed by: 6 -thickness loss of rocks of O_2hr-O_1lt, 7 -subsidence of bottom of Rägavere Formation, 8 - subsidence of bottom of Latorp Stage; 9 - numbers of revealed zones of disturbances with small amplitudes.

(5) to estimation of the rigidity of the rock mass, on the basis of the established characteristics of the formations (Shogenova, 1991a,1992); and

(6) allowed forecasting the possibility of the karsting in different carbonate formations.

The results and example of using the computer for the structural analysis in a sedimentary basin in Estonia permits recommendation of the application of such a technique for all of Estonia and also for other regions with similar geological conditions.

ACKNOWLEDGMENTS

The author would like to thank the colleagues from the Geological Survey of Estonia for making available geological and geophysical data, my supervisor Prof. Vaino Puura, my colleagues from Institute of Geology, Estonian Academy of Sciences who helped me in preparation of this paper and editors Andrea Förster and Daniel F. Merriam for helpful suggestions and remarks.

REFERENCES

Davis, J.C., 1986, Statistics and data analysis in geology: John Wiley & Sons Inc., New York, 646 p.

Bonham-Carter, G.F., 1989, Comparison of image analysis and geographic information systems for integrating geoscientific maps, *in* Agterberg, F.P., and Bonham-Carter, G.F., eds., Statistical analysis in the earth sciences: Geol. Survey Canada Paper 89-9, p. 141-155.

Merriam, D.F., and Jewett, D.G., 1988, Methods of thematic map comparison, *in* Current trends in geomathematics: Plenum Press, New York, p. 9-18.

Gazizov, M.S., 1971, Karst i ego vliyanie na gornye raboty (Karst and its influence on the mining works): Nauka, Moscow, 204 p.

Harff, J., Lange, D., and Olea, R., 1992, Geostatistics for computerized geological mapping: Geol. Jb., A 122, p. 335-345.

Heinsalu, U., and Andra, H., 1975, Treshtshinovatost' v rayone slancevyh shaht Estonii i geofizicheskie metody eyo issledovaniya. (Jointing in the Estonian oil shale basin and geophysical research methods for its study): Valgus, Tallin (In Russian with English summary), 116 p.

Hints L., Meidla T., Gailite L.-I., and Sarv L., 1993, Catalogue of Ordovician stratigraphical units and stratotypes of Estonia and Latvia: Estonian Acad. Sciences, Tallinn, 62 p.

Nikitin, A.A., 1986, Teoreticheskie osnovy obrabotki geofizicheskoy informacii. (Theoretical basis of the treatment of geophysical information): Nedra, Moscow, 342 p.

Phillip, G.M., and Watson, D.F., 1987, Neighborhood discontinuites in bivariate interpolation of the scattered observation: Math. Geology, v.19, no. 1, p. 69-74.

Puura, V., ed., 1987, Geology and mineral resources of the Rakvere phosphorite-bearing area: Tallinn Valgus Publisher (In Russian with English summary and explanation to the Figures), 211 p.

Puura, V., Vaher R., and Tuuling, I., 1986, Tectonics, *in* Puura, V., ed., Geology of the kukersite-bearing beds of the Baltic oil shale basin: Valgus Publishers, Tallinn (In Russian with English Summary), p. 55-62.

Puura, V., Vaher R., and Tuuling, I, 1987, Tektonika. (Tectonics), Puura, V., ed., Geology and mineral resources of the Rakvere phosphorite-bearing area: Valgus Publisher, Tallinn (In Russian with English summary and explanation to Figures), p. 90-104.

Shogenova, A., 1989a, Raschlenenie ordovikckoy karbonatnoy tolshtshi na Rakvereskom mestorozhdenii fosforitov po dannym skvazhinnoy geofiziki. (Detailed subdivision of the Ordovician carbonate beds in the Rakvere phosphorite deposit by geophysical logging): Proc. Academy Sciences Estonian SSR, Geology, v.38, no.1 (In Russian with English summary), p. 10-13.

Shogenova, A., 1989b, Ocenka narushennosti ordovikskoy karbonatnoy tolshtshi na Rakvereskom mestorozhdenii fosforitov po dannym skvazhinnoy geofiziki. (Fracturing estimation of the Ordovician carbonate beds on the Rakvere phosphorite deposit by geophysical logging): Proc. Academy Sciences Estonian SSR, v. 38, no. 3 (In Russian with English summary), p. 118-123.

Shogenova, A., 1991a, Ocenka ustoychivosti gornogo massiva Rakvereskogo mestorozhdeniya fosforitov po dannym skvazhinnoy geofiziki. (Rigidity estimation of the rock mass on the Rakvere phosphorite deposit by geophysical logging.) Proc. Estonian Academy Sciences, v. 40, no. 3 (In Russian with English summary), p. 104-111.

Shogenova, A., 1991b, Primenenie EVM dlya detal'nogo strukturnogo analiza Rakvereskogo slance-fosforitovogo mestorozhdeniya po dannym skvazhinnoy geofiziki. (Use of the computer for the detailed structural analysis of the Rakvere oil-shale and phosphorite deposit by geophysical logging): Primenenie EVM pri reshenii zadach geologii poleznyh iskopaemyh Estonii: AN Estonii, Tallinn (In Russian), p. 49-65.

Shogenova, A., 1992, Struktura Rakvereskogo mestorozhdeniya fosforitov po geologo-geofizicheskim dannym (s ispol'zovaniem modelirovaniya na EVM). (Structure of the Rakvere phosphorite deposit by geological and geophysical data by computer analysis), Abstract of candidate thesis: Sankt-Petersburg Mining Institute, Sankt-Petersburg, (In Russian), 22 p.

Suveizdis, P.I., ed., 1979, Baltic tectonics: Mokslas, Vilnius (In Russian), 92 p.

Vaher, R.M., 1983, Tektonika fosforito-slancevogo basseyna severo-vostochnoy Estonii. (Tectonic of the phosphorite and oil shale basin of North-Eastern Estonia): Abstract of candidate thesis, Institute of Geochemistry and Geophysics, Belorussian Academy Sciences, Minsk (In Russian), 22 p.

Vaher, R.M., 1986, Primenenie metoda soprotivleniy dlya vyyavleniya zon treshtshinovatosti v Severo-Vostochnoy Estonii. (Application of the resistivity method for revealing fracture zones in north-eastern Estonia): Proc. Academy Science, Estonian SSR, Geology, v. 35, no. 4 (In Russian with English summary), p. 146-155.

PAIRWISE COMPARISON OF SPATIAL MAP DATA

Daniel F. Merriam
University of Kansas, Lawrence, Kansas, USA

Ute C. Herzfeld
Universität Trier, Trier, Germany

Andrea Förster
GeoForschungsZentrum Potsdam, Potsdam, Germany

ABSTRACT

Map comparison/integration of spatial data has become important recently with the plethora of data being generated and accumulated. In many instances it is of interest to compare/integrate these data. The simplest comparative procedure is to overlay two maps and visually compare the areas of similarity and differences. This cursory examination is quick, but important aspects of the comparison may be overlooked. The simplest integration procedure is to overlay several maps and note the areas of correspondence on a resultant map. This type of comparison/integration is visual and subjective; quantatively, the degree of similarity may be expressed either (1) as a coefficient of overall similarity or (2) as a resultant map. The similarity coefficient gives an indication as to the goodness of correspondence and the resultant map shows where. Several techniques of pairwise map comparison have been utilized recently including bivariate and multivariate statistics, probability, AI/ES (Boolean logic), algebraic, and fuzzy set theory. A geothermal

Geologic Modeling and Mapping
Edited by A. Förster and D. F. Merriam, Plenum Press, New York, 1996

data set from southeastern Kansas is used to demonstrate pairwise map comparison with other spatial geological and geophysical data and to interpret local and regional conditions.

INTRODUCTION

The problem of quantifying map comparison/integration has been of interest to geologists for many years. The easiest way to compare two maps, of course, is to overlie them on a light table and note the areas that are similar and dissimilar based on some criterion (or set of criteria). To compare two maps where measured units are the same, one map can be subtracted from the other to produce a resultant map (e.g. an isopachous map), which then can be interpreted from the pattern that shows the similarity between the two maps. Where the two map units are different, a crossplot of the variables on each map can be used to compute a correlation coefficient. These first ways to accomplish the task were straightforward and simple.

With the advent of remote sensors and automatic data acquisition, these simplistic comparisons were not feasible anymore. Those working in remote sensing began to explore ways in which to analyze literally billions of items of data routinely collected in space from airborne and space platforms. Geophysicists also had to devise new ways of integrating diverse data obtained from the thousands of miles of land and sea profiles collected during exploration for mineral resources. Geographers were faced with organizing all types of spatial data and comparing and integrating them. Geologists, likewise, have been struggling with masses of data generated by automated equipment designed to provide adequate data for today's sophisticated three-dimensional studies.

There are several reasons to determine the relation of the map variables, not only to each other, but between maps. It may be desirable to know the (1) predictive value of one map to another, (2) evaluate the map similarity for classification, or (3) use the information for geological interpretation (Merriam and Jewett, 1988). For whatever reason to compare/integrate, the technique should be simple and fast. We suggest one such approach in this paper where we do a pairwise comparison of geological, geophysical, geothermal, and other spatial data to determine the predictive value of one map to the others.

PREVIOUS WORK

Many papers have appeared in the literature in the past several years related to map comparison/integration of geological data. A good summary on the subject is given by Merriam and Jewett (1988); other sources include Unwin (1981), Gold (1980), and Davis (1986).

An early work was by Mirchink and Bukharsev (1959) who took corresponding data points from maps and plotted them on a scatter diagram to determine the degree of accordance. Later, Ribeiro and Merriam (1979) used scatter diagrams and regression curves to determine the relationship of one map variable to another.

As early as 1966, Merriam and Sneath proposed using the coefficients from low-degree polynomial trend surfaces of structural surfaces as descriptors of those surfaces. The values were used as input into a clustering routine and plotted as dendrograms showing the overall similarity of one map to another. Merriam and Lippert (1966) noted the relation of residue values from trend analysis in determining map similarity. Thrivikramaji and Merriam (1976) used the trend-surface coefficients from thickness maps for input into a clustering routine in a study of the structural development of Kansas during the Pennsylvanian. Miller (1964) proposed comparing maps using the values of the computed trend-surface matrices, and Rao (1971) used the least-square equation.

Robinson and Merriam (1972) used values at grid nodes on spatially filtered maps to determine the relation of one map to another. Robinson, Merriam, and Burroughs (1979) reported on a technique to relate the location of oil production to geologic structure. Merriam and Robinson (1981) utilized a point-by-point comparison to obtain a coefficient of resemblance in comparing spatially filtered maps to each other. Eschner, Robinson, and Merriam (1979) and Robinson and Merriam (1984) constructed 'resultant' maps to compare spatially filtered maps to each other.

In 1988 Merriam and Sondergard published on a technique they devised to determine the relationship between one map and another by computing a spatial Reliability Index. Brower and Merriam (1990) used several multivariate algorithms, including principle component analysis (PCA), principal coordinates, Q-mode vector analysis, correspondence analysis, and clustering to determine the relationship between gridded structural maps. They followed up this initial work in 1992 (Brower and Merriam, 1992) with a method utilizing correlation coefficients to compare adjacent points on thematic maps. Merriam (1992) published an example of pairwise comparison utilizing similarity coefficients represented by dendrograms of standardized structural, stratigraphic, topographic, and geophysical data. Förster, Merriam, and Brower (1993) have compared geological and geothermal map data using cluster analysis.

A computer program, MAPCOMP, to perform a weighted comparison of two or more maps using a matrix algorithm was published in 1988 by Herzfeld and Sondergard. This algebraic algorithm, which produces a resultant map showing areas of similarity and dissimilarity, has been used successfully in several subsequent studies involving the use of geological, geophysical, and geothermal data (Herzfeld and Merriam, 1991; Merriam, Fuhr, and Herzfeld, 1993; Förster and Merriam, 1993; Merriam, Herzfeld, and Förster, 1993). Herzfeld and Merriam (1995) followed up this approach by suggesting optimizing techniques to automate the process for the selecting the 'best fit' for the algebraic algorithm.

THE DATA

The data for this study were compiled for an area in southeastern Kansas comprising Chautauqua County between T32S and T35S and R8E and R13E; an area of approximately 620 sq mi (1600 sq km). The subsurface data were collected from the files at the Kansas Geological Survey and consisted of petrophysical logs of all types, sample logs, drillers logs, and scout tops. A structural map on top of the Mississippian was constructed as well as bottom-hole temperature (BHT) maps for the Lower Pennsylvanian, Mississippian, and Cambro-Ordovician Arbuckle Group.

Geophysical data, aeromagnetic (Yarger and others, 1981) and gravimetric (Lamb and Yarger, 1985), were digitized on a 3-mi grid from published maps. Precambrian data were used from material by Cole (1962). Background information for the study was obtained from Merriam (1963) and other Kansas Geological Survey publications.

THE BACKGROUND

A variety of techniques have been utilized to compare spatial data - each practical for different conditions. These techniques are based either on a point-by-point comparison to obtain a coefficient of resemblance or some combination of spatial data to produce a resultant map. The approaches that have been used in map comparison, in addition to bivariate and multivariate statistics, include probability (Brower and Merriam, 1990), AI/ES (Boolean logic) (Bonham-Carter and Agterberg, 1990; Agterberg, Bonham-Carter, and Wright, 1991; Maslyn, 1991), algebraic (Herzfeld and Sondergard, 1988; Herzfeld and Merriam, 1991), and fuzzy set theory.

Similarity coefficients and resultant maps can be computed from the raw data, gridded data, or a set of descriptors derived from the original data (Fig. 1). Where the data are measured in different units, they first must be pretreated to standardize them. The similarity measure usually is a correlation or distance coefficient and has the general form (Unwin, 1981)

$$C_a = \frac{\textit{area over which phenomena are located together}}{\textit{total area covered by the two phenomena}}$$

The coefficients computed between each pair of maps then can be tabulated in a row,column (r,c) matrix and the relationship between the variables or samples displayed in a dendrogram for visual ease in interpretation.

Resultant maps can be constructed by noting the relation on a point-by-point correspondence between maps or using an interpolated grid. The values then are contoured for visual interpretation.

INTERPRETATION AND RESULTS

The geological, geophysical, and geothermal maps for our area in southeastern Kansas were analyzed by first computing correlation coefficients and constructing dendrograms and then by using the algebraic map-comparison program developed by Herzfeld and Sondergard (1988). One data subset consisted of seven maps, constructed just for Chautauqua County, digitized on a 3-mile grid. This data set consisted of two

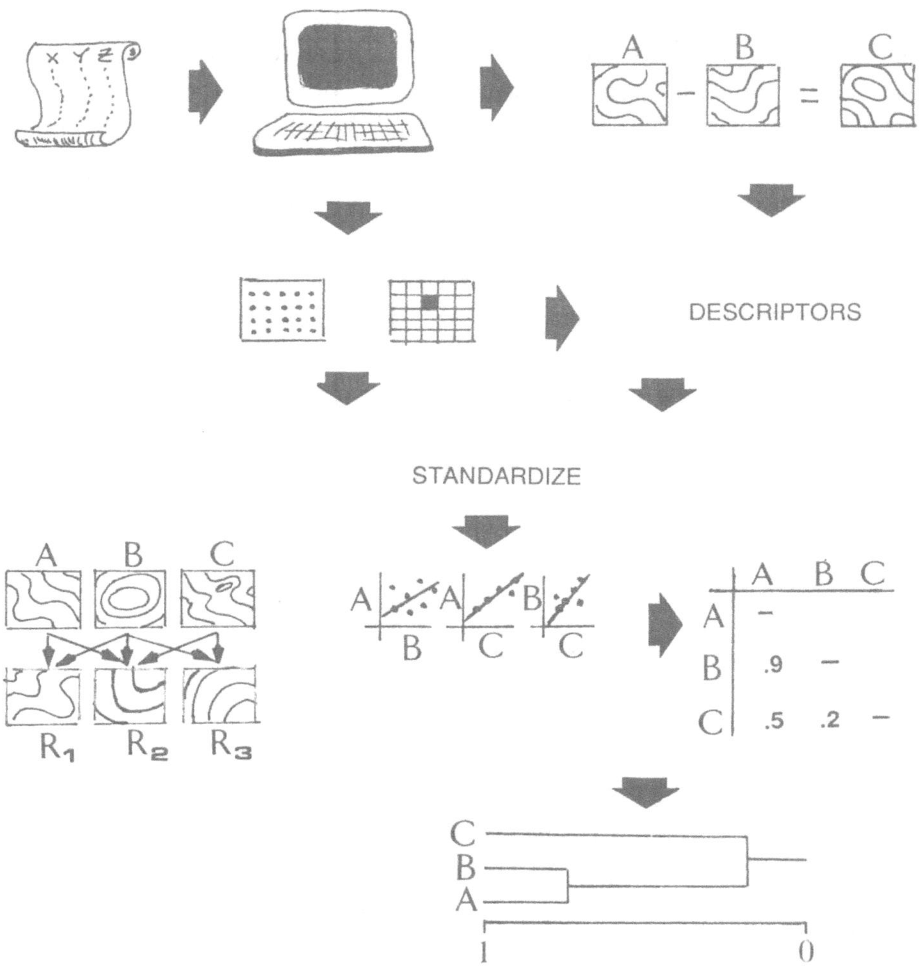

Figure 1. Flowdiagram of spatial analysis from data input to graphic output.

structural, two geophysical, and three geothermal maps. Regression analysis was performed pairwise on the data sets collected from the seven maps and the results summarized in a similarity matrix (Table 1). These data then were displayed graphically as a dendrogram (Fig. 2).

From the matrix (Table 1), it is obvious that the highest relationship (0.91) is between the Precambrian configuration and structure on top of the Mississippian. Gravity is related to the Precambrian configuration (0.70) and Mississippian structure (0.75). Although the

Table 1. Correlation coefficients computed for pairwise comparison of 7 maps of Chautauqua County based on gridded interpolated values. Best match is 0.91 between Precambrian configuration and Mississippian structure. Results displayed in Figure 2 as dendrogram.

	Penn. Temp.	Miss. Temp.	Arb. Temp.	Miss. Struct.	Gravity	Aero-mag.	Pre-Camb.
Pennsylvanian Temperature	--						
Mississippian Temperature	.34	--					
Arbuckle Temperature	.14	.31	--				
Mississippian Structure	.31	.30	.08	--			
Gravity	.33	.34	.16	.75	--		
Aeromagnetics	.03	.01	.10	.00	.03	--	
Precambrian Configuration	.21	.23	.05	.91	.70	.01	--

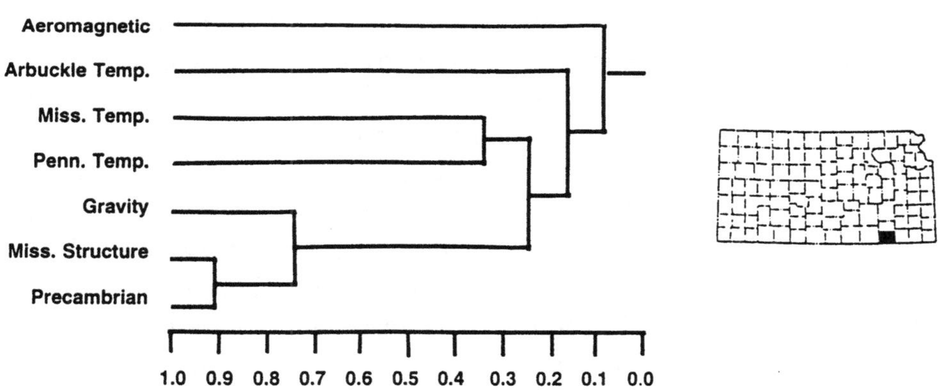

Figure 2. Dendrogram based on correlation coefficients determined between pairwise comparison of maps. 1.0 is perfect match; 0.0 is no match. Dendrogram based on the WVGM method (Sneath and Sokal, 1963). Index map shows location of Chautauqua County in southeastern Kansas.

gravity is related to structure, the aeromagnetics are not related to either the structure or geothermics. This confirms the interpretation that the aeromagnetics reflect anomalies within the Precambrian basement. This relationship has been noted in others studies in eastern Kansas (Merriam, Herzfeld, and Förster, 1993). The geothermal data for the three stratigraphic units are not related highly to each other nor to the structural and geophysical variables (Förster, Merriam, and Brower, 1993).

The original data sets (except for the aeromagnetic data) are contoured and shown in Figure 3. Casual inspection of the patterns displayed on these maps show little obvious similarity. The three geothermal maps have a series of 'highs' and 'lows' but not all in the same geographic location. There is a subtle increase in temperature to the west that corresponds to the increased depth as a result of the westward dip of the units. The gravity map also shows an overall increase in values from east to west that coincides with the westward dipping structural surfaces. Both the Precambrian and Mississippian surfaces, which dip to the west, have numerous small features superimposed on them. Thus, without a similarity coefficient, it would be difficult to determine the relationship of one map to another from just a visual inspection of the contour pattern.

Utilizing the algebraic map-comparison program developed by Herzfeld and Sondergard (1988), a pairwise comparison was made of the data sets. Results of this comparison of the six maps are shown in Figure 4. The most obvious feature of this analysis is the high similarity between the Precambrian-Mississippian structure maps (where the values are 0.10 or less) over most of the area. The next best relationship is that of the Precambrian-Gravity match. These results reenforce the results obtained from the computation of the correlation coefficients, but show spatially where the similarities are high. The similarities of the pairwise comparison of the geothermal maps to each other and their relation to the other map variables is spotty. However, the relation of Mississippian temperature and structure shows that the similarity is close locally, especially where the structure is positive and the temperature is higher. Because some oil and gas fields in southeastern Kansas are associated with positive structural features, it is reasonable to assume for these structures a higher temperature will occur on the anticlines. This relation of geothermal anomalies to the occurrence of petroleum in Kansas has been noted by Ball (1982) and Merriam and Förster (1995). The Pennsylvanian temperature map is difficult to interpret because of the considerable thickness of the unit and BHTs collected from different depth levels. The problems with the BHT maps are discussed in more detail in Förster and Merriam (1993).

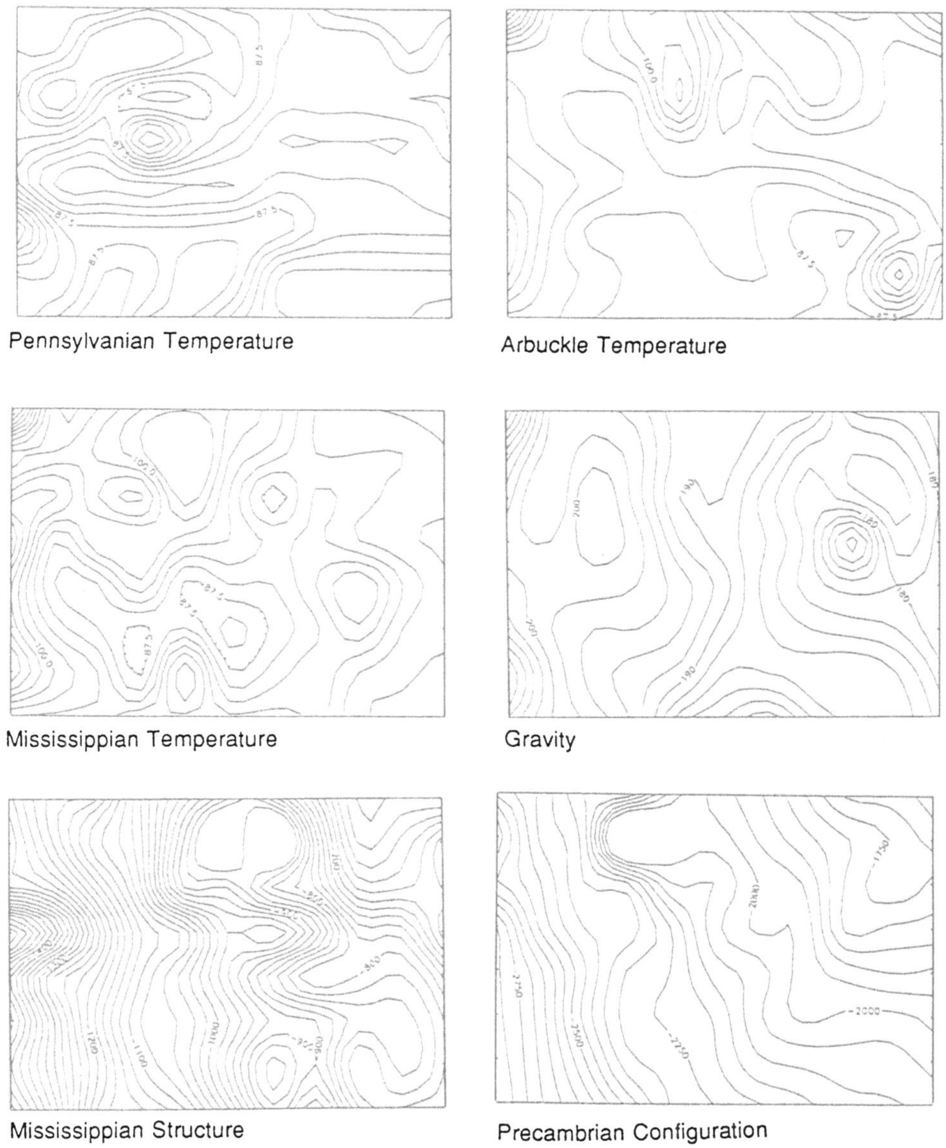

Figure 3. Contour of original 3-mi gridded data used for pairwise comparison. Contour Interval (CI) = 2.5°F for temperature maps; 2 mgs for gravity map; 20 ft for Mississippian structure map; and 50 ft for Precambrian configuration map. Area is Chautauqua County, Kansas (T.32 S. - T. 35 S., R. 8 E. - R. 13 E.), approximately 620 sq mi. Location of area shown on Figure 2 inset map. Contouring by SURFACE III (Sampson, 1988).

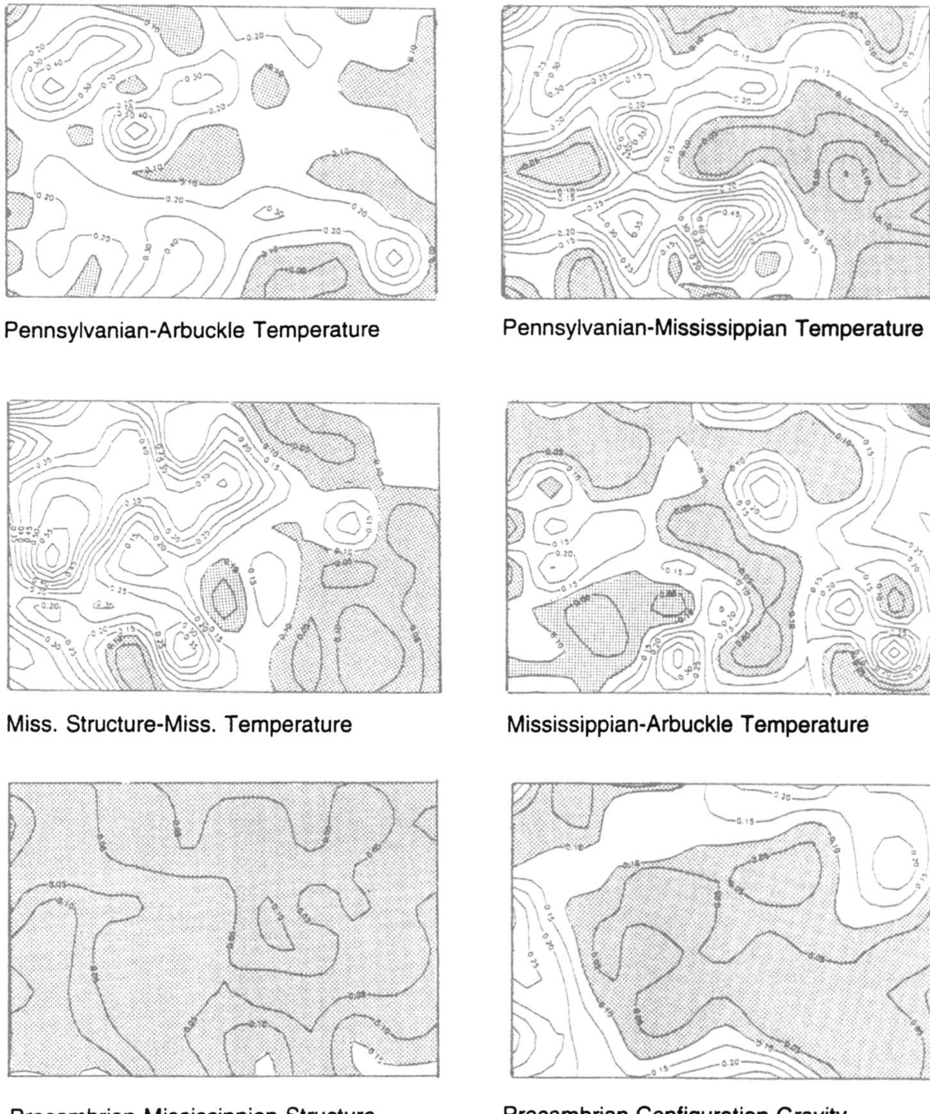

Figure 4. Contoured pairwise comparison of map variable using algebraic MAPCOMP (Herzfeld and Sondergard, 1988). Small values indicate high similarity; high values poor similarity. Areas of high similarity (0.10 or less) are outlined. Contour Interval (CI) is 0.10 and 0.05. Area is Chautauqua County, same as in Figure 3. Amount of correspondence is given in Table 1 by similarity coefficients - between Precambrian-Mississippian structure (0.91%); between Mississippian structure and gravity (0.75%); and between Precambrian and gravity (0.70%). Areas of high similarity shaded.

CONCLUSION

Pairwise comparison of spatial map data offers a good, quick method of analyzing the relationship between a series of maps of different parameters. By combining a quantitative assessment on the degree of similarity and a technique of outlining areas of high similarity, a good understanding on spatial relations can be gained to help in the geological interpretation.

ACKNOWLEDGMENTS

We would like to thank Rick Brownrigg of the Kansas Survey for help in processing the data. Partial funding for the study was provided by Conoco Oil Company and the Wichita State University Geology Research Fund.

REFERENCES

Agterberg, F.P., Bonham-Carter, G.F., and Wright, D.F., 1990, Statistical pattern integration for mineral exploration, *in* Gaal, G., and Merriam, D.F., eds., Computer applications in resource estimation, prediction, and assessment for metals and petroleum: Pergamon Press, Oxford, p. 1-21.

Ball, S.M., 1982, Exploration application of temperatures recorded on log headings - an up-the-odds method of hydrocarbon-charged porosity prediction: Am. Assoc. Petroleum Geologists Bull., v. 66, no. 8, p. 1108-1123.

Bonham-Carter, G.F., and Agterberg, F.P., 1990, Application of a microcomputer-based Geographic Information System to mineral-potential mapping, *in* Hanley, J.T., and Merriam, D.F., eds., Microcomputer applications in geology, II: Pergamon Press, Oxford, p. 49-74.

Brower, J.C., and Merriam, D.F., 1990, Geological map analysis and comparison by several multivariate algorithms, *in* Agterberg, F.P., and Bonham-Carter, G.F., eds., Statistical applications in the earth sciences: Geol. Survey Canada Paper 89-9, p. 123-134.

Brower, J.C., and Merriam, D.F., 1992, A simple method for the comparison of adjacent points on thematic maps, *in* Kürzl, H., and Merriam, D.F., eds., Use of microcomputers in geology: Plenum Press, New York and London, p. 227-240.

Cole, V.B., 1962, Configuration of top Precambrian basement rocks in Kansas: Kansas Geol. Survey Oil and Gas Invest. 26, map.

Davis, J.C., 1986, Statistics and data analysis in geology (2nd ed.): John Wiley & Sons, New York, 646 p.

Eschner, T.R., Robinson, J.E., and Merriam, D.F., 1979, Comparison of spatially filtered geologic maps: summary: Geol. Soc. America Bull., pt. 1, v. 90, p. 6-7.

Förster, A., and Merriam, D.F., 1993, Geothermal field interpretation in south-central Kansas for parts of the Nemaha Anticline and flanking Cherokee and Sedgwick Basins: Basin Research, v. 5, no. 4, p. 213-234.

Förster, A., Merriam, D.F., and Brower, J.C., 1993, Relationship of geological and geothermal field properties: Midcontinent area, USA, an example: Math. Geology, v. 25, no. 7, p. 937-947.

Gold, C., 1980, Geological mapping by computer, *in* Taylor, D.R.F., ed., The computer in contemporary cartography: John Wiley & Sons, Chichester, p. 151-190.

Herzfeld, U.C., and Sondergard, M.A., 1988, MAPCOMP - a FORTRAN program for weighted thematic map comparison: Computers & Geosciences, v. 14, no. 5, p. 699-713.

Herzfeld, U.C., and Merriam, D.F., 1991, A map-comparison technique utilizing weighted input parameters, *in* Gaal, G., and Merriam, D.F., eds., Computer applications in resource estimation, prediction, and assessment for metals and petroleum: Pergamon Press, Oxford, p. 43-52.

Herzfeld, U.C., and Merriam, D.F., 1995, Optimization techniques for integrating spatial data: Math. Geology, v. 27, no. 5, in press.

Lamb, C., and Yarger, H.L., 1985, Absolute gravity map of Kansas: Kansas Geol. Survey Open-File Rept. 85-15, map. Scale 1:500,000

Maslyn, R.M., 1991, Boolean grid logic - combining expert systems with grid nodes (abst.): Proc. GeoTech/GeoChautauqua '91, Denver GeoTech, Inc., Denver, CO., p. 316.

Merriam, D.F., 1963, The geologic history of Kansas: Kansas Geol. Survey Bull. 162, 317 p.

Merriam, D.F., 1992, Geological interpretation of integrated thematic spatial data: Geologisches Jahrbuch Reihe A, Hf. 122, p. 233-241.

Merriam, D.F., and Förster, A., 1995, Subsurface temperature anomalies as a key to petroleum-producing areas in the Cherokee and Forest City Basins, eastern Kansas? (abst.): Am. Assoc. Petroleum Geologists Regional Meeting (Tulsa, Oklahoma), 1 p.

Merriam, D.F., and Jewett, D.G., 1988, Methods of thematic map comparison, *in* Merriam, D.F., ed., Current trends in geomathematics: Plenum Press, New York and London, p. 9-18.

Merriam, D.F., and Lippert, R.H., 1966, Geologic model studies using trend-surface analysis: Jour. Geology, v. 74, no. 3, p. 344-357.

Merriam, D.F., and Robinson, J.E., 1981, Comparison functions and geological structure maps, *in* Future trends in geomathematics: Pion Ltd., London, p. 254-264.

Merriam, D.F., and Sneath, P.H.A., 1966, Quantitative comparison of contour maps: Jour. Geophysical Res., v. 71, no. 4, p. 1105-1115.

Merriam, D.F., and Sondergard, M.A., 1988, A reliability index for the pairwise comparison of thematic maps: Geologisches Jahrbuch, v. A104, p. 433-446.

Merriam, D.F., Fuhr, B.A., and Herzfeld, U.C., 1993, An integrated approach to basin analysis and mineral exploration, *in* Harff, J., and Merriam, D.F., eds., Computerized basin analysis; the prognosis of energy and mineral resources: Plenum Press, New York and London, p. 197-214.

Merriam, D.F., Herzfeld, U.C., and Förster, A., 1993, Integration of geophysical data for basin analysis: Proc. 3rd Intern. Brasilian Geophysical Soc. Congress (Rio de Janeiro), v. 2, p. 1025-1030.

Miller, R.L., 1964, Comparison-analysis of trend maps, *in* Computers in the mineral industries, pt. 2: Stanford Univ. Publ., Geol. Sci., v. 9, no. 2, p. 669-685.

Mirchink, M.F., and Bukhartsev, V.P., 1959, The possibility of a statistical study of structural conditions: Doklady Akad. Nauk SSSR (English translation), v. 126, no. 5, p. 1062-1065.

Rao, S.V.L.N., 1971, Correlations between regression surfaces based on direct comparison of matrices: Modern Geology, v. 2, no. 3, p. 173-177.

Ribeiro, J.C., and Merriam, D.F., 1979, Quantitative analysis of depositional environments (Aratu Unit, Reconcavo Series, Lower Cretaceous) in the Reconcavo Basin, Bahia, Brazil, *in* Gill, D., and

Merriam, D.F., ed., Geomathematical and petrophysical studies in sedimentology: Pergamon Press, Oxford, p. 219-234.

Robinson, J.E., and Merriam, D.F., 1972, Enhancement of patterns in geologic data by spatial filtering: Jour. Geology, v. 80, no. 3, p. 333-345.

Robinson, J.E., and Merriam, D.F., 1984, Computer evaluation of prospective petroleum areas: The Oil & Gas Jour., v. 82, no. 34, p. 135-138.

Robinson, J.E., Merriam, D.F., and Burroughs, W., 1979, A quantitative technique to determine relation of oil production to geologic structure in Graham County, Kansas, *in* Davis, J.C., and deLopez, S.L., eds., Computer mapping for resource analysis (Mexico City): Inst. de Geografia de la UNAM and Kansas Geol. Survey, p. 256-264.

Sokal, R.R., and Sneath, P.H.A., 1963, Principles of numerical taxonomy: W.H. Freeman and Co., San Francisco, 359 p.

Thrivikramaji, K.P., and Merriam, D.F., 1976, Trend analysis of sedimentary thickness data: the Pennsylvanian of Kansas, an example, *in* Merriam, D.F., ed., Quantitative techniques for the analysis of sediments: Pergamon Press, Oxford, p. 11-21.

Unwin, D., 1981, Introductory spatial analysis: Methuen, London and New York, 212 p.

Yarger, H.L. and others, 1981, Aeromagnetic map of Kansas: Kansas Geol. Survey Map M-16. Scale: 1:500,000

APPLICATIONS OF SPATIAL FACTOR ANALYSIS TO MULTIVARIATE GEOCHEMICAL DATA

E.C. Grunsky
Geological Survey Branch, Victoria, British Columbia, Canada

Q. Cheng
University of Ottawa, Ottawa, Ontario, Canada

F.P. Agterberg
Geological Survey of Canada, Ottawa, Ontario, Canada

ABSTRACT

Spatial Factor Analysis (SPFAC) is a technique used to determine multivariate relationships from the auto- cross-correlation relationships of spatially referenced data. The method derives linear combinations of variables with maximum autocorrelation for a specified lag distance. Eigenvectors are calculated from a covariance quotient matrix derived from auto-cross correlation estimates at lag 0 and distance d. SPFAC has been applied to regional geochemical (moss mat) sampling data collected in the southern part of Vancouver Island, British Columbia, Canada. The results of the analysis show that regional geochemical trends and interelement associations change with lag distance and orientation and can reflect the underlying spatially based lithological variations. The technique also has been applied to lithogeochemical data in the Sulphurets area of British Columbia, Canada. Zones of alteration associated with porphyry copper and gold mineralization have been shown to be distinct spatially from the background.

Geologic Modeling and Mapping
Edited by A. Förster and D. F. Merriam, Plenum Press, New York, 1996

INTRODUCTION

Traditional approaches to multivariate analysis such as principal components analysis do not consider the spatial relationships of the data. Multivariate relationships of spatially based data have been studied by Agterberg (1966, 1974), Myers (1982, 1988), Royer (1988), Switzer and Green (1984), Wackernagel (1988), Grunsky and Agterberg (1988, 1992), Grunsky (1989). Grunsky and Agterberg (1988, 1992) have shown that the application of a technique termed, SPFAC, produces linear combinations of variables that have maximum correlation based on auto-cross-correlation estimates at specific lag intervals.

Multivariate analysis has been applied to a wide range of geochemical studies and techniques for assessing geochemical data are well documented (Howarth and Sinding-Larsen, 1983; LeMaitre, 1982). Geochemical data usually are correlated and within a given sampling regime, several populations of data may represent different geochemical processes. The study of multielement geochemical data has been studied in a spatial context using a variety of techniques. Wackernagel (1988) contrasted the results of a principal components analysis between the variance-covariance matrix and a spatial covariance matrix based on variogram estimations using geochemical data from a regional soil survey. Bourgault and Marcotte (1991) used a spatial form of the Mahalanobis distance (multivariable variogram). The technique was applied to a lithogeochemical survey over a pluton. Bellehumeur, Marcotte, and Jebrak. (1994) have studied the spatial properties of principal components from two fractions of a stream sediment survey in Quebec.

The spatial factor technique employed here differs from the described approaches in the following ways:

(1) SFA determines linear combinations of variables and their significance to the spatial pattern from a given direction and some specified lag distance, d. Thus, spatial relationships can be evaluated as a function of lag distance and direction.

(2) The SPFAC technique uses the auto- cross-correlation relationships based on the signal at lag zero as opposed to the covariances of the variables.

The technique of SPFAC is described by Grunsky and Agterberg (1988, 1992). Details of the method will not be described here. The application of SPFAC as used in this study employs the use of the two point model only, that is, only one orientation is considered in the analysis.

Given a vector of variables, X, with lag 0 auto- cross-correlations of R_0, and lag d auto- cross-correlations of R_d where

$$R_0 = \begin{vmatrix} a_{11} & a_{12} & \cdots & a_{1m} \\ a_{21} & a_{22} & \cdots & a_{2m} \\ \cdot & & & \\ \cdot & & & \\ a_{m1} & a_{m2} & \cdots & a_{mm} \end{vmatrix} \qquad R_d = \begin{vmatrix} F_{d11} & F_{d12} & \cdots & F_{d1m} \\ F_{d21} & F_{d22} & \cdots & F_{d2m} \\ \cdot & & & \\ \cdot & & & \\ F_{dm1} & F_{dm2} & \cdots & F_{dmm} \end{vmatrix}$$

F_{dij} represents estimates of auto- cross-correlations of variables X_i and X_j for lag d using a function F. If α is a column vector with m unknown elements, then $\alpha'x$ has lag 0 auto- cross-correlations of $\alpha' R_0 \alpha$ and lag d auto- cross-correlations of $\alpha' R_d \alpha$. $\alpha R_d \alpha$ can be maximized subject to the constraint that $\alpha' R_0 \alpha = 1$.

The solution satisfies:

$$R_d \alpha = \lambda R_0 \alpha \qquad \text{or} \qquad R_0^{-1} R_d \alpha = \lambda \alpha$$

where λ is the Lagrange multiplier and α is an eigenvector of the transition matrix U such that

$$R_0^{-1} R_d = U_d$$

For m variables U_d can be decomposed into p separate spectral components. Estimates of the overall noise of the auto- cross-correlations, and a measure of the goodness of fit (squared multiple correlation coefficients, R^2) to estimate the predictive power of each component can be computed as outlined in Grunsky and Agterberg (1988, 1991b). For the approach to be valid, R_d and R_0 must be positive definite. A correction procedure can be applied to modify R_d, or R_0 such that they are positive definite (Grunsky and Agterberg, 1988, 1991b).

The two-point model is determined by a single vector that determines the relationship between locations using the underlying statistical model:

$Z'_i = Z'_j U_d + E'_i$

where Z'_i and Z'_j are row vectors consisting of standardized values of variables i and j; E_i is a row vector consisting of residuals. If S_j represents a column vector of the signal values corresponding to Z, then premultiplication of both sides of the given equation by S_j yields:

$S_j Z'_i = S_j Z'_j U_d + S_j E'_i$

Each column of U_d represents a set of regression coefficients by which the value of a variable at location i is predicted from the values of all variables at location j. If the residual variance for variable Z_k is written as $\sigma^2{}_k$, then the k-th column of U_d has $\sigma^2 R^{-1}_0$ as its variance-covariance matrix. The variances of the variables are proportional to the elements along the main diagonal of this matrix. Grunsky and Agterberg (1991a, 1991b) have provided computer programs for the implementation of the spatial factor technique.

MODELS OF AUTO- CROSS-CORRELATION

The spatial factor procedure requires that estimates of auto- cross-correlation be obtained for the variables over the spatial range of the data. Auto- cross-correlation relationships can be expressed such that a random variable X_k (k=1,...,m), with zero mean, at location x_i, in geographical space is related to every other location x_j by some function F where,

$x_i = F(d_{ij})x_j + y_i$

Both i and j go from 1 to N where N denotes total number of observations (Agterberg, 1970). $F(d_{ij})$ is a function of distance d_{ij} between these two points. The function is considered to decay to zero at the maximum distance of the "neighborhood" for which it is defined. The residual y_i is the realization of a random variable Y_i at point i. It satisfies $E(Y_i) = 0$ and Y_i is assumed to be independent of X_j.

The selection of functions used to estimate auto- cross-correlation is dependent on the nature of the spatial relationships. A number of models can be used such as the spherical, Gaussian, or exponential models. These functions have specific characteristics that are useful for describing the behavior of regionalized variables (Journel and Huijbregts, 1978). A specific requirement of autocorrelation functions is that they satisfy the following conditions. Given the autocorrelation of a random variable, r_x, and lag, d: a) $|r_x(d)| \leq 1$; b) $r_x(d) = r_x(-d)$. As well, functions that obey the properties required of regionalized variables within defined limits also can be applied. Quadratic functions were used as estimates of spatial auto- and cross-correlation by Agterberg (1970) and Grunsky, and Agterberg (1988, 1989).

In this study, three functions of autocorrelation (quadratic, exponential, and Gaussian) were used to define the spatial variation of the irregularly spaced multielement geochemical data sets. For each variable and pair of variables, estimates of auto- cross-correlation were computed for specified lag intervals representing the lag coefficients of a correlogram. From these experimental correlograms, the parameters for each function then were calculated using the downhill simplex method of multidimensional minimization (Press and others, 1986, p. 289) whereby successive estimates of the function parameters are selected based on the sum of squares between the observed autocorrelations and the function estimates. The parameter estimates converge as the sum of squares converges to a minimum.

SPFAC can make use of any designated function specified to create values for R_0 and R_d. In this study SPFAC was run using four types of estimates of auto- cross-correlation. Three estimates were based on the exclusive use of quadratic, exponential, and Gaussian functions. A fourth estimate was made using a mix of functions (mixed model) based on the minimum sum of squares criteria. For each pair of variables the function whose parameters satisfied the minimum sum of squares was selected. For most variables, the exponential function usually met the minimum sum of squares criterion

STUDY AREAS

Two areas were selected for study. Southern Vancouver Island was selected as an area to evaluate the use of SPFAC for interpreting regional geochemical survey data. The Sulphurets area in the northern part of the province was selected to evaluate the use of SPFAC in a mineralized area.

The minimum distance between sample sites on southern Vancouver Island ranges from 0.07 to 9.65 kilometers with a median distance of 3.65 kilometers. The average minimum distance between sample sites in the Sulphurets area ranges from 1 to 670 meters with a median minimum distance of 55 meters.

The data from regional geochemical sampling program in the southern part of Vancouver Island contains information about regional geological processes such as major lithologic differences and structural trends of major lithologies. Anomalous samples that reflect local processes were not included in this study because of the poor autocorrelation of such data as a result of low-sampling densities.

Data were analyzed using FORTRAN programs prepared by Grunsky and Agterberg (1991a, 1991b). In addition data from the southern part of Vancouver Island were analyzed and displayed using Splus® (Statistical Sciences, 1993). Data from the Sulphurets area were displayed using SPANS™ (INTERA TYDAC, 1993).

GEOLOGY OF SOUTHERN VANCOUVER ISLAND

The southern part of Vancouver Island lies at the southern end of the Insular Belt of the Canadian Cordillera (Fig. 1). It comprises three fault-bounded terranes which were assembled in the late Eocene (N.W.D. Massey, pers. comm., 1994). The bulk of the island is underlain by the Wrangellia terrane. This thick Paleozoic to Mesozoic sequence is made up of three major volcanic cycles - the oceanic island-arc volcanics of the Devonian Sicker Group, the oceanic flood-basalts of the Upper Triassic Karmutsen Formation and the Lower Jurassic arc volcanics of the Bonanza Group. Coeval with the Bonanza Group are the plutons of the Island Intrusive Suite and the mid-crustal level metamorphic rocks of the West Coast Complex. The volcanic sequences are separated by the clastic sediments and limestones of the upper Paleozoic Buttle Lake Group and limestones and shales of the Quatsino and Parson Bay Formations. Sandstones, shales, and conglomerates of the Upper Cretaceous Nanaimo Group lie unconformably on the older rocks. Wrangellia has undergone a complex structural history, with frequent rejuvenation of previous structures, imparting a prominent northwesterly aligned structural fabric. The present map pattern is dominated by the northwesterly trending contractional faults of the Tertiary Cowichan fold and thrust system.

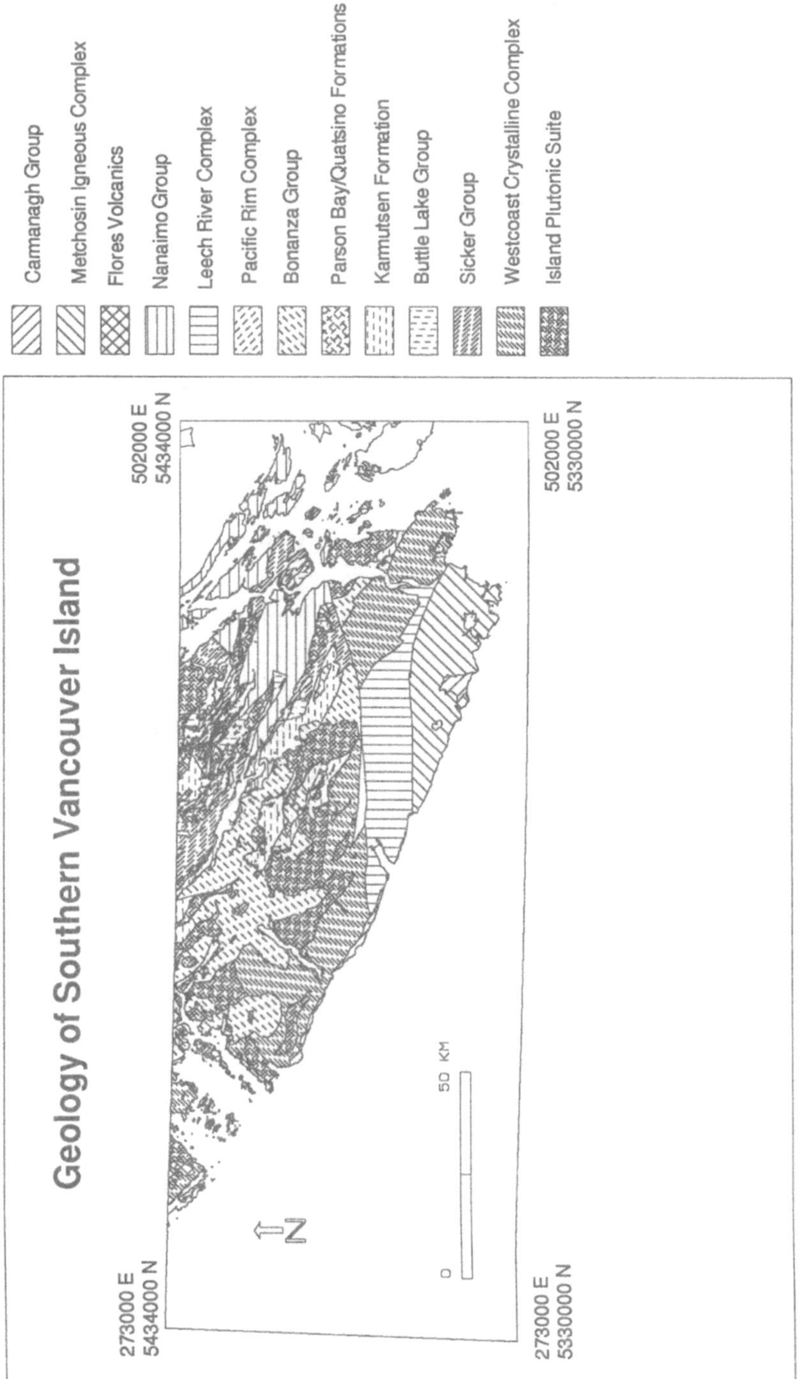

Figure 1. Geology of southern part of Vancouver Island, British Columbia. See text for detailed explanation of map codes.

The Leech River Complex is separated from Wrangellia by the San Juan - Survey Mountain Faults. It comprises Jura-Cretaceous marine sediments and basaltic rocks, regionally metamorphosed and deformed to schists and amphibolites. The ophiolitic basalts and gabbros of the Metchosin Igneous Complex have been thrust beneath the southern end of Vancouver Island along the northerly dipping Leech River Fault. The post-accretion, clastic sediments of the Eocene-Oligocene Carmanagh Group unconformably overlie all older rocks along the southwestern coast.

GEOLOGY AND AU-ASSOCIATED ALTERATION OF THE SULPHURETS AREA

The Mitchell-Sulphurets mineral district in northwestern British Columbia is about 120 km^2 in size. It is noted for extensive alteration zones associated with porphyry copper and molybdenum systems as well as other types of gold and silver mineralization. The area is underlain by Jurassic volcanics and sedimentary rocks, primarily from the Hazelton and Stuhini Groups. The central part of the area is underlain by so-called pyritic altered rocks (Fig. 2) that host most of the potassic, sulphidic, and silicic alteration zones. The contacts of the pyritic altered rocks with the Stuhini Group to the west and the Hazelton Group to the east consist of faults striking NS to NE-SW and NS to NNW-SSE, respectively. Most of the geochemical anomalies in the area are spatially related to the alteration (Cheng, Agterberg, and Ballantyne, 1994).

Vein-stockwork systems occurring in the pyritic altered rocks (Fig. 2) show both brittle and ductile deformation which probably took place at relatively great depths. Most volcanic and sedimentary rocks in the study area are cut by subvolcanic porphyritic intrusions of dioritic, monzonitic, syenitic, and low silica granitic composition. Relatively large granitic intrusions are located in the western part of the study area and smaller dioritic intrusions in the southern and eastern parts. The pyritic alteration, extensive quartz vein stockworks and several types of gold mineralization zones (Cu-Au, Mo-Au, Cu-Mo-Au, and As-Au-Ag) probably are related genetically to the emplacement of these porphyry alkalic intrusions. Some Au mineralization zones may be related to younger faults (Alldrick, 1987; Alldrick and Britton, 1988; Kirkham, 1990; Kirkham, Ballantyne, and Harris, 1989, 1990, 1992).

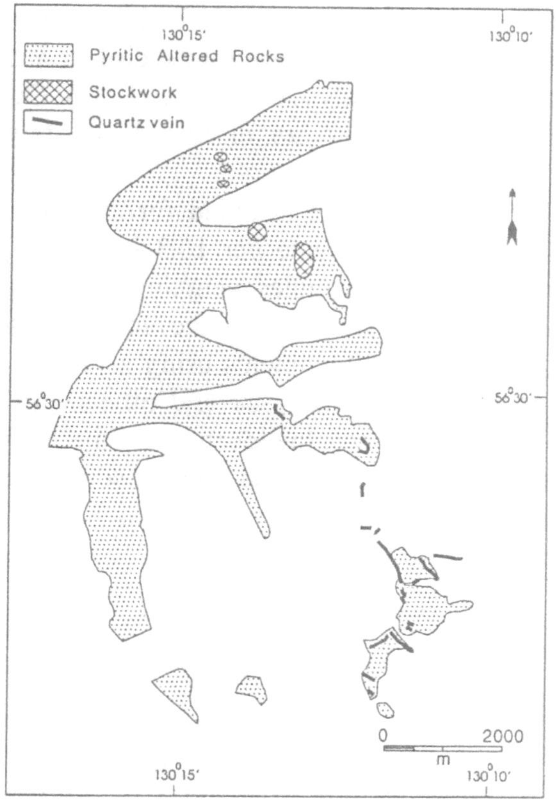

Figure 2. Alteration zones in Mitchell-Sulphurets district.

Gold and Au-associated elements and oxides show spatial correlation with the intrusive rocks and the faults. The main purpose object of applying SPFAC method in the Sulphurets area is to study these relationships.

REGIONAL GEOCHEMICAL SAMPLING PROGRAM - SOUTHERN VANCOUVER ISLAND

The southern part of Vancouver Island, (National Topographic Series Maps 92B and 92C) was sampled as part of a regional geochemical reconnaissance sampling program (Matysek and others., 1990). Moss mats were sampled and analyzed for their geochemical responses in place of traditional stream sediments. Moss mat sampling is an alternative to sampling stream sediments in areas where fine sediment material generally is lacking in stream beds, as is the situation for Vancouver Island. Moss mats capture fine sediment and are plentiful in Vancouver Island streams (Matysek and Day, 1987). Quality control and sample preparation procedures are discussed in Matysek and others. (1990).

A total of 599 sites were sampled for moss-mat sediment at an average sampling density of 1 site per 10.3 km^2. The sample sites are shown in Figure 3. Moss mat samples weighing 1-2 kg were scraped from logs and boulders which occur within the active stream channel. The samples were analyzed for 22 elements using a variety of determination techniques as documented in Matysek and others. (1990). Because the sample sites follow stream beds, the locations of the samples do not cover the map area in a uniform manner. As a result some lithologies have not been sufficiently sampled.

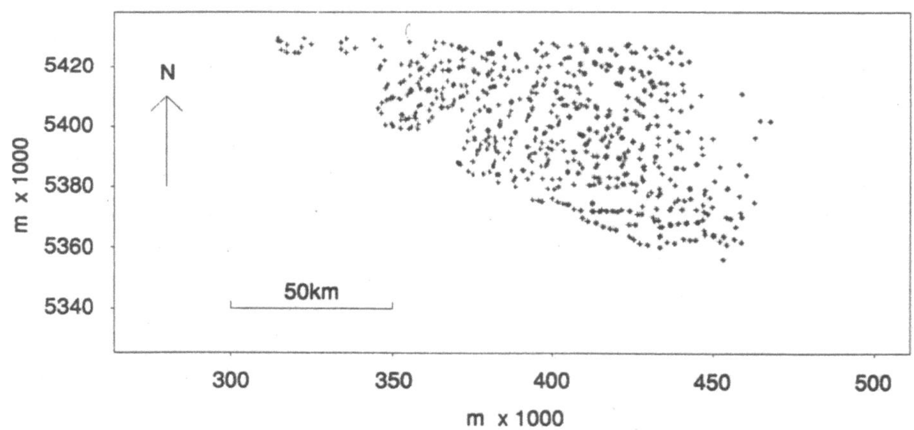

Figure 3. Sample locations, southern part of Vancouver Island. Sample sites follow drainage network.

PRELIMINARY ASSESSMENT OF DATA SOUTHERN VANCOUVER ISLAND

Regional geochemical survey data can be difficult to interpret. A significant difficulty in the interpretation of geochemical data is the nonnormal nature of their distributions. Meaningful estimates and interpretations of nonnormal distributions is unlikely if statistical measures of the distributions are made under the assumption of normality.

In many distributions, the skewed nature of the data can be overcome by applying a suitable transformation that shifts the values of the distribution such that it becomes normally distributed. The method selected for the Southern Vancouver Island study was that of applying Box-Cox generalized power transformations using a method outlined by Howarth and Earle (1979). The Box-Cox family of transformations are defined as:

$$y=(x^{\lambda} - 1)/^{\lambda} \quad \text{for } \lambda > 0$$
$$y=\ln(x) \text{ for } \lambda=0$$

The presence of outliers in the data poses a problem particularly for determining an appropriate value of λ. Campbell (1986) has developed a method for determining a suitable value of λ based on the presence of outliers but this method was not used in this study. One way to minimize the effect of outliers, is to trim the data at the 99, 98, and 95 percentile rankings and compute values of λ for each group of data. This approach was taken for this study. Table 1 lists the values of λ for a selected group of elements calculated and the subsequent value that was used. The selection of an appropriate value of λ in the situation of outliers is problematic.

Table 1. Values of λ and replacement values for selected elements.

Element	L.L.D	n<L.L.D	Trim	λ	Replacement Value
Pb	2.00	256	0.98	Log	1.00
Cu	2.00	0	0.98	0.2	
Ni	2.00	0	0.98	0.25	
Co	2.00	0	0.98	0.5	
Mn	5.00	0	0.98	Log	
Fe	0.02	0	0.98	Log	
U	0.50	13	0.98	0.3	0.04
As	1.00	216	0.98	Log	0.39
V	5.00	0	0.98	0.2	
Cr	5.00	0	0.98	0.3	

L.L.D. - Lower Limit of Detection
n<L.L.D. - number of samples less than the detection limit.
Trim - the maximum percentile used for determining the value of λ
λ - Value for Box - Cox power transformation
For λ=0, the natural logarithm is used for the transformation.
Replacement value determined only for censored distributions.

Another problem is the occurrence of censored distributions. Censored distributions result when a portion of the sample population is less than some specified value (i.e., the detection limit). The problem of censored data becomes more apparent when means and covariances are required. The use of a single substituted value biases the computation of the moments of the distribution. However, if the distribution is assumed to be normal, then the replacement value of the censored data and parameters of the distribution (mean, variance) can be estimated from the portion of the distribution that is not censored. Estimates of the distribution parameters can be obtained using the EM algorithm (Dempster, Laird, and Rubin, 1977) and is discussed by Campbell (1986), and Chung (1985, 1988). Chung (1985) and Campbell (1986) have published computer programs for the statistical treatment of geochemical data with observations below the detection limit. The assumption of normality is essential for the EM algorithm to work. Campbell (1986) invokes an algorithm to transform the data to normality using Box-Cox transformations as described in the previous section. More recently, Sanford, Pierson, and Crovelli (1993) have developed a method that allows the calculation of a suitable replacement value based on a maximum likelihood approach.

Q-Q plots were made for 22 of the elements for both the untransformed and power transformed data. The plots were compared for assurance that the power transform was appropriate and to evaluate the degree of censoring of the data. Elements including Ag, Cd, and Bi are highly censored which results in difficulties for meaningful estimates of auto- cross-correlations. An example of the effects of the Box-Cox power transformation for the element, Cu, is shown in Figure 4.

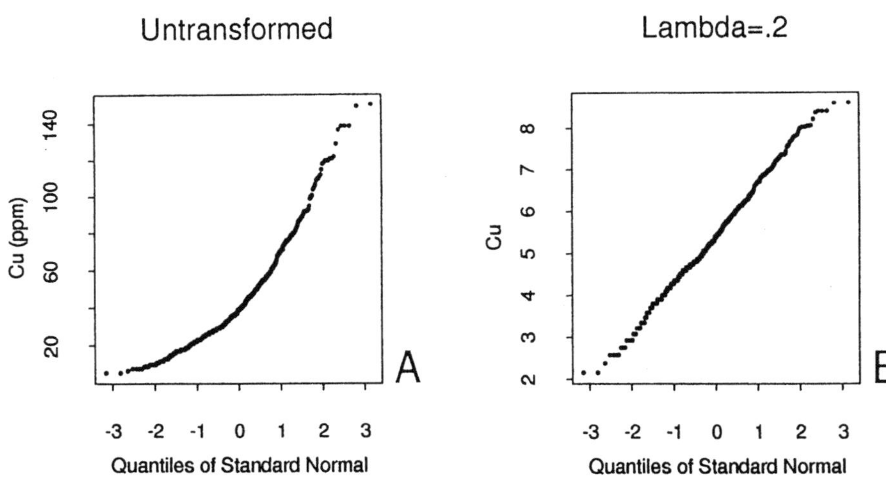

Figure 4. Effect of Box-Cox power transformation for Cu, southern Vancouver Island. A, Q-Q plot of original data for Cu. Distribution is not normal and can affect calculations of means and variances. B, Q-Q plot of transformed data for Cu. Note straightline indicating distribution is near normal.

In the situation of constant sum data, the problem of transformations and outlier detection are more problematic (Barcelo, Pawlowsky, and Grunsky, 1994). The use of Box-Cox power transformations can achieve normality for the marginal distributions; however, the multivariate distribution may not be multivariate normal. The application of log-ratios as outlined by Aitchison (1986) overcomes the problem of the interpretation of constant sum data, however the resulting log-ratio distributions may themselves not be normal and thus may be difficult to interpret.

Subsequent to determining the most appropriate transformation for each element, the technique outlined by Chung (1985) was used for determining suitable estimates of the means and standard deviations of the transformed elements. From the estimates of the distribution parameters, replacement values were calculated as described by Sanford, Pierson, and Crovelli (1993). Table 1 lists the values of λ, and the subsequent replacement values that were used for the censored elements, Pb, U, and As.

During the initial evaluation of the data, the elements were examined for their potential usefulness for deciphering geochemical trends. The use of scatterplot diagrams with all possible pairs of elements indicated that some elements show poor correlation with others. This was the situation for the elements such as Sn and Ag. These elements were eliminated for any further consideration.

Principal components analysis was applied to the transformed data. The results indicated that there is a clear distinction between the mafic igneous rocks, the felsic plutonic rocks and the metasedimentary sequences. Statistical summaries of the data by Matysek and others (1990) show that many of the elements display distinct differences in population distributions as a function of lithology. Elements that typically reflect the mafic volcanic lithologies of the Sicker Group volcanics, mafic rocks of the Coast Plutonic Complex, the Karmutsen basalts, and the Metchosin Group pillow basalts (Fig. 1) include Co, Cr, Cu, Ni, Fe, and V. Elements that reflect felsic volcanic and intermediate to felsic intrusive lithologies include, As, Pb, and U. These elements represent lithologies such as the felsic Island Plutonic Complex. Sedimentary rocks which include the Sooke, Carmanah, Nanaimo, Leech River, Pacific Rim, and Buttle Lake groups are less obvious to distinguish solely on the basis of 10 selected elements that were used for the study. Differences between the mafic lithologies are described by associations of Co, Cr, Cu, Ni, Fe, and V. Iron and Cu usually are associated with mafic volcanic and intrusive rocks of the area including some phases of the Westcoast Crystalline Complex, probably representing assimilated mafic protolith. Elements such as Ni, Co, and Cr usually are restricted to the Sicker, Karmutsen and Bonanza volcanic groups and the Metchosin Igneous Complex.

ESTIMATES OF SPATIAL AUTO-CROSS-CORRELATION

Subsequent to the correction procedures outlined previously, auto-cross-correlation values were determined for a range of lag intervals and orientations. Examination of the spatial distribution of the data suggested that a lag interval between 10 and 40 kilometers would reflect the most of the spatial processes. From Figure 1 it can be seen that the dominant tectonic and lithologic trends in the area are in the NW-SE direction for the northern and eastern parts of the area. An E-W trend predominates in the southwestern part of the area and some N-S trends are observed for some units in the eastern part of the area. Thus, the orientations were subdivided into N-S, NE-SW, E-W, and NW-SE directions. Experimental auto- cross-correlation correlograms were determined for the four orientations with an angle of tolerance of 15° and lag intervals ranging from 10 to 40 kilometers. The data generated from these correlograms then were used to determine the parameters for the three models of autocorrelation as discussed earlier.

An examination of correlograms and signal to noise ratios of the elements indicated that the following elements would be the most suitable for study: Cu, Pb, Ni, Co, Mn, Fe, U, As, V, Cr. As described previously, these elements reflect mafic and felsic igneous processes to a significant extent at the scale of sampling and size of the area. Other elements such as Ag, Au, Bi, W, Sn, and Sb tend to be censored highly or have such high nugget effects at the scale of sampling, that it is difficult to use these elements to detect spatial geochemical processes. The elements, Sb, Au, Ag, Cd, Mo, Sn, Bi, Hg, W, F, and L.O.I. were not included in the study because the low-sampling density resulted in large nugget effect values. Auto-cross-correlograms for these elements were difficult to interpret.

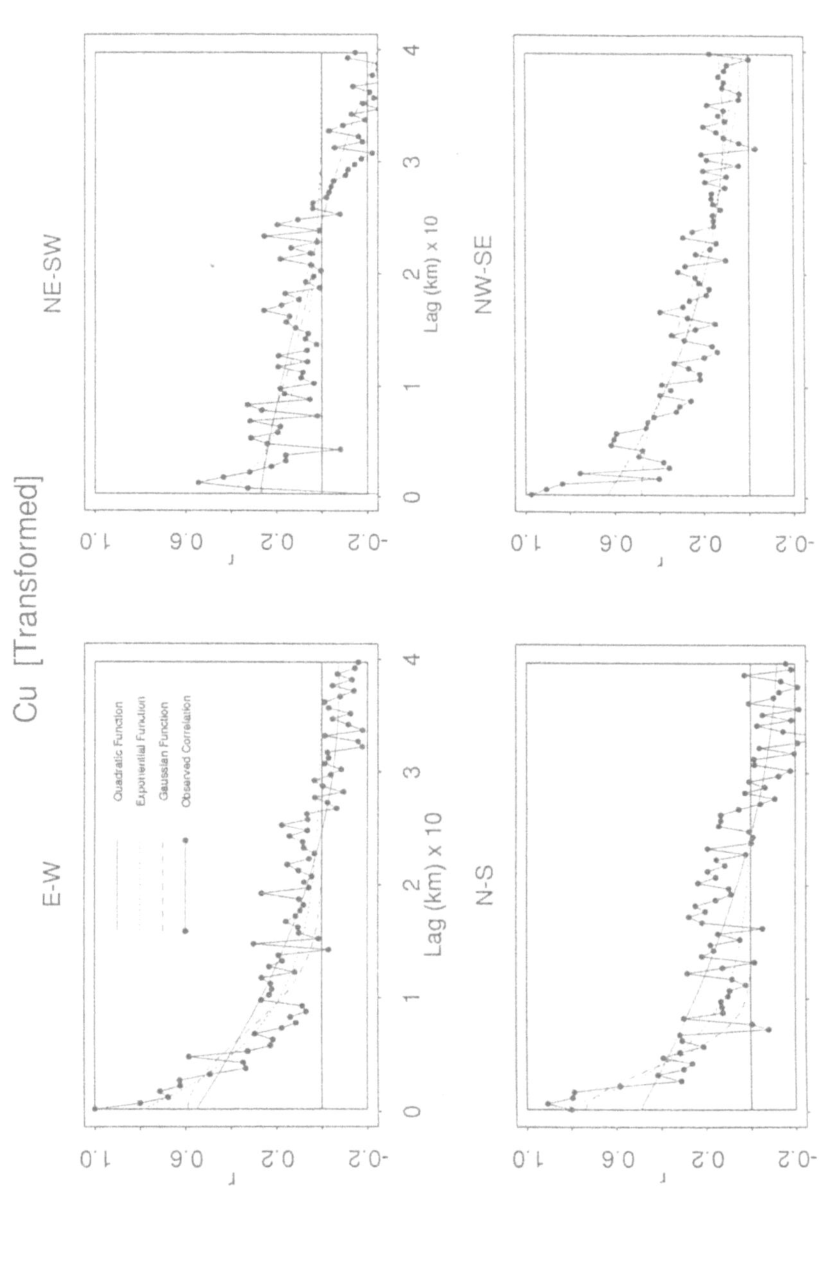

Figure 5. Experimental correlogram and 3 function model fits for Cu, southern Vancouver Island. Note changing signal to noise ratio and decay rate for different orientations. 5a: Correlogram for E-W direction with signal to noise ratio of ~0.6. 5b: Correlogram for NE-SW direction with signal to noise ratio of ~0.3. 5c: Correlogram for N-S direction with signal to noise ratio of ~0.5. 5d: Correlogram for NW-SE direction with signal to noise ratio of ~0.7.

An examination of the autocorrelations of the 10 selected elements showed that correlations were strong in some directions and weak in others. Elements that show strong autocorrelation signals in the NW-SE direction include Cu, Ni, Co, Mn, Fe, and V. Elements that show strong autocorrelation in the N-S direction include U, As, and Cr. Figure 5 displays plots of the auto-correlation functions derived for Cu over the four different orientations. The figure shows that Cu is distributed anistropically with the longest decay in the NW-SE direction and the shortest decay in the NE-SW direction. The figure also shows plots of the three function fits (quadratic, Gaussian, and exponential) to the correlogram. The correlograms and functions indicate a decay to zero in the range of 25 km for the E-W, NE-SW, and N-S directions. The correlograms decay to values below zero for the these three directions and indicate that there is a lack of stationarity. For the NW-SE orientation the correlogram and functions decay to zero at a distance greater than 40 km.

LITHOGEOCHEMICAL DATA FROM THE SULPHURETS AREA

About 1233 surface outcrop samples (Fig. 6) were collected along mapping traverses for mineral deposit research by R.V. Kirkham, S.B. Ballantyne, and D.C. Harris of the Geological Survey of Canada during the period 1986 to 1990. Both altered and unaltered rocks were sampled. The samples were analyzed chemically for approximately 30 trace elements and 13 oxides (Ballantyne, 1990; Ballantyne and others, 1992). The lithogeochemical data were stored using the Geographical Information System SPANS (INTERA TYDAC, 1993) for preparation of contour maps and integration with other types of data.

Statistical characteristics of the lithogeochemical data were studied using various multivariate analysis procedures including factor analysis and cluster analysis showing that As, Sb, Ag, Pb, Zn, Cu, Mo, F, Nb, S, Cd, Se, Eu, K_2O, and SiO_2 are correlated positively whereas Cr, V, Ce, Co, Tb, Fe, T, Sr, Sm, W, La, MnO, MgO, and CaO are correlated negatively with Au. Contour maps for these elements show that most Au-associated elements have higher concentration values in two separate zones. The first zone strikes SN to NE-SW and lies close to the western contact between the pyritic altered rocks and the Stuhini Group, and the other one strikes SN to NW-SE along the eastern contact with the Hazelton Group. Au-associated elements in the first zone include Au, Cu,

Mo (to the north), K_2O, and Al_2O_3 , and those in the second zone As, Sb, Ag, and SiO_2. The higher temperature element association in the first zone may be the result of being in closer proximity to intrusive rocks and greater depth below the topographic surface at the time of origin.

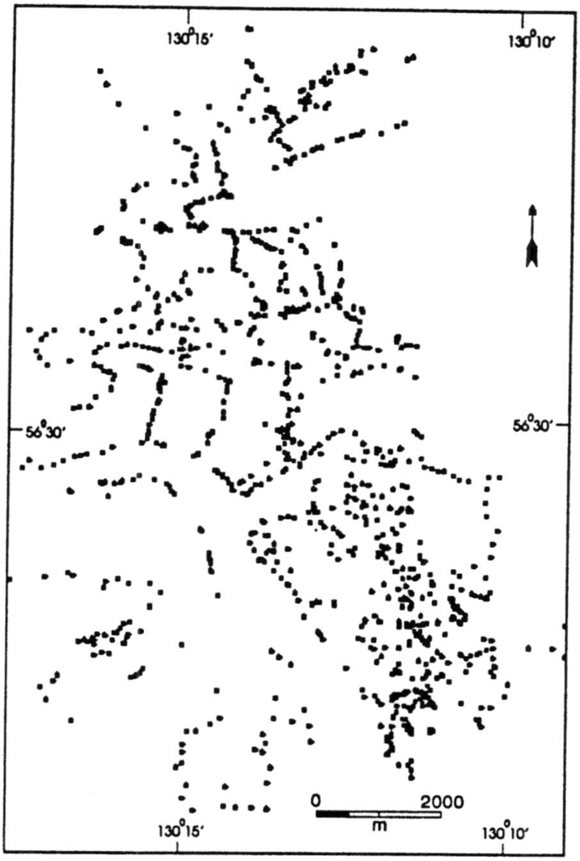

Figure 6. Sample location map, Sulphurets area.

APPLICATION OF SPATIAL FACTOR ANALYSIS

Southern Vancouver Island

SPFAC was applied to the 10 selected elements for the four orientations and at lag distances from 10 to 40 km at 5 km intervals using the quadratic, exponential, Gaussian, and mixed models. This resulted in 96 spatial factor results for subsequent interpretation. From these analyses only the mixed model results are presented here.

The results of the analysis indicated that the significance and relationships of the elements change as a function of lag distance and orientation. As discussed in Grunsky and Agterberg (1988), the squared multiple correlation coefficient (R^2) provides a measure of significance of each variable. Values of R^2 can be calculated for the variables over the transition matrix U and for each of the spectral components U_i. Ideally, the values of R^2 should not exceed a value of 1. Figure 7 shows the values of R^2 for the transition matrix U over the 7 lag distances and 4 orientations. The figure shows that as the lag distance increases, there is generally, a corresponding decrease in the value of R^2. Of particular interest is the change in relative significance of variables as a function of orientation.

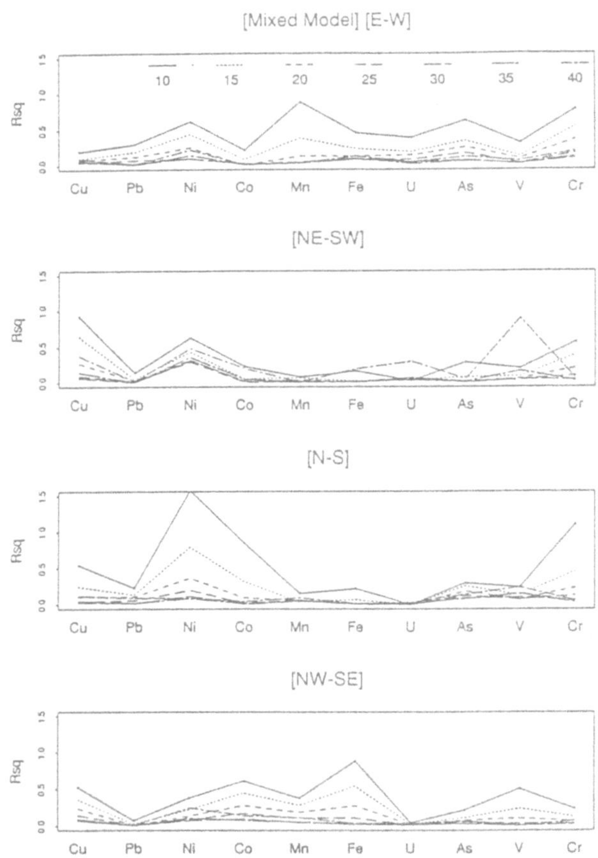

Figure 7. Multiple R^2 correlation coefficients for mixed model SPFAC. Each line represents R^2 coefficients for lag distance (in kilometers) indicated at top of figure. Note that for N-S orientation, Ne, Co, and Cr exceed values of 1.0. This suggests that collinearity may interfere in determination of meaningful results.

Comparison of the R^2 coefficients in Figure 7 illustrates that as orientation changes, the relative significance of the variables changes. In the E-W and NW-SE directions the relative significance of the elements are closer to each other than in the N-S and NE-SW directions. In the NW-SE and E-W directions Cu, Ni, Co, Mn, Fe, V, and Cr share a similar relative significance. In the N-S and NE-SW directions, this is not the situation. In the N-S direction Ni, Co, and Cr have significantly greater R^2 values than the other elements and in the NE-SW direction Cu, Ni, and V are substantially greater than the other elements. The associations of these elements reflect the orientations of various lithologies with geochemical responses in preferred directions.

The spectral components U_i, explain which elements are associated in a manner similar to principal components analysis. The components generally are fewer than the number of the elements. In this study the number of components seldom exceeded three. An examination of the spectral components as a function of lag distance and orientation indicate that several geochemical associations with defined spatial structures can be distinguished.

The geochemical sampling pattern over the area does not reflect adequately the distribution of the lithologies that exist on the geological map. Thus, the interpretation of the geochemical patterns must be made with the knowledge that the sampling does not represent the complete geological picture. The results of the SPFAC indicated that as lag distance and orientation change, specific associations of elements were obtained. Not all of the element associations derived from the analysis can be interpreted in a geological context. This is most probably the result of the nature of the moss mat sediment that may be a mix of one or more lithologies, making interpretation difficult as well as an insufficient sampling density to adequately represent the spatial extent of the underlying lithologies. An evaluation of the most significant component for each of the results for the 7 lag distances and 4 orientations provided the following interpretations.

From the application of the spatial factor methodology several patterns and elemental associations have been identified in addition to the two examples as described. These associations generally are described for each orientation as follows.

As shown in Figure 7, from the R^2 coefficients, the dominant elements for the E-W direction are Mn, Fe, As, Cr, V, U, and Ni. Generally, as lag distance increases, the overall significance of the variables decreases however the relative significance of the variables remains approximately the same. Examination of the most significant component for the 7 lag distances indicated that the dominant lithologies highlighted by this orientation are the felsic intrusive rocks of the Island Plutonic Suite and the West Coast Crystalline Complex that contains a mix of felsic to mafic plutonic rocks with varying amounts of assimilated supracrustal material. At the lag distance of 40 km, the association does not seem to be present. This most likely reflects the spatial extent of these rocks in terms of the samples used to derive the auto- cross-correlation coefficients.

In the NE-SW direction, Cu, Ni, Cr, and As are the dominant elements for all of the lag distances with the exception of the 40 km distance. Typically the plotted scores of these components highlighted areas of mafic provenance including the Karmutsen, Bonanza, and Sicker volcanics as well as sediments with mafic components such as the Leech River Complex containing metavolcanics and mafic derived metasediments, and parts of the West Coast Crystalline Complex intrusions. The elements usually associated with these components are Ni, As, and U. At the 40 km lag distance the dominance of V highlights the distinction between the Leech River Complex and the Nanaimo Group sediments with Co, Fe, V-bearing mafic volcanic rocks. Figure 8A graphically displays the R^2 coefficients for the dominant second component at the 25 km lag distance. Nickel, Cr, and U are the most significant elements. Figure 8B shows the corresponding amplitude vectors. Nickel and Cr contribute to the positive scores and U, V, Cu, and Pb contribute to the negative scores. Figure 9 shows a plot of interpolated scores for this component. Positive scores of the samples outline the Leech River metasedimentary rocks, and the Bonanza and Sicker volcanics. The negative scores outline the Westcoast Crystalline Complex and portions of the Metchosin Igneous Complex.

From Figure 7, Ni, Co, Cr, and Cu are shown to be the most significant elements for the N-S direction. A comparison of scores for the 7 lag distances shows that at shorter lag distances (< 30 km) mafic volcanics of the Karmutsen and Sicker volcanic assemblages, and mafic derived metasediments of the Leech River Complex are highlighted. For lag distances greater than 35 km, the West Coast Crystalline Complex is highlighted.

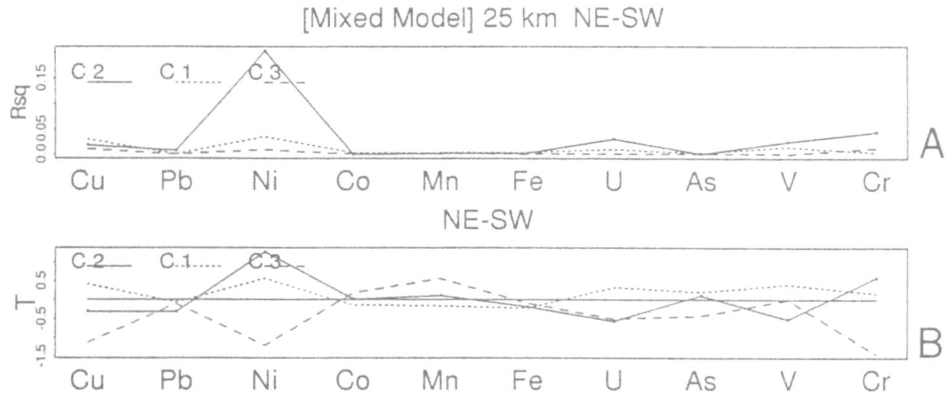

Figure 8. A, Multiple R^2 correlation coefficients for mixed model SPFAC in NE-SW orientation at lag 25 km. For dominant second component, Ni is most significant element for this orientation and lag distance. Components 1 and 3 are insignificant relative to second component. B, Amplitude vector loading for mixed model SPFAC in NE-SW orientation at lag 25 km. Loadings indicate that positive scores are associated with samples containing Ni and Cr.

For the NW-SE direction, Fe, Co, Cu, Ni, V, and Cr are the most significant elements (Fig. 7). The components for this orientation and the 7 lag distances contrast the mafic volcanic and igneous rocks from the sedimentary and felsic plutonic suite. Distinctions between the various volcanic rocks also is highlighted at the 20 km lag distance, whereby Mn-Fe enriched volcanics are distinguished from Ni-Cr enriched volcanics. The NW-SE direction is the dominant direction for many of the volcanic lithologies in the area. This also is demonstrated in Figure 7 where most of the elements associated with the mafic volcanics (Sicker, Karmutsen, and Bonanza Groups) have similar R^2 values. These lithologies also are dominated by Ni, Co, Cr, Cu, Fe, and V. There are 5 components associated with this lag distance and orientation. Figures 10A and 10B show the R^2 and amplitude vectors for the results at the 20 km lag distance. Figure 10A shows that Co, Cu, Fe, and V are the dominant elements and Figure 10B shows that the negative loadings are associated with the elements of mafic associations (Cu, Co, Fe, and V). The first component describes the dominant NW-SE trends of the Karmutsen, Bonanza, and Sicker Group volcanics; the slight positive loading of Pb is displayed in Figure 11 where positive scores highlight areas of Pb

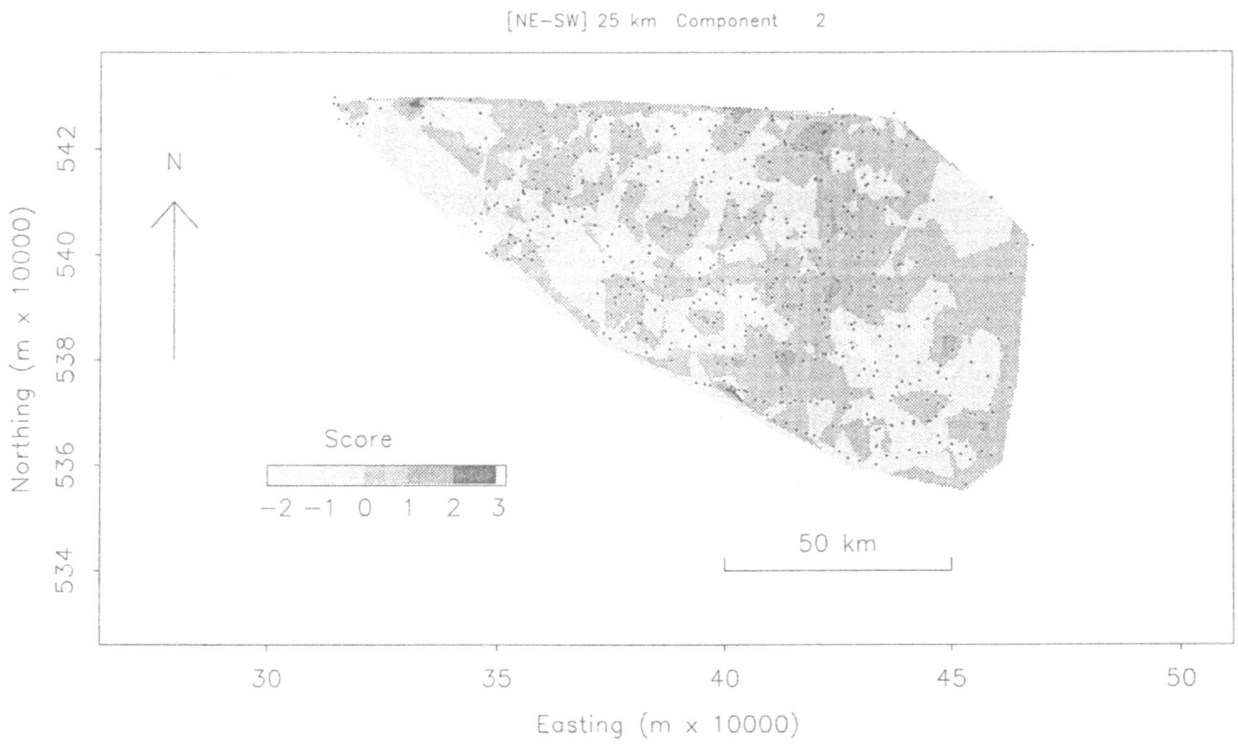

Figure 9. Contour map of first component for NE-SW direction at lag 25 km. For this orientation positive scores associated with Ni and Cr enriched samples outline Leech River metasedimentary complex, Bonanza Group volcanics and parts of Sicker Group volcanics are outlined. Samples with negative scores outline portions of Westcoast Crystalline Complex and parts of Metchosin Igneous Complex.

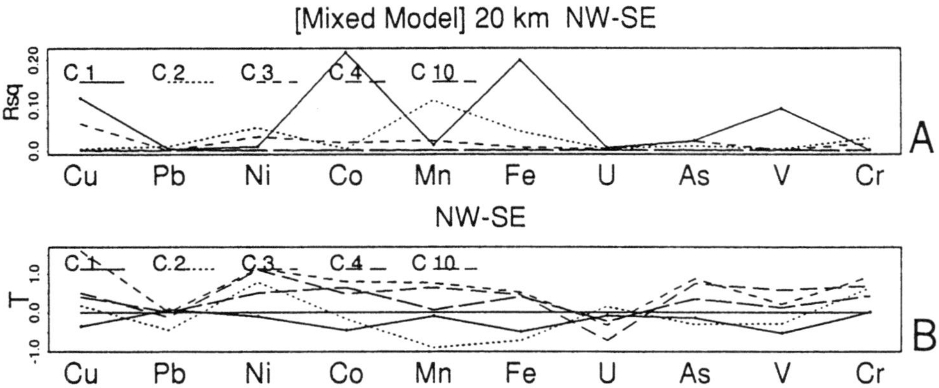

Figure 10. A, Multiple R^2 correlation coefficients for mixed model SPFAC in NW-SE orientation at lag 20 km. For this lag and orientation there are 5 components and at least three of them indicate significant contribution from at least one element. First component indicates that Co, Fe, and Cu are most significant elements at this lag and orientation. Third component shows that Mn, Ni, and Cu are dominant elements and second component indicates that Mn and Ni are significant. B, Amplitude vector loadings for mixed model SPFAC in NW-SE orientation at lag 20 km. Note varying associations of As, Co, and V which describe differences between volcanic assemblages, metasedimentary assemblages and plutonic complexes.

enrichment (felsic intrusives) and the negative scores outline the volcanics of the Karmutsen and Sicker Groups and the Metchosin Igneous Complex. The second component outlines associations, for positive loadings of elements that correspond with the Leech River metasediments, and the Westcoast Crystalline Complex. Negative loadings are associated with felsic igneous rocks. The third components represents a unique association of Ni, Co, Cr, Cu Mn, V, Fe, and As which describes most of the volcanic rocks and mafic derived sediments in the Leech River Complex and Bonanza Group.

APPLICATION OF SPATIAL FACTOR ANALYSIS IN THE SULPHURETS AREA

Because of anisotropy of the spatial distribution of Au and Au-associated elements in the study area, auto- cross-correlation coefficients were estimated for different directions (E-W, NE-SW, NW-SE, and N-S) with lags extending to 1000m. Because of relatively large differences in

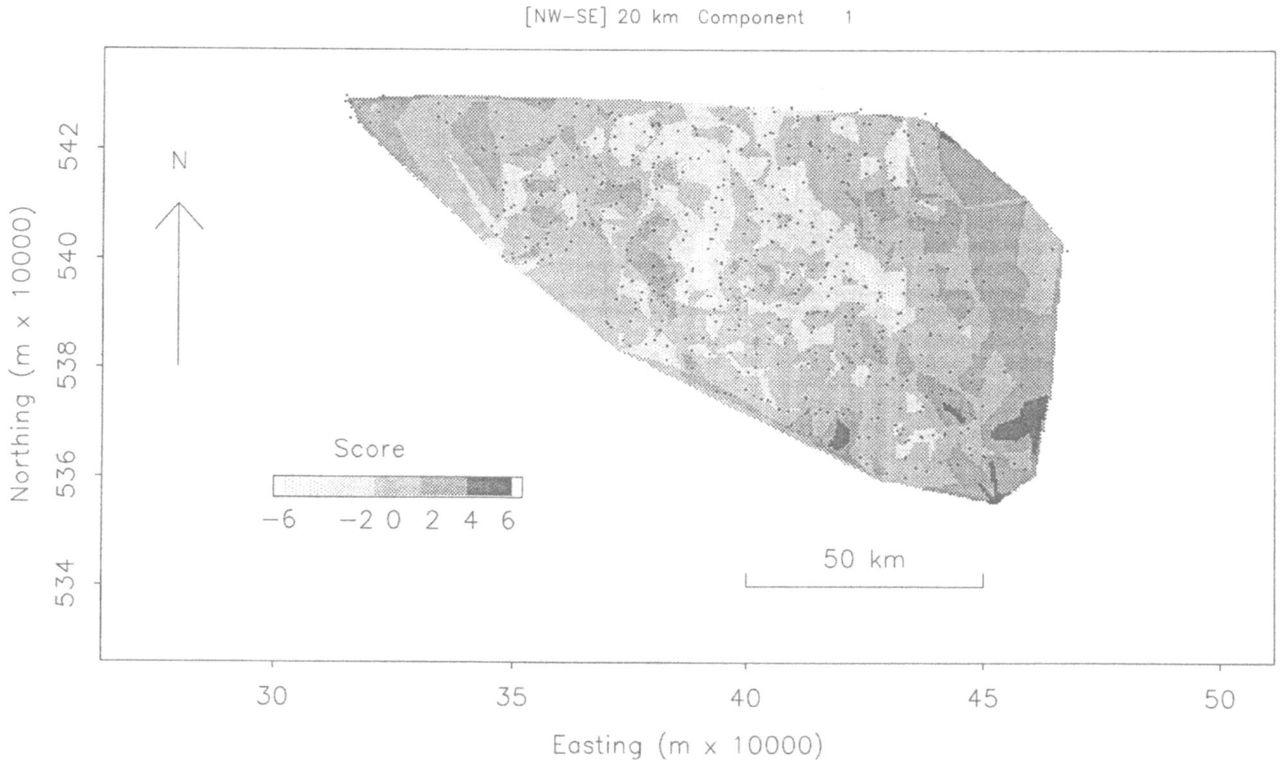

Figure 11. Contour map of first component for NW-SE direction at lag 20 km. Components outline element associations that distinguish major volcanic groups; mafic derived metasedimentary assemblages and igneous complexes.

elevation, true distances between sample locations were computed from digital elevation data and these were used instead of their horizontal projections. The unit of lag was set equal to 12.5 meters with a tolerance of 12.5m. The search angles were 0°±90° (isotropic case), 0°±15° (E-W), 45°±15° (NE-SW), 90°±15° (NS), and 135°±15° (NW-SE). Before computation of the auto- cross-correlation coefficients, the trace-element determinations were log-transformed and the oxides subjected to Aitchison's log-contrast transformation. Three models (quadratic, exponential, and Gaussian) were used for fitting auto- and cross-correlation functions. The elements and oxides have auto- cross-correlation functions that differ for different orientations as shown for Au in Figure 12 where the exponential model is used. In general, the spatial correlation remains relatively strong for longer distances in the trend direction than across trends. For example, Au, Ag, As, Sb, and SiO_2 show relatively strong spatial correlation in the SN to NW-SE directions, Mo and W in the E-W to NE-SW directions, and K_2O and Cu in the SN to NE-SW directions.

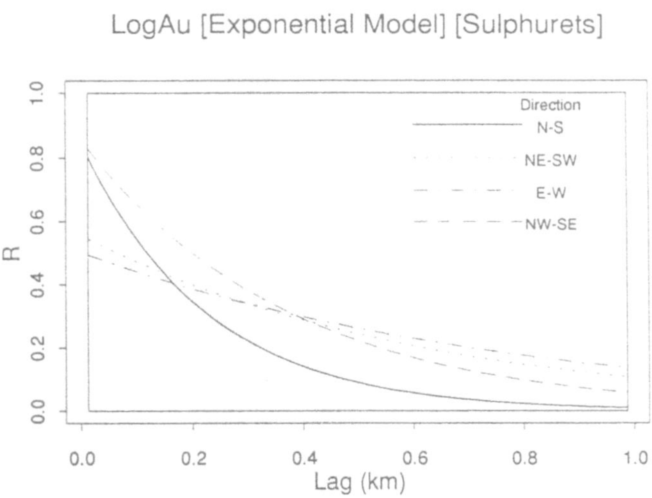

Figure 12. Exponential correlogram functions for log Au in four different directions.

Spatial factors were calculated for groups of 10 trace elements (Zn, As, Mo, Ag, Cd, Sb, Ba, W, Au, and Cu) and 11 oxides (SiO_2, TiO_2, Al_2O_3, total iron, MnO, MgO, CaO, Na_2O, K_2O, CO_2, and P_2O_5). These calculations were carried out for the isotropic case and the four different directions (E-W, NE-SW, NS, and NW-SE) with lag distances increasing from 100m to 900m in steps of 100m. For each orientation and lag distance two to four major spatial factors could be distinguished. Examples are given in Figures 13 and 14 and Tables 2 and 3. Figure 13 is a contour map of the scores for oxides in the NW-SE direction) with lag distance equal to 400m. The coefficients of the corresponding amplitude vector are given in column 4 (A-NW400) of Table 2. Silica and (negative) sodium have the largest amplitudes in this situation. These two oxides also have the largest squared correlation coefficients (R^2) (column 5, R-NW400, Table 2). The pattern of Figure 13 outlines primary areas of silicic alteration and shows spatial correlation with the occurrences of quartz veins and vein stockwork systems (see Fig. 2). After Si and -Na in Table 2 (NW400), -Mg and CO_2 have the largest amplitudes and correlation coefficients for the pattern of Figure 13.

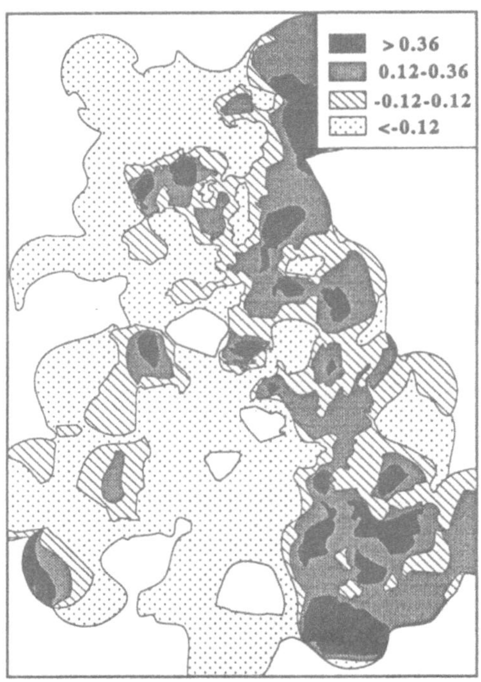

Figure 13. Contour map of third spatial factor, 11 oxides, NW-SE direction, 400m lag, delineates silicic alteration.

To illustrate the dependence of the preceding results on direction and distance, amplitudes and squared correlation coefficients are shown for similar silicification factors with 400m lag in the north-south direction

(N400 in Table 2) and 600m lag in the NW-SE direction (NW600). This indicates that the negative loading of Mg increases in magnitude with clockwise rotation toward the north whereas the relative contribution of CO_2 increases with longer distances in the NW-SE direction.

Table 2. Amplitude (A-NW400) and squared correlation (R-NW400) coefficients for spatial factor of Figure 12 in comparison with coefficients for similar factors with 400m lag in NS direction (columns 2 and 3) and 600m lag in NW-SE direction (last two columns). Largest values are highlighted.

	A-N400	R-N400	A-NW400	R-NW400	A-NW600	R-NW600
Si	**1.150**	**0.677**	**1.089**	**0.035**	**0.782**	**0.016**
Ti	0.201	0.063	-.239	0.003	-.244	0.003
Al	-.020	0.001	0.139	0.002	-.365	0.009
Fe	0.043	0.001	0.311	0.006	0.031	0.000
Mn	0.381	0.279	0.344	0.012	0.570	0.030
Mg	**-.918**	**0.901**	**-.602**	**0.017**	**-.654**	**0.018**
Ca	-.238	0.034	-.067	0.000	0.395	0.007
Na	**-.343**	**0.106**	**-.918**	**0.047**	**-.918**	**0.042**
K	-.076	0.007	0.038	0.000	-.275	0.004
CO_2	0.157	0.020	0.588	0.024	**1.053**	**0.068**
P	-.122	0.323	-.115	0.001	-.361	0.011

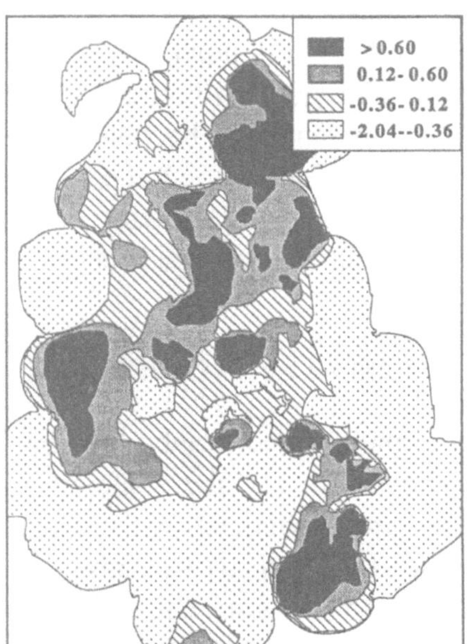

Figure 14. Contour map of third spatial factor, 10 trace elements, NS direction, 400m lag, delineates gold mineralization.

Table 3. Amplitude (A-NW400) and squared correlation (R-NW400) coefficients for spatial factor of Figure 13 in comparison with coefficients for similar actors with 400m lag in NW-SE direction (columns 2 and 3) and 600m lag in NS direction (last two columns). Largest values are highlighted.

	A-NW400	R-NW400	A-N400	R-N400	A-N600	R-N600
Zn	-.040	0.000	0.035	0.000	0.369	0.004
As	**1.015**	**0.100**	**0.557**	**0.032**	**0.799**	**0.014**
Mo	0.537	0.030	0.169	0.003	0.479	0.004
Ag	0.240	0.004	0.076	0.000	-.098	0.000
Cd	**0.876**	**0.073**	**0.352**	**0.012**	**0.949**	**0.018**
Sb	0.788	0.051	0.308	0.010	0.093	0.000
Ba	-.019	0.000	0.110	0.005	0.200	0.003
W	-.112	0.003	-.073	0.001	-.188	0.001
Au	**1.452**	**0.235**	**0.521**	**0.034**	**1.467**	**0.058**
Cu	-.449	0.031	-.181	0.003	-.080	0.000

A second example for trace elements is illustrated in Figure 14. It shows the spatial factor for regional Au mineralization with lag of 400m in the NS direction. This pattern is correlated positively with As and Au (Table 3). The same type of spatial factor could be distinguished for 400m lag in the NW-SE direction and 600m lag in the NS direction. In the latter two situations, Cd is positively associated with Au and As.

The examples of Figures 13 and 14 illustrate that SPFAC is helpful for determining associations between trace elements and oxides. In this respect its role is similar to that of factor analysis. However, SPFAC has the added advantage of allowing these associations to change as a function of distance as well as direction.

CONCLUDING REMARKS

Results of the spatial factor method (SPFAC) applied to Southern Vancouver Island show that for the scale of sampling, multielement geochemical patterns are associated with a particular spatial process related to underlying lithological variation, or variation in stream sediment composition. The interpretation is dependent on the spatial continuity of the elements and a suitable sampling density to ensure minimum nugget effects. The mafic igneous volcanic and intrusive lithologies, felsic intrusive rocks, and metasediments have characteristic geochemical signatures that are highlighted at various orientations and lag distances. Generally, lithological patterns that are indicated by the spatial factor scores are consistent in shape for similar orientations, however the extent of the patterns change with changing lag distance. Lithologies such as the Leech River metasediments show to their maximum spatial extent at the shorter lag intervals. As the lag distance increases the pattern decreases due to the decreasing value of auto- cross-correlations.

The analysis of the Sulphurets area shows that specific zones of alteration can be distinguished readily using the spatial factor technique. Specific geochemical processes that have preferred directions can be isolated using this methodology and can assist in mapping favorable horizons.

The application of SPFAC provides more detail on the multielement relationships of lithologies than can be described using more conventional methods. In the application of the technique to a regional sampling program, multielement associations of regional trends can be factored providing that they have been adequately sampled. In areas of more detailed sampling, as shown in the Sulphurets area, significant geochemical trends that are associated with alteration and mineralization can be enhanced using the spatial factor technique.

ACKNOWLEDGMENTS

This work has been supported by the Geological Survey Branch, British Columbia Ministry of Energy Mines and Petroleum Resources. We are grateful for the assistance of Nick Massey and Steve Sibbick of the Geological Survey Branch, for information about the geology and geochemistry of the Southern Vancouver Island area.

REFERENCES

Agterberg, F.P., 1966, The use of multivariate Markov schemes in petrology: Jour. Geology, v. 74, no. 5, pt. 2, p. 764-785.

Agterberg, F.P., 1970, Autocorrelation functions in geology, *in* Merriam, D.F., ed., Geostatistics: Plenum Publ. Co., New York, p. 113-142.

Agterberg, F.P., 1974, Geomathematics: Elsevier, Amsterdam, 596 p.

Aitchison, J., 1986, The statistical analysis of compositional data: Chapman and Hall, New York, 416p.

Alldrick, D.J., 1987, Geology and mineral deposits of the Salmon River Valley, Stewart area, NTS 104A and 104B: British Columbia Ministry of Energy, Mines and Petroleum Resources, Geology Survey Branch, Open File Map 1987-22, 1:50,000.

Alldrick, D.J., and Britton, J.M., 1988, Geology and mineral deposits of the Sulphurets area: British Columbia Ministry of Energy, Mines and Petroleum Resources, Open File Map, 1988-4, 1:50,000.

Ballantyne, S.B., 1990, Geochemistry of Sulphurets area, British Columbia (abst.): Program with Abstracts, Geol. Survey Canada, Minerals Colloquium, 1 p.

Ballantyne, S.B., Shives, R.B.K., Harris, D.C., Plouffe, A., Judge, A., and Pilon, J.A., 1992, An integrated approach and model for the discovery of blind Cu-Au porphyry systems: Geol. Survey Canada, Minerals Colloquium, poster.

Barcelo, C., Pawlowsky, V., and Grunsky, E.C., 1994, Outliers in compositional data: a first approach, *in* Proc. Intern. Assoc. Math. Geology, Ann. Meeting (Mount Tremblant, Quebec, Canada), p. 21-28.

Bellehumeur C., Marcotte. D., and Jebrak, M., 1994, Multielement relationships and spatial structures of regional geochemical data from stream sediments, southwestern Quebec, Canada: Jour. Geochemical Exploration, v. 51, no. 1, p. 11-35.

Bourgault, G., and Marcotte, D., 1991, Multivariable variogram and its application to the linear model of coregionalization: Math. Geology, v. 23, no. 7, p. 899-928.

Campbell, N.A., 1986,. A general introduction to a suite of multivariate programs: CSIRO Division of Mathematics and Statistics, Perth, Australia, unpubl. rept., unpaginated.

Cheng, Q., Agterberg, F.P., and Ballantyne, S.B., 1994, The separation of geochemical anomalies from background by fractal methods: Jour. Geochemical Exploration, v.51, no. 2, p.109-130.

Chung, C.F., 1985, Statistical treatment of geochemical data with observations below the detection limit, *in* Current Research, Part B: Geol. Survey Canada Paper 85-1B, p. 141-150.

Chung, C.F., 1988, Statistical analysis of truncated data in geosciences: Science de la Terre, Inform. Geol. (Nancy), v. 27, p. 157-180.

Dempster, A.P., Laird, N.M., and Rubin, D.B., 1977, Maximum likelihood from incomplete data via the EM algorithm: Jour.Roy. Statist. Society, Ser. B, v. 39, p. 1-38.

Grunsky, E.C., and Agterberg, F.P., 1988, Spatial and multivariate analysis of geochemical data from metavolcanic rocks in the Ben Nevis Area, Ontario: Math. Geology, v. 20, no. 7, p. 825-861.

Grunsky, E.C., 1989, Spatial Factor Analysis: a technique to assess the spatial relationships of multivariate data, *in* Agterberg, F.P, and Bonham-Carter, G.F. eds., Statistical applications in the earth sciences: Geol. Survey Canada Paper 89-9, p. 329-347.

Grunsky, E.C., and Agterberg, F.P., 1989, The application of spatial factor analysis to unconditional simulations with implications for mineral exploration, *in* Proc. Applications of Computers and Operations Research in the Mineral Industry (APCOM) 1989 Chapter 20, p. 194-208.

Grunsky, E.C., and Agterberg, F.P., 1991a, FUNCORR, a FORTRAN 77 program for computing multivariate spatial autocorrelation: Computers & Geosciences, v. 17, no. 1, p. 115-131.

Grunsky, E.C., and Agterberg, F.P., 1991b, SPFAC: a FORTRAN 77 program for spatial multivariate analysis: Computers & Geosciences, v. 17, no. 1, p. 133-160.

Grunsky, E.C., and Agterberg, F.P., 1992, Spatial relationships of multivariate data: Math. Geology, v. 24, no. 6, p. 731-758.

Howarth, R.J., and Sinding-Larsen, R., 1983, Multivariate analysis, (Chapter 6), *in* Howarth, R.J., ed., Statistics and data analysis in geochemical prospecting, v. 2, *in* Govett, G.J.S., ed., Handbook of exploration geochemistry: Elsevier, Amsterdam, 437 p.

Howarth, R.J., and Earle, S.A.M., 1979, Application of a generalized power transformation to geochemical data: Math. Geology v. 11, no. 1, p. 45-62.

INTERA TYDAC Technologies Inc. 1993, Reference manual of SPANS Version 5.2

Journel, A.G., and Huijbregts, C.J., 1978, Mining geostatistics: Academic Press, London, 600 p.

Kirkham, R.V., 1990, Sulphurets area, British Columbia: Geology (abst.): Geol. Survey Canada, 1990.

Kirkham, R.V., Ballantyne, S.B., and Harris, D.C., 1989, Sulphurets area, British Columbia: geology, geochemistry, and mineralogy (abst.): Geol. Survey, Canada, Minerals Colloquium, 1989.

Kirkham, R.V., Ballantyne, S.B., and Harris, D.C., 1990, Sulphurets, British Columbia: Geology, geochemistry, and mineralogy of a deformed porphyry copper, molybdenum and precious metal system (abst.): Geol. Survey Canada, Minerals Colloquium, 1990.

Kirkham, R.V., Ballantyne, S.B., Harris, D.C., Henderson, J.R., Henderson, M.N., and Wright, T.O., 1992, Lower Jurassic Sulphurets porphyry Cu-Au system, northwest British Columbia. (abst.): Geol. Surv. Canada, Minerals Colloquium, 1992.

LeMaitre, R.W., 1982, Numerical petrology, statistical interpretation of geochemical data: Elsevier, New York, 281.p.

Matysek, P.F., and Day, S.J., 1987, Geochemical orientation surveys: northern Vancouver Island, fieldwork and preliminary results, *in* Geological fieldwork, a summary of field activities and current research: British Columbia Ministry of Energy Mines and Petroleum Resources Paper 1988-1, p.493-502.

Matysek, P.F., Gravel, J.L., Jackaman, W., and Feulgen, S., 1990, British Columbia regional geochemical survey, stream sediment and water geochemical data, Victoria/Cape Flattery - NTS 92B / 92C: British Columbia Ministry of Energy Mines and Petroleum Resources RGS Report 24, Geol. Survey Canada, Open File Rept. 2182, 1 Map Booklet, 1 Data Booklet

Myers, D.E., 1982, Matrix formulation of co-kriging: Math. Geology: v. 14, no., 3, p. 249-257.

Myers, D.E., 1988, Some new aspects of multivariate analysis, *in* Fabbri, A.G., Chung, C.F., and Sinding-Larsen, R., eds., Statistical treatments for estimation of mineral and energy resources: Proc. NATO Conf (Il Ciocco, Italy), Reidel, Dordrecht, p. 669-687.

Press, W.H., Flannery, B.P., Teukolsky, S.A., and Vetterling, W.T., 1986, Numerical recipes, the art of scientific computing: Cambridge Univ. Press, Cambridge, 818p.

Royer, J.-J., 1988, Geochemical data analysis, Fabbri, A.G., Chung, C.F., and Sinding-Larsen, R., eds., *in* Statistical treatments for estimation of mineral and energy resources: Proc. NATO Conf. (Il Ciocco, Italy), Reidel, Dordrecht , p. 89-112.

Sanford, R.F, Pierson, C.T., and Crovelli, R.A., 1993, An objective replacement method for censored geochemical data: Math. Geology, v.25, no. 1, p. 59-80.

Statistical Sciences Inc., 1993, SPLUS Reference Manuals, Version 3.2: Seattle, Washington.

Switzer, P., and Green, A.A., 1984, Min/Max autocorrelation factors for multivariate spatial imagery: Stanford University, Dept. Statistics, Tech. Rept. No. 6, 14 p.

Wackernagel, H., 1988, Geostatistical techniques for interpreting multivariate spatial information, *in* Fabbri, A.G., Chung, C.F., and Sinding-Larsen, R., eds., Statistical treatments for estimation of mineral and energy resources: Proc. NATO Conf. (Il Ciocco, Italy), Reidel, Dordrecht, p. 393-409.

GEOSTATISTICAL SOLUTION FOR THE CLASSIFICATION PROBLEM WITH AN APPLICATION TO OIL PROSPECTING

Jan Harff
Institut für Ostseeforschung Warnemünde, Warnemünde, Germany

Ricardo A. Olea, John C. Davis, and Geoffrey C. Bohling
Kansas Geological Survey, Lawrence, Kansas, USA

ABSTRACT

Mathematically modeled thematic maps can relate productivity to geology and display the probability of occurrence of mineral deposits. Stochastic simulation, a geostatistical technique that regards a surface as an outcome of a random function, is used for nonparametric characterization of the uncertainty involved in the estimation of the Mahalanobis' distances at unsampled locations. This avoids overly smooth maps and precludes the need for distributional assumptions to assess misclassification probabilities, problems encountered in earlier studies that described the relations between multivariate observations and rock type by distances and treated them as univariate variables.

In the stochastic approach, discriminant analysis is used to simultaneously partition the multidimensional space occupied by the observations and select the most informative geological variables. Clusters are identified by regarding each observation as a point in space with as many dimensions as variables. Mahalanobis' distances to the cluster centroids at all geographic locations are estimated using sequential Gaussian simulation. By generating multiple realizations of the distances, the most likely distances and their estimation uncertainties can be assessed. Finally, using Bayesian relations,

Geologic Modeling and Mapping
Edited by A. Förster and D. F. Merriam, Plenum Press, New York, 1996

probabilities are adjusted for misclassification and sampling bias. The technique is used to demonstrate that results of stochastic simulation agree well with actual distributions of fields in an oil-producing area in western Kansas, USA; the map of probability of drilling a producing well corresponds with previous results using kriging to map Mahalanobis' distances.

INTRODUCTION

The interpolation between sampling sites irregularly distributed in an area of interest is a spatial estimation problem in the construction of geological maps in mineral exploration and in marine geology. In these situations the problem reduces to extracting the maximum information from a minimal number of observations by preparing maps covering the entire area of investigation. Lately mathematical models and computers have become widely used for the construction of such geological maps.

Geostatistics, based on the theory of regionalized variables, has provided valuable tools for handling univariate and multivariate geological variables (Journel and Huijbregts, 1978; Myers, 1982). Instead of interpolating the multivariate geological feature fields, Harff and Davis (1990) proposed to take advantage of the relations between geological features. This is done by describing the relations between multivariate observations and types or classes of rocks or rock sequences using distance measures which are inversely proportional to the similarity to these types. The problem is reduced to the handling of these distances as univariate variables. Regionalized classification involves a unification of the theory of classification of geological objects developed in the former Soviet Union (Rodionov, 1981) with the theory of regionalized variables (Matheron, 1970). Harff, Davis, and Olea (1992) used kriging for the interpolation of the distances. This method, however, produces overly smooth maps and requires that distributional assumptions be made in order to assess misclassification probabilities.

The main purpose of this paper is to present the use of stochastic simulation for a nonparametric characterization of the uncertainty involved in the estimation of the Mahalanobis' distances at unsampled locations (Journel, 1989; Deutsch and Journel, 1992). In addition to the construction of maps, we address the problem of data interpretation related to a special geological task. In Harff, Davis, and Olea (1992) mapping was done to assess exploration risk. Although the present case study focuses on hydrocarbon exploration, the methodology is completely general and can be applied equally well to other geological mapping problems and interpretation of geological data.

STOCHASTIC SIMULATION FOR NONPARAMETRIC CHARACTERIZATION OF UNCERTAINTY

Thematic maps conveniently display the likelihood of occurrence of mineral deposits by relating a productivity variable P with information provided by geological variables G. The prediction of P by G is based on probabilistic techniques and the existence of a sample subset comprising measurements of both P and G. Regression is not adequate because the correlation between P and G typically is weak. Instead, if each observation is regarded as a point in G-dimensional space, clusters can be identified that have a one-to-one correspondence with a partition of P.

The first step in our procedure is a systematic selection of the most informative variables in G concurrently with optimal partitioning of both P and the multidimensional space defined by the retained variables in G. The search is based on discriminant analysis and in a first stage ignores those observations without production information. The best possible partition of P and G is one in which all the observations in one cluster in G belong to one and only one partition of P.

Once the partitions are decided, the correlation matrix and mean vector for each cluster in G are calculated. At this point those observations without production information enter the analysis. The correlation matrix and mean vector are used to calculate the Mahalanobis' distances from all observations in G to the centroids of the clusters. Observations without production are assigned to the closest G class and the calculation of the correlation matrix and mean vector is repeated using this time all observations in G. The calibration stage of the procedure ends by recording, for each G cluster, the proportion of observations that belong to each P partition.

The next stage is estimation of Mahalanobis' distances at all geographic locations. The estimation is done by stochastic simulation, a geostatistical technique that regards a surface as an outcome of a random function. By generating multiple realizations of the Mahalanobis' distances, the most likely distances and their estimation uncertainties can be assessed. At any location, the likelihood of membership in any cluster G is given by the proportion of times the simulated distance to cluster G is smaller than the distance to the alternative cluster.

Using Bayesian relationships, the final step is adjustment of the probability of belonging to any P class for misclassification and for sampling bias.

A CASE STUDY

The methodology is easier to understand if each step is illustrated with an example. We will use data on petroleum geology in the western part of the State of Kansas, USA. The study relates to oil deposits in the Lansing and Kansas City Groups of the Pennsylvanian System. These consist of cyclic marine limestones with interbedded dark gray, marine shales. A geological overview is given by Merriam (1963). The most prolific deposits are along the Central Kansas Uplift (Fig. 1), where production has reached a mature stage. This tectonic unit has influenced the entire sedimentary process during Paleozoic time. To the west of the Central Kansas Uplift lies the Hugoton Embayment, a part of the Anadarko Basin. For more information about development of hydrocarbon exploration and exploitation in Kansas, see Watney (1984) and Watney and others (1989).

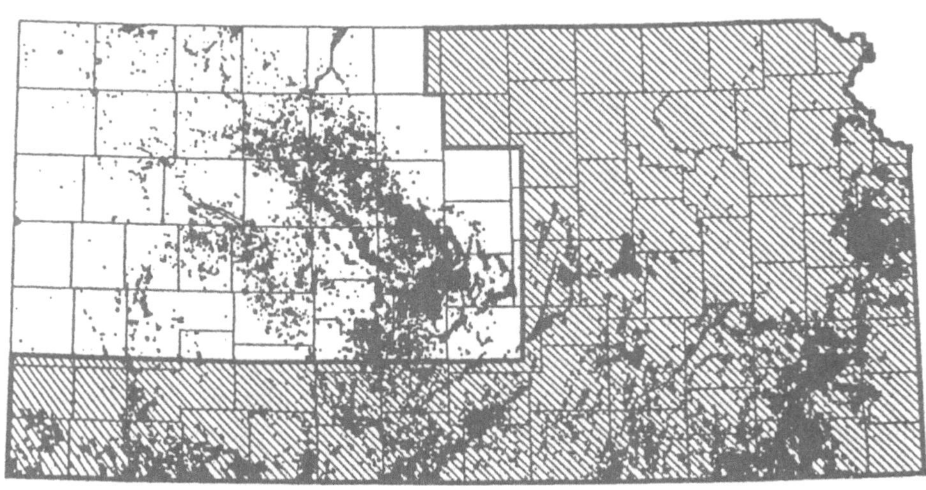

Figure 1. Oil fields in Kansas and study area (nonshaded).

For this study, we employed a data set containing stratigraphic, structural, and petrophysical information from 1046 wells penetrating the Lansing and Kansas City Groups. Because productivity was available as an indicator that does not give the amount produced, we were forced to partition the productivity variable P into only two classes: productive and nonproductive. Ninety-five of the wells were classified as oil wells (class "oil") and 753 were classified as dry holes (class "dry"). The balance are wells with unknown production status. Figure 2 shows the area of investigation with the locations of oil, dry, and unclassified wells. These data can be regarded as a censored sampling. The purpose of the case study is to reproduce the more detailed portrait of the reality shown in Figure 1 by using only the censored data. The successful completion of the exercise should encourage readers to use the method when a sparse sample is all that is available to characterize a prospective area.

Each observation has information about 16 variables. Remember that we want to have as many partitions in the geological variable space as we have production classes. Thus in this case study we are interested in a partition into two classes. After running a discriminant analysis, we concluded that ten of the variables were either redundant or irrelevant, reducing the number of prediction variables of interest to six.

Although a six-dimensional space is smaller than a 16-dimensional space, the dimensions are still too numerous to permit graphical display of six-dimensional clusters. An illustration such as Figure 3 is an alternative to the display of the six-dimensional space; it shows the distances to the two centroids after an anamorphosis that forces each of the sets of distances to follow a normal distribution with mean zero and a variance of one. Even though the relative locations of the points in the figure are different from those in the six-dimensional space, the display is instructive because it is still possible to keep the points from each partition separated. The boundary between the two partitions is the S-shaped curve in the middle of the display. The curvature of the boundary is the result of anamorphosis and the small displacement is an effect of differences in the covariance matrices of the two clusters.

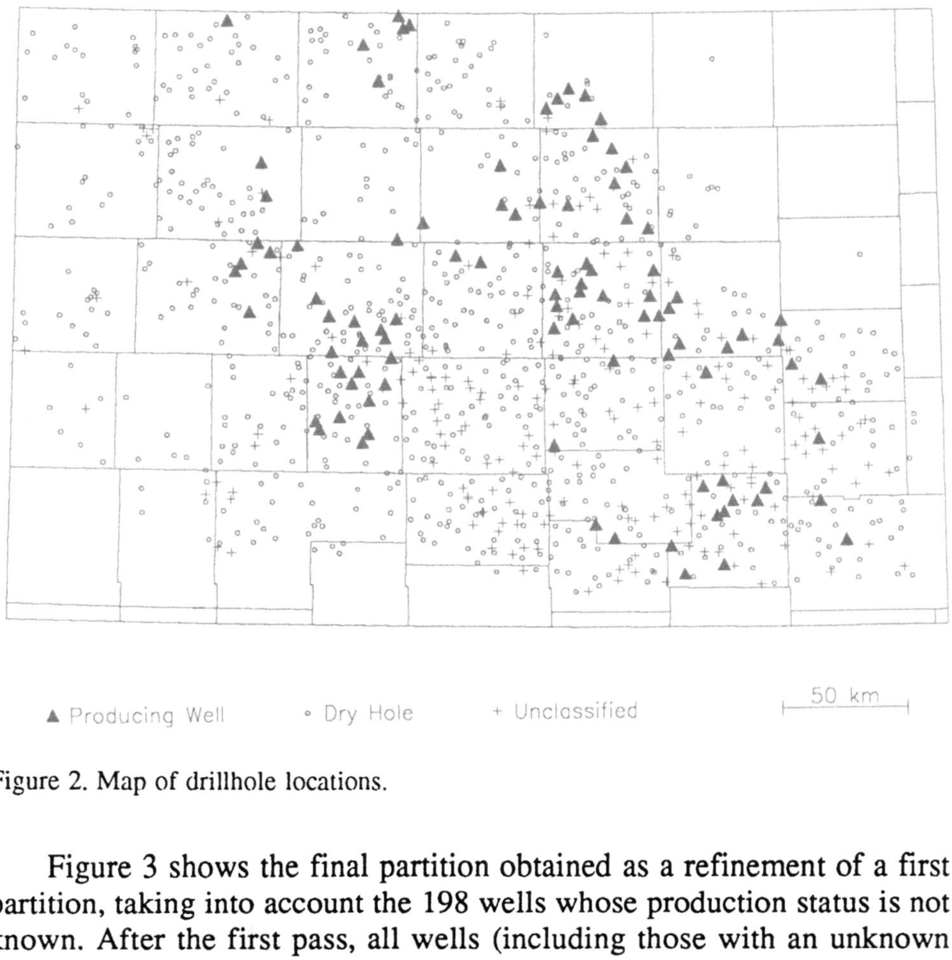

Figure 2. Map of drillhole locations.

Figure 3 shows the final partition obtained as a refinement of a first partition, taking into account the 198 wells whose production status is not known. After the first pass, all wells (including those with an unknown production status) were assigned to the closest cluster, the locations of the centroids were recalculated using all 1046 wells, and the wells were reassigned to the closest new centroids. The final count of oil, dry, and unclassified wells on each of the two partitions in the geological variable space is given in the lower part of Figure 3. Partition G_2 contains 72% of the producing wells, thus is the predictor for the target class "oil." By default, partition G_1—containing 63% of the dry wells—is the predictor partition for the target class "dry." The 95 unclassified wells in G_1 were considered as dry wells and the 103 in G_2 as oil wells. Figure 3 can be regarded as a graphical display of the calibration stage.

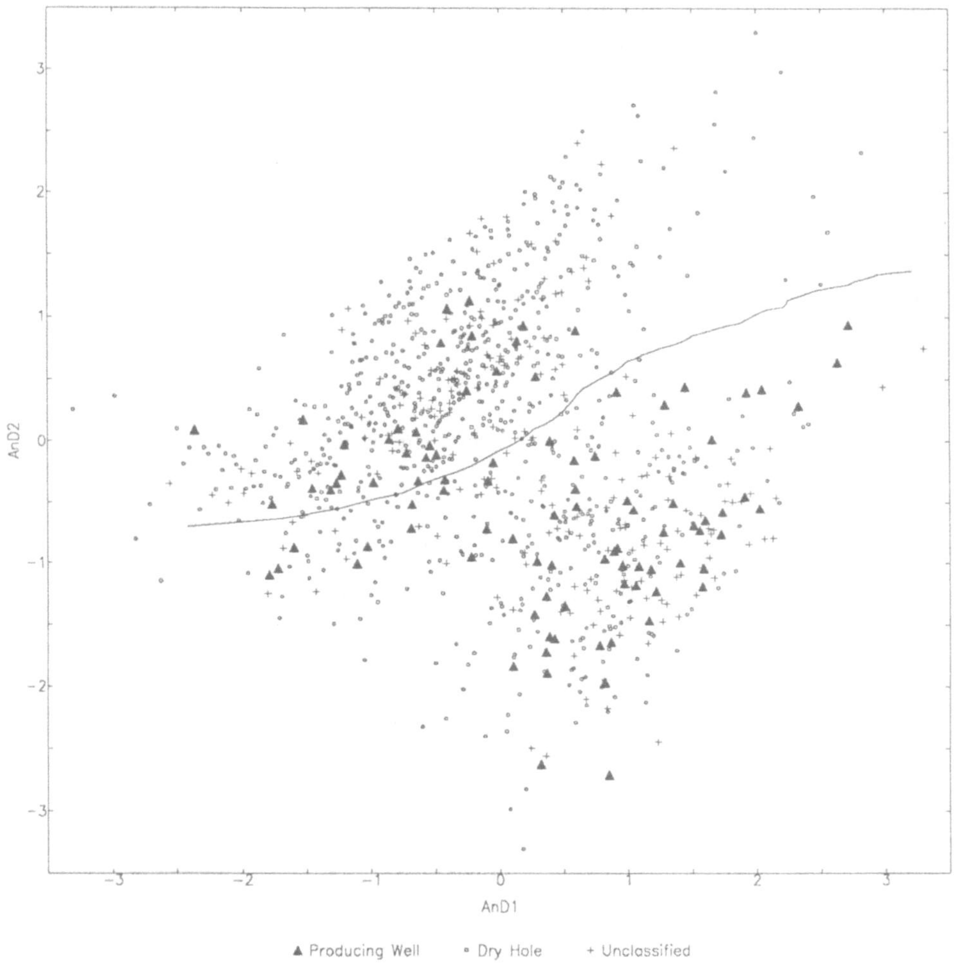

Partition P	Partition G 1	Partition G 2
P	27	68
D	474	279
U	95	103

cor. coef. (AnD1, AnD2) = −0.04

Figure 3. Observations plotted in anamorphosed distance space.

Table 1 contains the means and the covariance matrices for both clusters in the six-dimensional geological space after recomputing the location of the centroids of G_1 and G_2 using all 1046 observations. Changes in the assignments, means, and covariances due to the changes in the centroids obtained in the first pass were of minor importance.

Table 1. First and second moments.

Variable	Mean G_1	Mean G_2
HEEB, Surface elevation of the Heebner Shale, m	-349.7	-353.2
K-BP, Thickness base of the K-zone-top Pennsylvanian, m	106.2	41.08
I, Thickness of the I-zone, m	5.354	4.307
PorH, Thickness of porous carbonate in H-zone, m	0.7814	0.5804
PorI, Thickness of porous carbonate in H-zone, m	0.5912	0.3901
GaJ, Maximum gamma-ray radiation in J-zone, API units	172.9	165.7
Total number of observations	596.0	450.0

Covariance matrix for cluster G_1

	HEEB	K-BP	I	PorH	PorI	GaJ
HEEB	9721.317	2323.5028	-85.2216	-19.2278	-8.8930	837.647
K-BP	2323.503	1037.2876	2.8061	-3.9420	-0.8282	234.769
I	-85.222	2.8061	9.4949	0.3457	1.1737	30.839
PorH	-19.228	-3.9420	0.3457	0.9378	0.1284	-1.422
PorI	-8.893	-0.8282	1.1737	0.1284	0.9323	3.901
GaJ	837.647	234.7694	30.8390	-1.4217	3.9011	3279.531

Covariance matrix for cluster G_2

	HEEB	K-BP	I	PorH	PorI	GaJ
HEEB	1670.551	125.899	-21.70941	-9.15971	-4.61448	659.331
K-BP	125.899	738.311	14.32330	-4.13676	-3.55293	839.535
I	-21.709	14.323	2.48585	0.02744	0.14306	17.223
PorH	-9.160	-4.137	0.02744	0.66596	0.07759	-5.858
PorI	-4.614	-3.553	0.14306	0.07759	0.40483	-1.535
GaJ	659.331	839.535	17.22314	-5.85837	-1.53457	4916.712

The next stage is mapping the two sets of Mahalanobis' distances. Because the univariate distribution of the distances is strongly skewed, a normal score transformation is in order for more accurate geostatistical estimation. As shown in Figure 3, there is no correlation between the anamorphosed distances, indicating that a simpler estimation of each anamorphosed distance separately is as accurate as the simultaneous estimation of both variables. Here we resorted to geostatistical simulation to explore the potential of the technique to assess the probability of misclassification. The appeal of simulation is the avoidance of distributional assumptions about the estimation errors that are required when using universal kriging to map the anamorphosed distances (Harff, Davis, and Olea, 1992). Structural analysis of the variables produced the semivariograms in Table 2 and Figure 4.

Table 2. Semivariogram models.

Variable	Model
$AnD_1(x)$	Exponential (9,0.26) + Exponential (435,0.74)
$AnD_2(x)$	Exponential (6,0.16) + Exponential (315,0.84)

The tectonic pattern of the Central Kansas Uplift causes a drift in the data along the NE–SW direction. The semivariograms are calculated for the NW–SE direction perpendicular to the trend. Using the sequential Gaussian method (Deutsch and Journel, 1992), we simulated a set of equally probable realizations of distances $AnD_1(x)$ and $AnD_2(x)$. Figures 5 and 6 show the first five realizations per anamorphosed distance out of a total of 101 realizations. Instead of a continuous coverage each realization is made from a grid of 111 by 76 nodes. The seed used to simulate both sets of anamorphosed distances is the same. For comparative purposes the figures also include the result of an interpolation using universal kriging (Journel and Huijbregts, 1978). We apologize for the black and white illustrations and refer the reader to Harff and others (1993) for color renditions.

Figure 7 illustrates the calculation of the probability of membership in partition G_2 for a specific node in the lower-left corner of the study area, after back-transforming all realizations to the original distance space. To eliminate any possible effect of the simulation seed, the realizations were compared with an offset of one simulation. The value of $D_1(x)$ for the first simulation and the value of $D_2(x)$ for the second simulation provided the first point for Figure 7, the value of $D_1(x)$ for the second simulation was paired with the value of $D_2(x)$ in the third simulation to provide the second point in the crossplot, and so on until pairing $D_1(x)$ from the 100th simulation with $D_2(x)$ from the 101st simulation for a total of 100 pairs.

The proportion of pairs for which $d2 < d1 + 3.84$ provides the value of the probability $p(G_2(x))$ at the lower left node, 24% in this example. The value 3.84 accounts for the difference in covariance matrices of the clusters used in the calibration. By repeating this process for the remaining 8435 nodes, we obtained a 111 by 76 grid for $p(G_2(x))$ which is displayed in Figure 8. The NW–SE striking zone of elevated probabilities clearly outlines the Central Kansas Uplift where most of the oil deposits of western Kansas are located.

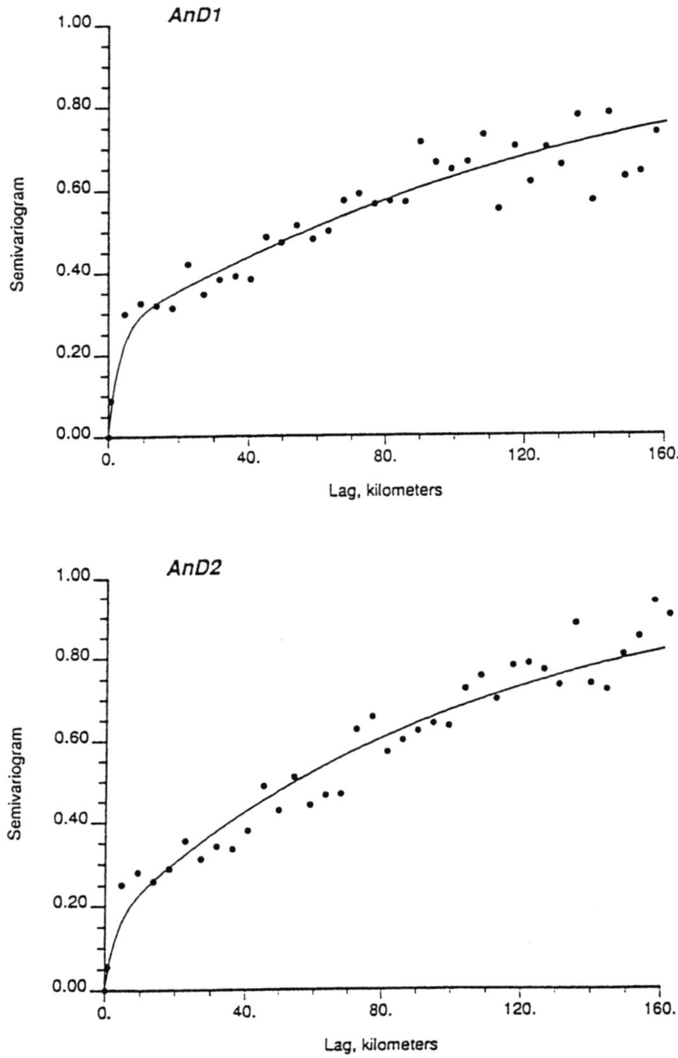

Figure 4. Semivariogram models.

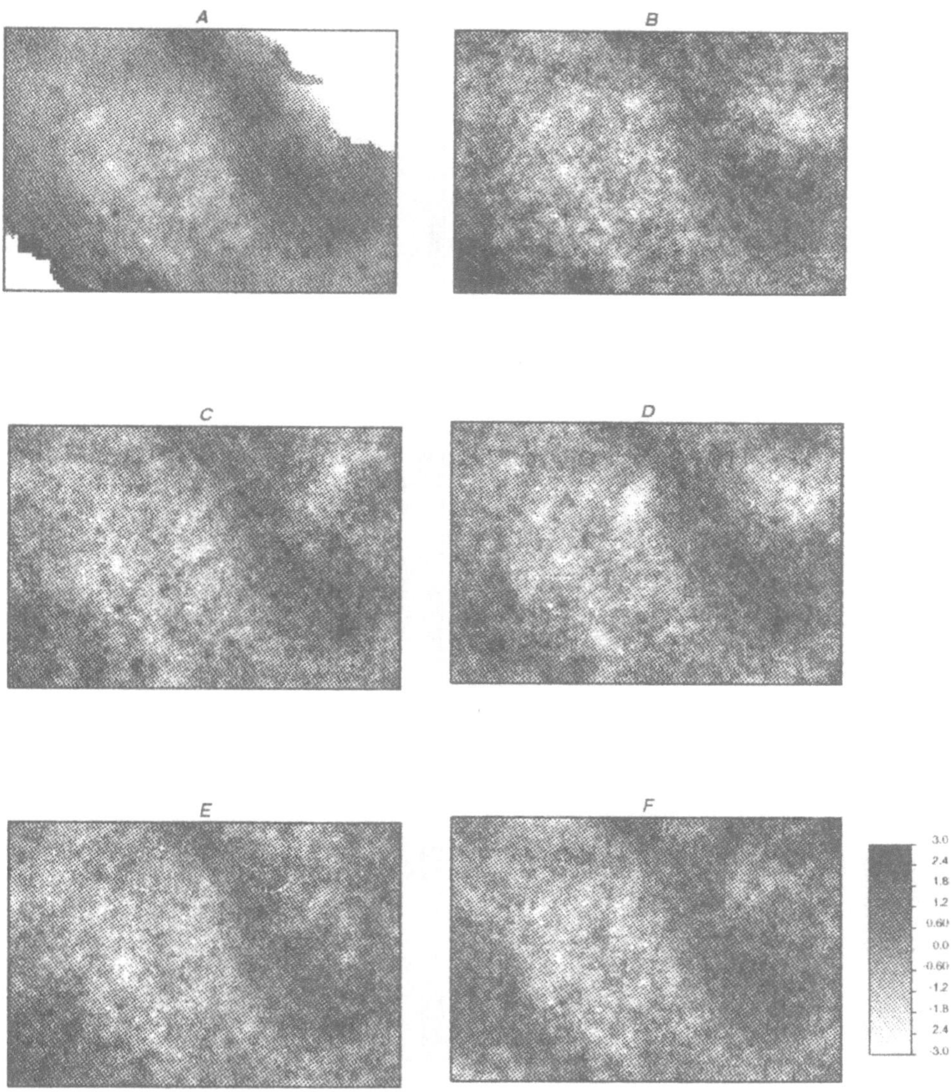

Figure 5. Anamorphosed distance $AnD_1(x)$ to centroid of dry partition G_1.
A, Universal kriging; and B through F, realizations.

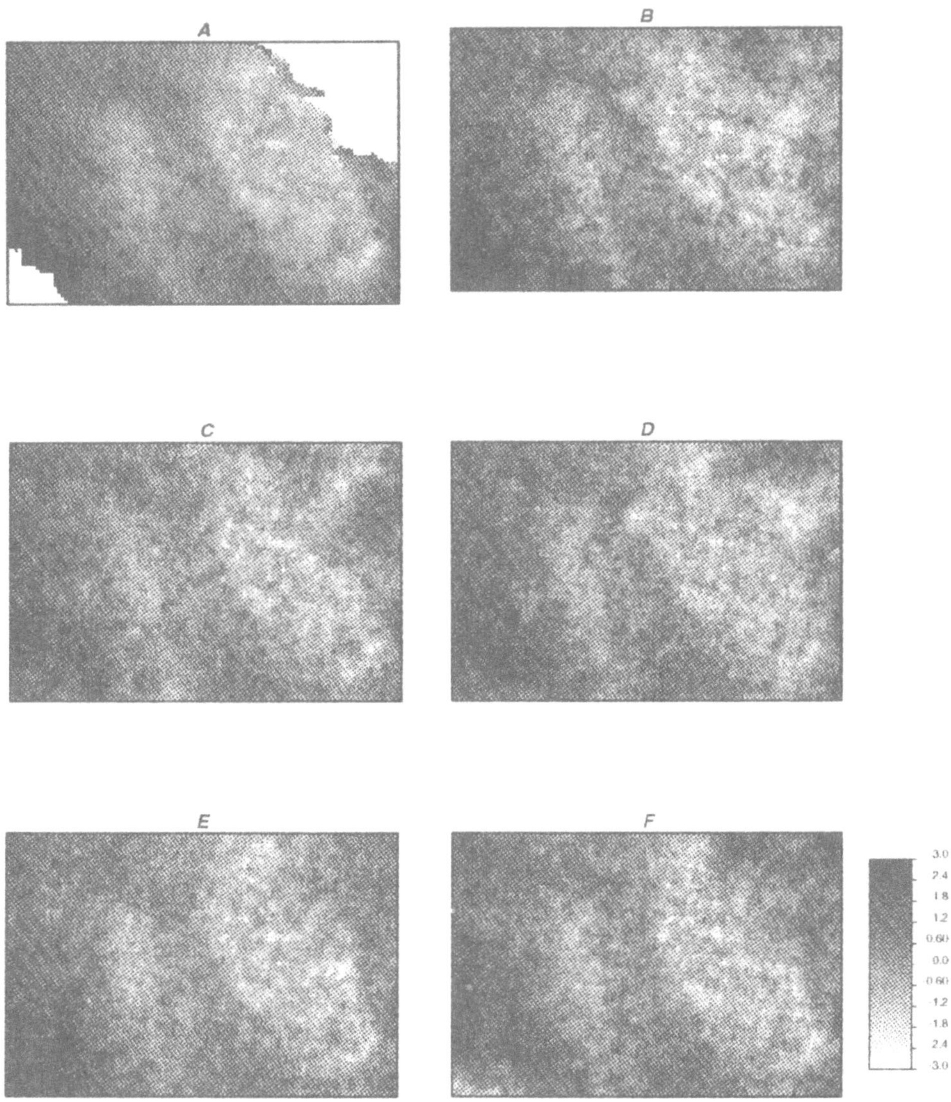

Figure 6. Anamorphosed distance $AnD_2(x)$ to centroid of oil partition G_2. A, Universal kriging; and B through F, realizations.

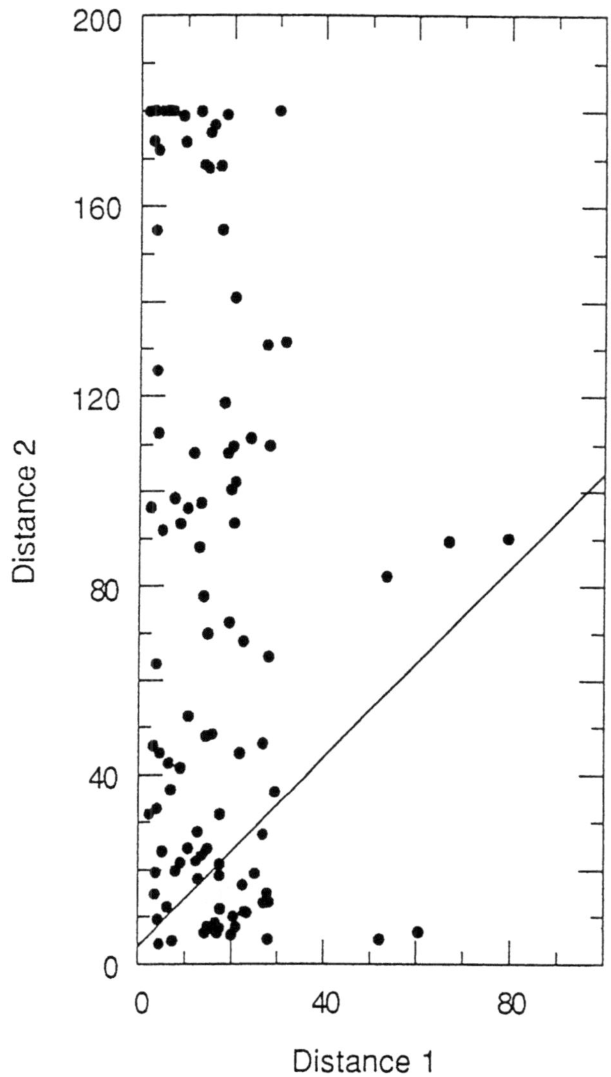

Figure 7. Determination of probability of belonging to oil partition for lower left node of area of study.

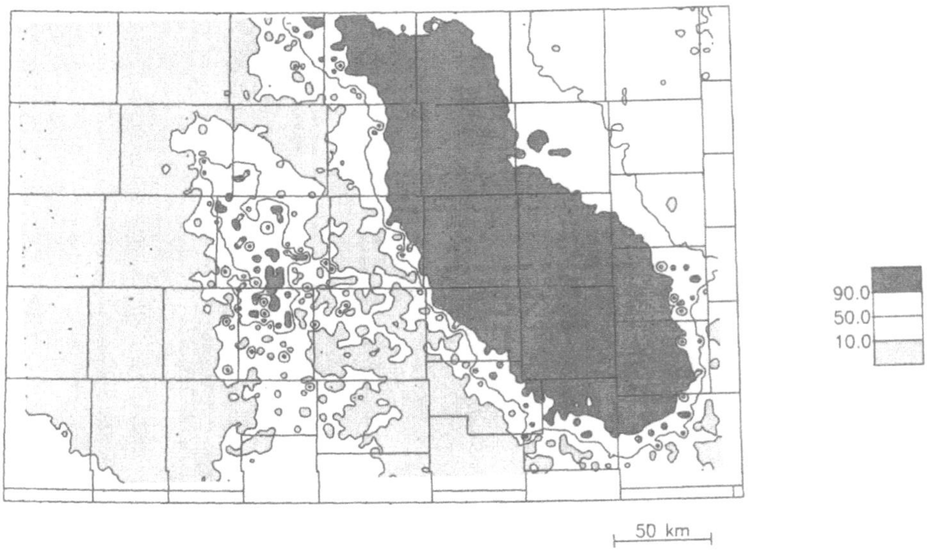

Figure 8. Percentage probability of belonging to oil partition G_2.

To estimate the probability of encountering oil in a well, the formula of the total probability can be used:

$$p(oil(\boldsymbol{x})) = p(G_1(\boldsymbol{x}))p(oil|G_1) + p(G_2(\boldsymbol{x}))p(oil|G_2)$$
$$= p(oil|G_1) + \{p(oil|G_2) - p(oil|G_1)\}p(G_2(\boldsymbol{x}))$$

which is a linear transformation of $p(G_2(\boldsymbol{x}))$. The coefficients $p(oil|G_i)$ can be estimated using Bayes' formula:

$$p(oil|G_i) = \frac{p(oil)p(G_i|oil)}{p(oil)p(G_i|oil) + (1 - p(oil))p(G_i|dry)}$$

Table 3 contains the conditional probabilities $p(G_i|oil)$ and $p(G_i|dry)$, which are computed directly as proportions in the tabulation in Figure 3. Consequently, the calculation of a map with the probability of hitting oil depends ultimately on a knowledge of the global probability of finding production. In the absence of this information, a sensitivity analysis can be performed. Figure 9 is a map of the probability of drilling a producing well obtained using the value $p(oil) = 0.42$ taken from a publication by Watney and others (1989). Note that Figure 9 is simply a rescaling of Figure 8.

An actual count of all wells reaching the Lansing-Kansas City in Barton County gives a value of $p(oil) = 0.69$. Barton, approximately 100 Km to the northwest of the lower right corner of Figure 9, is the only county entirely inside the area with $p(oil) \geq 0.55$. A comparison of Figures 8 and 9 with the map in Figure 1 shows that the general trend in the distribution of hydrocarbon deposits is reflected by the current study. The deposits linked with post-Mississippian–pre-Middle Pennsylvanian structures of the Central Kansas Uplift are reasonably recognizable. The oil favorability in the Hugoton Embayment in the central part in the region of investigation is marked by elevated probabilities in this zone.

Table 3. Conditional probabilities.

	Geologic partition	
	G_1	G_2
Production class		
Oil	0.284	0.716
Dry	0.629	0.371

CONCLUSIONS

The study provides a contribution to computerized geological mapping and interpretation of geological data. In order to make assumptions about the oil favorability of an entire area of investigation it is necessary: (1) to study empirically the relationship between oil-bearing (target variable) and easily measurable geological features (predictor variables) using a relatively small random sample of wells drilled in the region, and (2) to provide data vectors of the predictor variables assigned to drilling or measurement points covering the area of investigation so that a reasonable interpolation is possible. Stochastic simulation is suggested for the interpolation. The method is designed as an iterative procedure: by including more information in a stepwise manner, the results become more reliable. The method can be

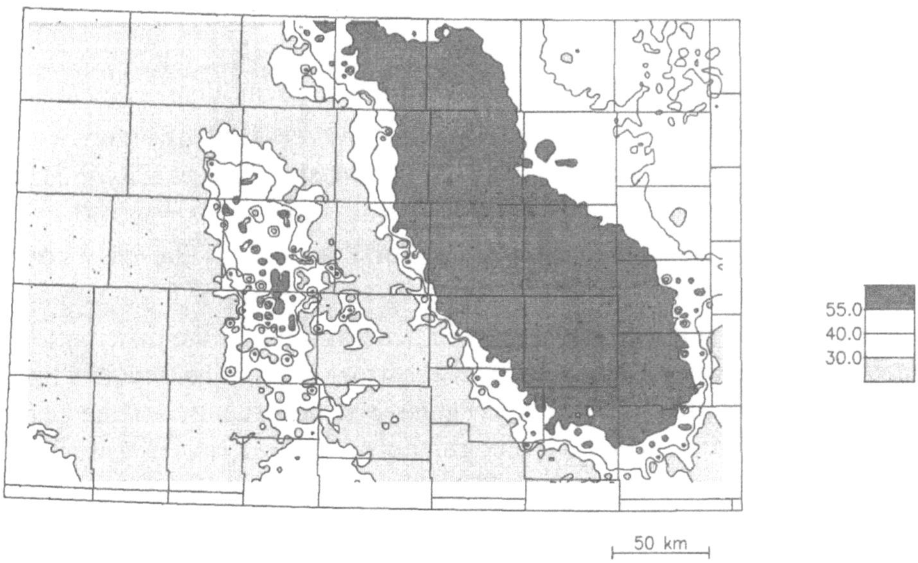

Figure 9. Percentage probability of belonging to productive class.

used for planning of exploration activities because it provides the geologist with information about favorable targets, although it cannot replace geological thinking and modeling.

ACKNOWLEDGMENTS

The authors thank Prof. A. Journel, Stanford University, USA, for valuable scientific advice and discussions; and express their appreciation to the Deutsche Forschungsgemeinschaft, the U.S. National Science Foundation (INT-9111646) and the Kansas Geological Survey for the support of this research project.

REFERENCES

Deutsch, C.V., and Journel, A.G., 1992, GSLIB—Geostatistical Software Library and user's guide: Oxford University Press, New York, 314 p.

Harff, J., and Davis, J.C., 1990, Regionalization in geology by multivariate classification: Math. Geology, v. 22, no. 5, p. 573–588.

Harff, J., Davis, J.C., and Olea, R.A., 1992, Quantitative assessment of mineral resources with an application to petroleum geology: Nonrenewable Resources, v. 1, no. 1, p. 74–84.

Harff, J., Davis, J.C., Olea, R.A., and Bohling, G.C., 1992, Computergestützte geologische Kartierung und Rohstoffperspektivität: Die Geowissenschaften, Oktober/November 1993, p. 375-379.

Journel, A.G., 1989, Fundamentals of geostatistics in five lessons: American Geophysical Union, Short Course in Geology 8, Washington, D.C., 40 p.

Journel, A.G., and Huijbregts, C. J., 1978, Mining geostatistics: Academic Press, London, 600 p.

Matheron, G., 1970, La Théorie des Variables Régionalisées et ses Applications: Les Cahiers du Centre du Morphologie Mathématique de Fontainebleau, Fascicule 5, Ecole Nationale Supérieure des Mines, Paris, 211 p.

Merriam, D.F., 1963, The geological history of Kansas: Kansas Geol. Survey Bull. 162, 317 p.

Myers, D.E., 1982, Matrix formulation of cokriging: Math. Geology, v. 14, no. 3, p. 249–257.

Rodionov, D.A., 1981, Statistističeskie Rešenija v Geologii: Nedra, Moscow, 211 p.

Watney, W.L., 1984, Recognition of favorable reservoir trends in Upper Pennsylvanian cyclic carbonates in western Kansas, *in* Hyne, N.L., ed., Limestones of the Mid-Continent: Tulsa Geol. Society Spec. Publ. 2, p. 201–246.

Watney, W.L., Newell, K.D., Collins, D.R., and Skelton, L.H., 1989, Petroleum exploration in Kansas—past trends and future options: Proc. Eighth Tertiary Oil Recovery Conference, TORP Contr. 10, Lawrence, Kansas, p. 4–35.

TRANSITION PROBABILITY APPROACH TO STATISTICAL ANALYSIS OF SPATIAL QUALITATIVE VARIABLES IN GEOLOGY

Junfeng Luo
Gesellschaft für wasserwirtschaftliche Planung und Systemforschung mbH, D-12526 Berlin, Germany

ABSTRACT

A spatial qualitative variable can be expressed by $\{z(\mathbf{x}) \in \{S_1, S_2 \ldots S_m\} \mid \mathbf{x} \in \mathbf{R}\}$, where $\{S_i\}$ represents a set of discrete states and $\mathbf{x}$ the coordinate vector in a two- or three-dimensional space $\mathbf{R}$. The descriptive aspect of geology as well as the qualitative information from basic geological investigations force the community of mathematical geology inevitably to handle such qualitative variables as rock types, sedimental phases, ore types, etc. when modeling geological phenomena connected with them. A spatial Markov chain, or transition probability approach, provides an appropriate formalism to model geological spatial phenomena characterized by qualitative and discrete state variables. Methods of Markov conditional simulation (estimation) as well as case studies are given to demonstrate the proposed approach.

INTRODUCTION

The analysis and modeling of geologic phenomena described by qualitative variables for example rock types, sedimentary facies, ore, or contaminant types, etc. are of special interest and importance to such practical problems as:

Geologic Modeling and Mapping
Edited by A. Förster and D. F. Merriam, Plenum Press, New York, 1996

- spatial distributions of certain rock types in underground (for related questions in environment geology, hydrology, or petroleum geology)

- spatial distribution patterns of different ore types as a basis for mine planning;

- environment monitoring of the distribution of contaminants, for example certain heavy metals.

In the context of mathematical modeling of geological phenomena Krumbein and Graybill (1965) distinguish geological data or variables as nominal, ordinal, and numeric type, that is three types, which correspond to qualitative (for example rock types), semiquantitative (hardness scale of minerals), and quantitative (geochemical survey) variables respectively. A spatial qualitative variable z(x) can be defined as a type of variable which takes, in nature, a set of nonnumeric discrete states as values, that is

$$\{ z(\mathbf{x}) \in \{S_1, S_2 \ldots S_m\} \mid \mathbf{x} \in \mathbf{R} \} \tag{1}$$

where $\mathbf{x}$ represents the coordinate (vector) in a two- or three-dimensional region $\mathbf{R}$ and $\{S_i\}$ is a finite set of m distinct discrete states considered, for instance, {dolomite, sandstone, limestone, shale} describes a simple geologic section of four sediment types. It is apparent that among the discrete states there exists generally no such order relation as $S_i > S_j$ $(i \neq j)$ or vice versa. That implies the difference $z(\mathbf{x}_1)-z(\mathbf{x}_2)$ $(z(\mathbf{x}_1), z(\mathbf{x}_2) \in \{S_i\}$ and $\mathbf{x}_1, \mathbf{x}_2 \in \mathbf{R})$ has no numerical meaning. This important property of qualitative variables distinguishes them from (measured) quantitative variables.

In the last years, geostatistical methods based on the theory of regionalized variables are used widely to model geological spatial phenomena. In geostatistics a regionalized variable $z'(\mathbf{x})$ refers usually to a measured quantitative variable given by (cf. Journel and Huijbregts, 1978; or Olea, 1991)

$$\{ z'(\mathbf{x}) \in (a, b) \text{ or } [a, b] \mid \mathbf{x} \in \mathbf{R} \} \tag{2}$$

where (a, b) or [a, b] represents a real value interval with $a < b$. In the situation of a spatial phenomenon described by a set of discrete states, for example rock types, or lithofacies, geostatistical methods can be employed only under certain previsions and restrictions. Whereas Markov chain models provide a straightforward formalism to handle such discrete variables.

This paper suggests that a transition probability approach or a spatial Markov chain model provide an appropriate formalism to describe as well as to model those geological spatial phenomena characterized by discrete variables $z(\mathbf{x})$.

First, it is argued that (k-step) transition probabilities are appropriate dependence measures for the spatial auto- and cross-correlations among the m states. The analysis of the transition matrices provides the information about *spatial dependence* and *expected homogeneous range*, *discontinuity*, *ergodic property* of states considered. Furthermore, the transition probabilities can be used straightforward for conditional simulation or estimation of unknown $z(\mathbf{x})$. Methods of Markov conditional simulation (estimation) as well as case studies are given to demonstrate the proposed approach.

DESCRIPTION OF SPATIAL VARIABILITY OF QUALITATIVE VARIABLES

A quantative description of the spatial variability of different discrete states is the first step towards modeling. In following discussions, let $z(\mathbf{x})$ represent some geological phenomenon, for example, the spatial distribution of rock types in a certain region **R**. $\mathbf{B}=\{ z(\mathbf{x}_\beta) \mid \beta=1...N\}$ will be used to denote a initial set of samples (observations) of $z(\mathbf{x})$: $z(\mathbf{x}_\beta) \in \{ S_1...S_m \}$, $\mathbf{x}_\beta \in \mathbf{R}$. $\mathbf{S}(\mathbf{h})=\{ z(\mathbf{x}_\alpha), z(\mathbf{x}_\alpha+\mathbf{h}) \mid \alpha=1...N(\mathbf{h})\}$ refers to a set of spatial sample pairs from the basic sample set **B**, that is $z(\mathbf{x}_\alpha)$, $z(\mathbf{x}_\alpha+\mathbf{h}) \in \{ z(\mathbf{x}_\beta) \mid \beta=1...N\}$ and their locations are separated by direction vector **h**.

Now we suppose that $z(\mathbf{x}_\alpha)=S_i$, $z(\mathbf{x}_\alpha+\mathbf{h})=S_j$, without loss of generality. To describe the **qualitative** variability between $z(\mathbf{x}_\alpha)$ and $z(\mathbf{x}_\alpha+\mathbf{h})$, an intuitive way is to interpret it directly as a state transition of $z(\mathbf{x})$ from S_i at $\mathbf{x}_\alpha$ to S_j at $\mathbf{x}_\alpha+\mathbf{h}$, and along direction **h** and with step length $|\mathbf{h}|$. Denote $n_{ij}(\mathbf{h})$ the total transition number from S_i to S_j observed in the spatial sample set $\mathbf{S}(\mathbf{h})$, a m by m transition count matrix $\mathbf{N}^{(\mathbf{h})}=(n_{ij}(\mathbf{h}))$ can be obtained

$$\mathbf{N}^{(\mathbf{h})} = (n_{ij}(\mathbf{h})) \qquad i,j=1...m \tag{3a}$$

$$n_{i\cdot}(\mathbf{h}) = \sum_{k=1}^{m} n_{ik}(\mathbf{h}) \ , \qquad n_{\cdot j}(\mathbf{h}) = \sum_{k=1}^{m} n_{kj}(\mathbf{h}) \qquad i,j=1...m \tag{3b}$$

$$n_{\cdot\cdot}(\mathbf{h}) = \sum_{i,j}^{m} n_{ij}(\mathbf{h}) = \sum_{i}^{m} n_{i\cdot}(\mathbf{h}) = \sum_{j}^{m} n_{\cdot j}(\mathbf{h}) = N(\mathbf{h}) \tag{3c}$$

Equations (3) summarize complete state transitions in $\mathbf{S}(\mathbf{h})=\{z(\mathbf{x}_\alpha+\mathbf{h}), z(\mathbf{x}_\alpha) \mid \alpha=1...N(\mathbf{h})\}$. Because a transition from S_i to S_j in direction **h** implies also a transition from S_j to S_i in the opposite direction **-h**, two estimated transition matrices can be obtained from Equations (3) as follows

$$\mathbf{P}^*(\mathbf{h}) = (\ p^*_{ij}(\mathbf{h})\ ) = (\ n_{ij}(\mathbf{h})/n_{i\cdot}(\mathbf{h})\ ) \qquad i,j=1...m \tag{4a}$$

$$\mathbf{P}^*(-\mathbf{h}) = (\ p^*_{ji}(-\mathbf{h})\ ) = (\ n_{ij}(\mathbf{h})/n_{\cdot j}(\mathbf{h})\ ) \quad j,i=1...m \tag{4b}$$

with

$$0 \le p^*_{ij}(\pm\mathbf{h}) \le 1, \qquad \Sigma_j\, p^*_{ij}(\pm\mathbf{h}) = 1 \quad \text{for } i,j=1...m \tag{4c}$$

Suppose that $z(\mathbf{x})$ is a realization of a spatial random function with Markov property along lines in direction **h** (Switzer, 1965, p.1860; Luo, 1993, p.18; see also the Appendix). Under the hypothesis of homogeneity, $p^*_{ij}(\mathbf{h})$ and $p^*_{ji}(-\mathbf{h})$ are valid estimates of transition probabilities $p_{ij}(\mathbf{h}) = p(\ Z(\mathbf{x}+\mathbf{h})=S_j \mid Z(\mathbf{x})=S_i\ )$ and $p_{ji}(-\mathbf{h}) = p(\ Z(\mathbf{x})=S_i \mid Z(\mathbf{x}+\mathbf{h})=S_j\ )$ respectively. Whereas $p_{ii}(\mathbf{h})$ describes the spatial auto-correlation of state S_i, $p_{ij}(\mathbf{h})$ $(i \neq j)$ reveals the cross-correlations between any two different states S_i and S_j. The complete spatial structure of $z(\mathbf{x})$ under consideration then can be studied through the structure analysis based on the so-called Markov diagrams of k-step transition probabilities $p_{ij}(k\mathbf{h})$ for different step k as well as for diverse directions **h** (see the following case study).

Remarks: The basic idea of geostatistical approachs to characterize the spatial variability of a regionalized variable $z'(\mathbf{x})$ given by Equation (2), is to define some spatial continuity measures based upon the ***quantitative*** difference between $z'(\mathbf{x})$ und $z'(\mathbf{x}+\mathbf{h})$ (cf. Journel, 1988). A widely used measure is the semivariogram $\gamma(\mathbf{h}) = 1/2\ E\ \{\ (z'(\mathbf{x}+\mathbf{h}) - z'(\mathbf{x})\ )^2\ \}$ under the intrinsic hypothesis, which is estimated by the experimental semivariogram (Journel and Huijbregts, 1978, p.12; Luo, 1984, p.130; Olea, 1991, p.25)

$$\gamma^*(\mathbf{h}) = \frac{1}{2N(\mathbf{h})} \sum_{\alpha=1}^{N(\mathbf{h})} (z'(\mathbf{x}_\alpha+\mathbf{h}) - z'(\mathbf{x}_\alpha))^2 \tag{5}$$

If a qualitative variable $z(\mathbf{x})$ is specified by Equation (1), it is apparently difficult as well as inappropriate to apply Equation (5) directly, unless some transformation f on $z(\mathbf{x})$ is made. In other words, the discrete state space $\Omega = \{S_1...S_m\}$ should be mapped by f into another space of numeric type Ω', so that Equation (5) can be applied to $f(z(\mathbf{x}))$. For this purpose Gaussian Anamorphosis (e.g. de Fouquet and others, 1989) and indicator function (Journel, 1983) usually are used. Whatever transformation is applied, a question remains, namly, to what extent $f(z(\mathbf{x}_\alpha+\mathbf{h})) - f(z(\mathbf{x}_\alpha))$ represents the purely ***qualitative*** difference between $z(\mathbf{x}_\alpha+\mathbf{h})=S_j$ and $z(\mathbf{x}_\alpha)=S_i$.

MODELING BASED ON THE TRANSITION PROBABILITIES

A modeling of spatial variability, for example *conditional simulation* or *estimation* of the spatial distribution of different states is possible, as soon as the transition matrices ($p_{ij}(k\mathbf{h})$) for different lags and directions $\mathbf{h}$ have been estimated from available data. There are several algorithms for *conditional Markov simulation* or *estimation* of the spatial phenomena (Luo, 1993a). In general, conditioning indicates that modeling procedures are directly controlled by known states and corresponding state-transition probabilities. Principly, a modeling procedure of an unknown $z(\mathbf{x}_0)$ ($\mathbf{x}_0$ for a point, a cell or a block in $\mathbf{R}$) is performed in the following sequences

(1) Detect n known neighboring cells of $z(\mathbf{x}_0)$, $z(\mathbf{x}_a)$, a=1...n. Let $p_{iaj}(\mathbf{h}_a)=p(z(\mathbf{x}_0)=S_j \mid z(\mathbf{x}_a)=S_{ia})$, $z(\mathbf{x}_a)=S_{ia} \in \{S_1...S_m\}$ and $\mathbf{h}_a=\mathbf{x}_a-\mathbf{x}_0$. specifies then the probability of the occurrence of S_j at $\mathbf{x}_0$, conditioned by S_{ia} at $\mathbf{x}_a$;

(2) Calculate $\mathbf{V}_s = (p_{s1}...p_{sj}...p_{sm})$, the *average transition vector* of the transition probabilities $p_{iaj}(\mathbf{h}_a)$ corresponding to n known $z(\mathbf{x}_a)$, a=1...n:

$$p_{sj} = 1/n \sum_{ia=1...n} p_{iaj}(\mathbf{h}_a) \text{ (for j = 1...}m) \tag{6}$$

$\mathbf{V}_s$ provides an approximation to p($z(\mathbf{x}_0) = S_j \mid z(\mathbf{x}_a) = S_{ia}$, a=1...n), the conditional distribution, upon which *conditional simulation* or *estimation* can be achieved by

(3a) Simulating state S_j for $z(\mathbf{x}_0)$, in accordance with the Monte-Carlo Method, or by

(3b) Assign S_{j0} to $z(\mathbf{x}_0)$, such that p_{sj0} is maximal among $\{p_{sj}, j=1...m\}$ (estimation).

A CASE STUDY

To illustrate the performence of the transition probability model, a case study is given with the emphasis on the modeling spatial pattern of a discrete variable $z(\mathbf{x})$ defined by Equation (1). Figure 1A shows a known geological profile section of four glacial deposits of Quaternary age. The four deposits: gravel, sand, mixture of sand and clay, and clay are subsequently represented by S_i i=1...4. For the sampling, the 480m long section with 60m in depth is represented by a regular grid of 96 x 60 cells of size 5 x 1m^2 in such a way that each cell contains one and only one of the 4 defined states S_i (cf. Fig.1B).

In this way we have a gridding representation of the geological section considered and a basic sample set $\{ z(\mathbf{x}_\beta) \mid \beta=1...N=5700\}$ on the 96 x 60 grid nodes, where $\mathbf{x}_\beta$=(x, y) is the coordinate vector of middle point of corresponding cells. Table 1 summarizes the observed frequencies of each state, that is n_i, along with estimated state probability p^*_i. To describe quantitatively the deposit (state) alternations along the vertical (y-) and the horizontal (x-) direction, transition count matrices [Eq. (3a)] along with the corresponding transition probability matrice [Eq. (4)] are calculated. Table 1b gives only the 1-step transition count and probability matrices, wheras k-step transition matrices $\mathbf{P}^{(k\mathbf{h})}$, k=2,3... can be derived easily by recursive using $\mathbf{P}^{(k\mathbf{h})}=\mathbf{P}^{(\mathbf{h})}*\mathbf{P}^{((k-1)\mathbf{h})}$ because of the Markov property and Chapman-Kolmogorov Equation (e.g. Feller, 1968; Ferschl, 1970).

Spatial Structure Analysis of the Markov Model

Based on k-step transition probabilities, we now can quantify the spatial variability of each deposit (state) as well as spatial correlations between

Table 1a. Estimates of state probability of S_i (5 x 1m^2-grid)

	S_1	S_2	S_3	S_4	
n_i	1320	2738	308	1394	$\Sigma_i n_i = 5760$
$p^*(S_i)$	.2292	.4753	.0535	.2420	$p^*(S_i) = n_i / \Sigma_i n_i$

Table 1b. Transition probability matrice (5 x 1m^2-grid)

$P(h_x) = (p_{ij}(h_x))$				$1\ \lvert h_x \rvert = 5m$	$P(-h_x) = (p_{ij}(-h_x))$			
.9603	.0191	.0053	.0153		.9640	.0153	.0115	.0092
.0074	.9841	.0074	.0011		.0092	.9809	.0070	.0029
.0502	.0635	.8863	.0000		.0240	.0685	.9075	.0000
.0087	.0058	.0000	.9855		.0145	.0022	.0000	.9834
$P(h_y) = (p_{ij}(h_y))$				$1\ \lvert h_y \rvert = 1m$	$P(-h_y) = (p_{ij}(-h_y))$			
.7720	.1076	.0530	.0674		.7720	.0667	.0235	.1379
.0333	.9107	.0106	.0454		.0537	.9107	.0254	.0102
.1006	.2175	.6818	.0000		.2273	.0909	.6818	.0000
.1306	.0194	.0000	.8501		.0638	.0861	.0000	.8501

Table 2. Structure parameter of Markov-Model (5 x 1m^2)

	S_1(gravel)	S_2 (sand)	S_3 (mixture)	S_4 (clay)
$E[D_j(x)]$	132m	286m	49m	321m
$E[D_j(y)]$	4.4m	11.2m	3.1m	6.7m
$d_j(x)$	650m	1300m	325m	1500m
$d_j(y)$	16m	32m	16m	25m
$p(S_j)$	.2292	.4753	.0535	.2420
π_j (x)	.2262	.4837	.0541	.2359
π_j (y)	.2330	.4664	.0544	.2461
$1-p_{jj}(x)$	.0379	.0175	.1031	.0156
$1-p_{jj}(y)$	.2280	.0893	.3182	.1499

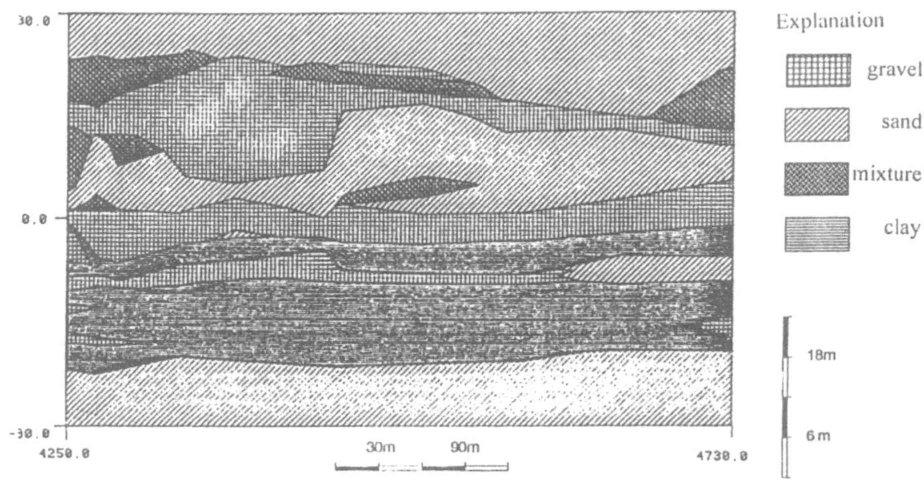

Figure 1A. Geologic profile section of four glacial deposits of Quaternary age.

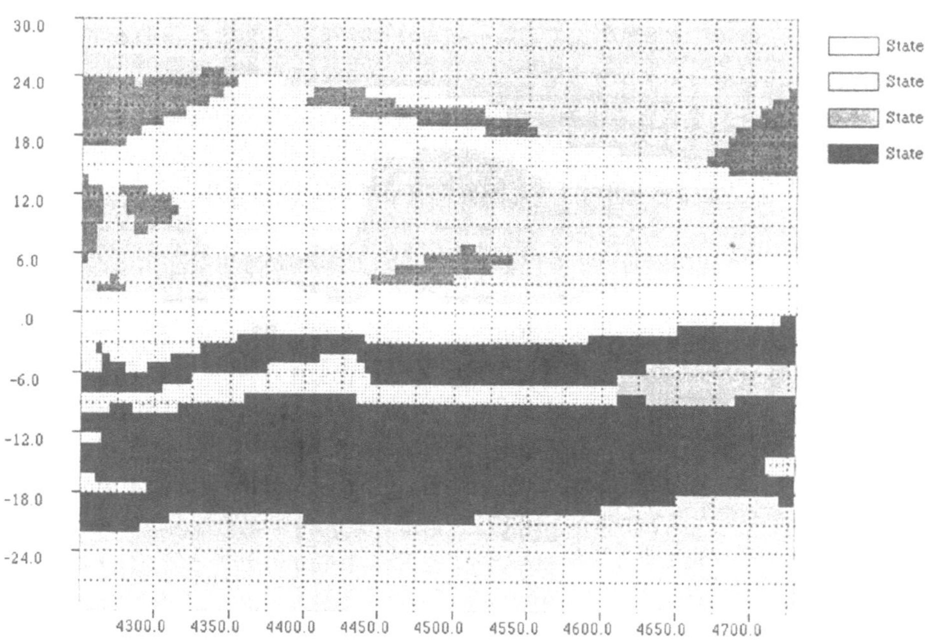

Figure 1B. Gridded representation of geologic section (96 x 60 cells of size 5x1m^2).

them. Let us look at the Markov diagrams (Fig.2A-D), where k-step transition probabilities $p_{ij}(k\mathbf{h})$ against corresponding steps k are graphically presented. The two diagrams in Figure 2A, for instance, document the variation of transition probabilities $p_{1j}(k\mathbf{h})$, j = 1,2,3,4 along the horizontal (x-) and vertical (y-) direction respectively. Here we observe:

(1) as step k (distance) increases, transition probability $p_{12}(k\mathbf{h})$ and $p_{14}(k\mathbf{h})$ increase, whereas $p_{11}(k\mathbf{h})$ decreases. Opposite to that, $p_{13}(k\mathbf{h})$ shows few variation. This implies that state 1 (gravel) has a high spatial correlation with state 2 and 4 (sand and clay), but few with state 3 (mixture);

(2) if step k remains small, then $p_{11}(k\mathbf{h}) > p_{12}(k\mathbf{h}) > p_{14}(k\mathbf{h}) > p_{13}(k\mathbf{h})$. But as soon as step k exceeds certain values that is $k_x < 24$ (= 120m) $k_y < 5$ (= 5m), the diagram shows that $p_{11}(k\mathbf{h}) < p_{12}(k\mathbf{h})$ at first, then $p_{11}(k\mathbf{h}) < p_{14}(k\mathbf{h})$. This phenomenon implies that if we observe state 1 occurring at position $\mathbf{x}$ [i.e. $z(\mathbf{x})$ = gravel], then, for relative smaller distance $k\mathbf{h}$, that again state 1 occurs at position $\mathbf{x}+k\mathbf{h}$ [i.e. $z(\mathbf{x}+k\mathbf{h})$ = gravel] is more possible than that other states occur. Generally, the larger the distance, the less the possibility of occurrence of state 1 at position $\mathbf{x}+k\mathbf{h}$ unless there is a period. If the distance $k\mathbf{h}$ exceed certain values, then it is more possible that state 2 (sand) occurs at position $\mathbf{x}+k\mathbf{h}$ [i.e. $z(\mathbf{x}+k\mathbf{h})$ = sand] because $p_{11}(k\mathbf{h}) < p_{12}(k\mathbf{h})$;

(3) furthermore, if step k is large enough such that $k_x > 130$ corresponding to 650m, and $k_y > 16$ corresponding to 16m, then the four transition probabilities $p_{1j}(k\mathbf{h})$, j = 1,2,3,4 become almost invariant and $p_{1j}(k\mathbf{h}) \rightarrow p_j$. In this situation, the spatial auto- and cross-correlation among the four states disappear;

(4) as $\mathbf{h} \rightarrow 0$, there occurs $p_{11}(k\mathbf{h}_x) \rightarrow 1$ as it should be, but $p_{11}(k\mathbf{h}_y)$ does not. This implies there exists some discontinuity of state 1 (gravel) along the y-axis (vertical) for smaller distance.

Similarly, we can quantify and analyze variations of the other states based on transition probabilities $p_{2j}(k\mathbf{h})$, $p_{3j}(k\mathbf{h})$, and $p_{4j}(k\mathbf{h})$ and the corresponding Markov-diagrams (Fig.2B-D). To quantify the spatial continuity of each deposit, the probability of k-time continuous occurrence of state S_i, $p(T_i=k)$, can be used (e.g. Krumbein and Dacey, 1969)

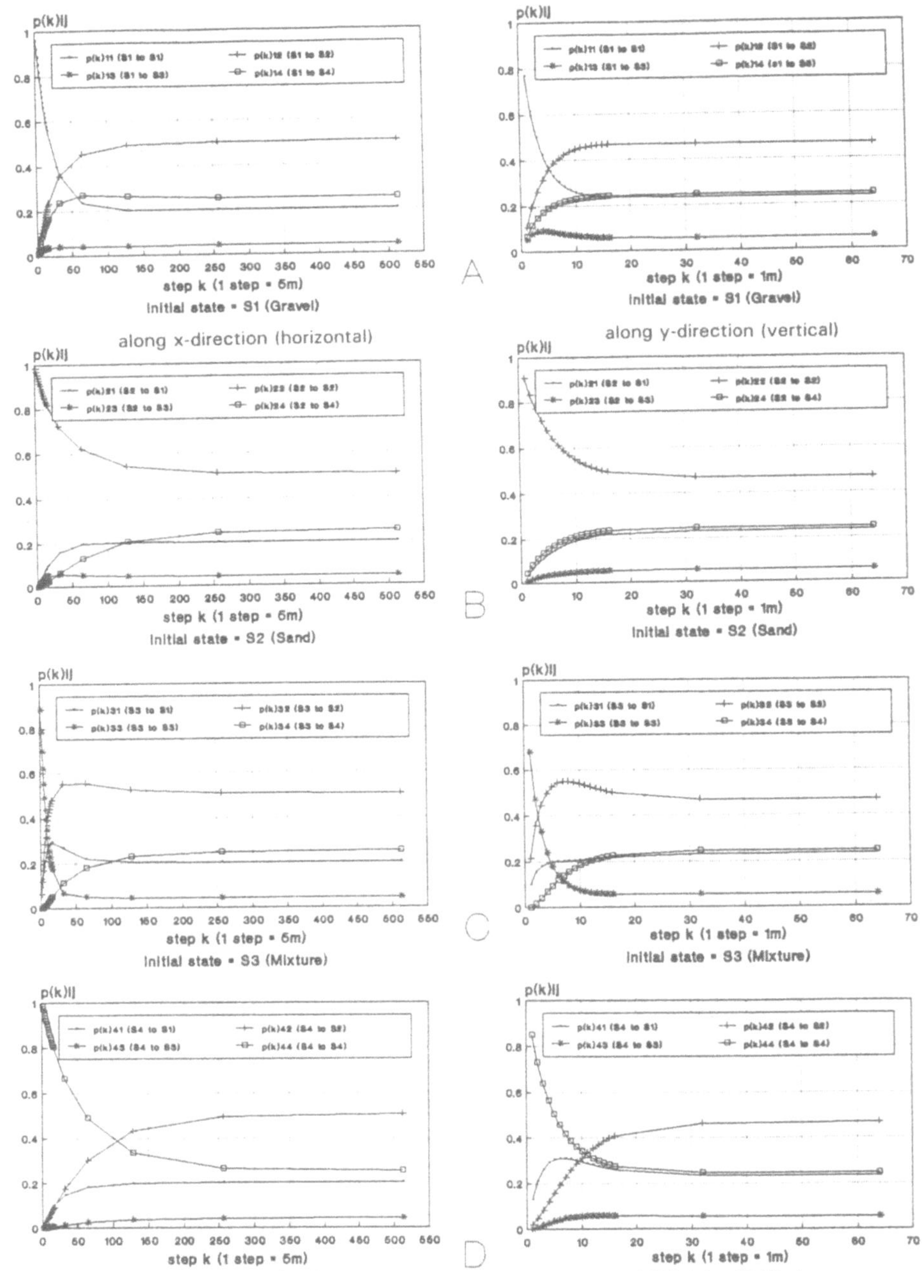

Figure 2. Markov diagrams of A, state 1 (gravel); B, State 2 (sand); C, state 3 (mixture); and D, state 4 (clay).

$$p(T_i=k) = (1 - p_{ii}(\mathbf{h})) * (p_{ii}(\mathbf{h}))^{k-1} \quad (\text{for } i = 1...4) \tag{7}$$

$p(T_i=k)$ describes practically homogeneous spatial extension of state S_i and also can be against distance k graphically presented in Figure 3. There we observe, in general, as distance k increases, $p(T_i=k)$ for i=1,2,3,4 decrease. But the decreasing rates for each deposit S_i are different. For instance, for given probability value of 0.6, it seems that clay and sand have larger spatial extension than gravel as well as that for the mixture of sand and clay [for example along the horizontal, $p(T_4=170\text{m})=p(T_2=160\text{m})=p(T_1=55\text{m})=p(T_3=20\text{m})=0.6$]. In comparision of $p(T_i=k)$ along the x-axis (horizontal) and the counterpart along the y-axis (vertical), it is easy to see that the anisotropic variability of the deposits is quantified [e.g. for deposit gravel S_1 $p_x(T_1=55\text{m})=p_y(T_1=2\text{m})=0.6$]. Furthermore, we can calculate the expected value of the extension of state S_i, $E(D_i)$ by

$$E(D_i) = (1 - p_{ii}(\mathbf{h})) \sum_{k=1...\infty} k(p_{ii}(\mathbf{h}))^{k-1} \quad (\text{for } i = 1...4)' \tag{8}$$

As summary, we have a set of structure parameters (similar to those in geostatistics) given in Table 2, where index j stands for deposit S_j, x and y in parenthesis represent along the x- and the y-axis respectively. $E[D_j()]$ is the expected value of the extension of S_j. $\pi_j() = \lim_{k\mathbf{h}\to\infty} p_{ij}(k\mathbf{h})$, the limited value of transition probability $p_{ij}(k\mathbf{h})$ and $\pi_j()$ is comparable directly to the corrsponding state probability $p(S_j)$. $d_j()$ is the so-called memory duration such that $\pi_j() = p_{ij}(\, d_j() \,)$ is satisfied at the first time, and hence $d_j()$ quantifies the correlation range of S_j. $1\text{-}p_{jj}()$ provides the information on the spatial continuity or variability of S_j as $k\mathbf{h} \to 0$ (similar to the so-called "nugget-effect").

Conditional Simulation of Spatial Distribution of Deposits

Now, based on the transition matrices ($p_{ij}(k\mathbf{h})$) mentioned previously, the spatial distribution pattern of different deposits can be modeled. For the purpose of demonstration, we make use of seven drillholes in the geological section (Fig.1) to condition or control the modeling procedure. This is known from drillholes which are used as starting states for the simulation. The modeling procedure extends sequentially from the bottom to the top of the section: the lowest "layer" is conditioned by every known deposit state at the seven drillhole positions and modeled first. The modeling of the second lowest "layer" then is controlled by

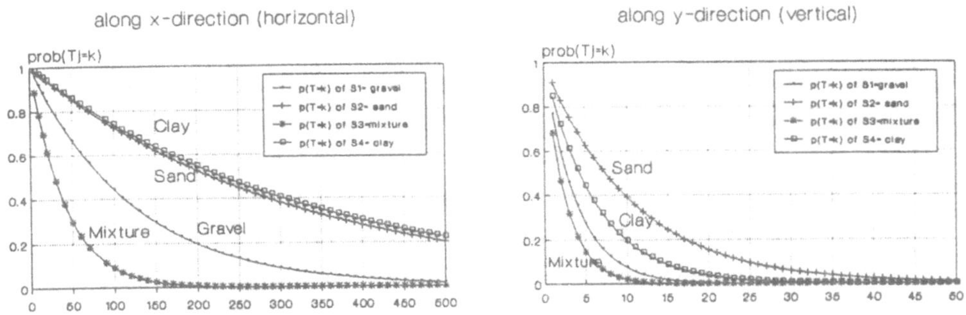

Figure 3. Diagram of probability of k-time continuous occurrence of S_i, $p(T_i=k)$

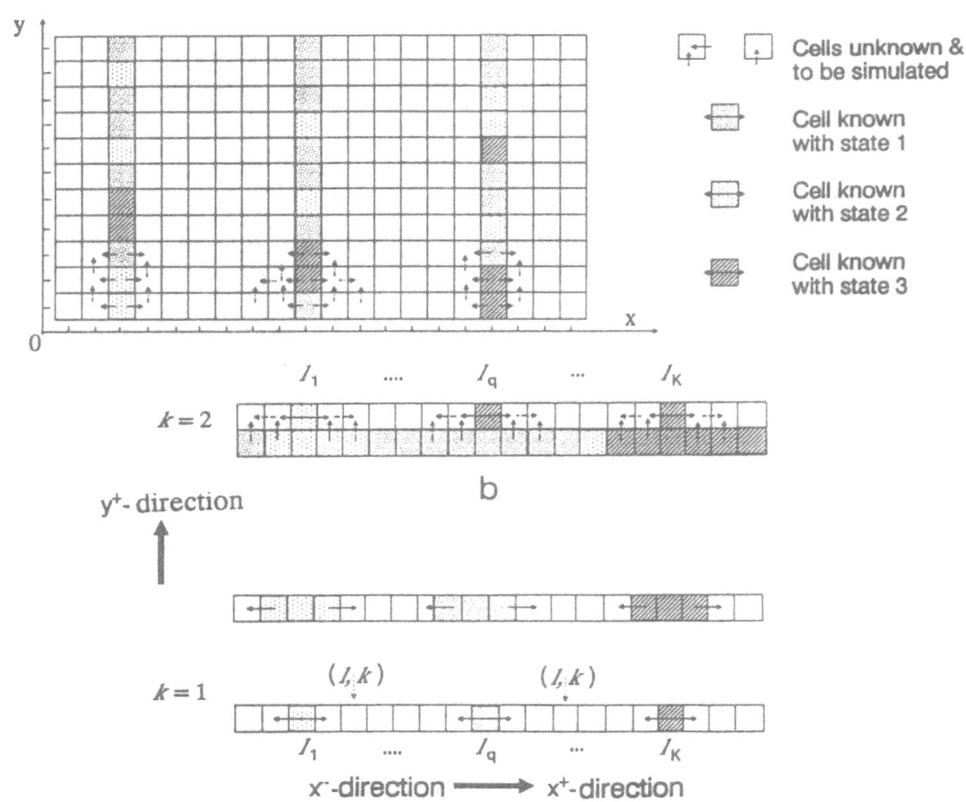

Figure 4. Description 2D-Markov simulation conditioned by drillholes

every known deposit state at the seven drillhole positions of the second lowest "layer" as well as by those "known" (simulated) states of the lowest "layer". Such modeling procedure continues until the entire section will have been simulated (Fig.4).

Figure 5A and 5B show a simulated section (stochastic modeling using Monte-Carlo method) and an "estimated" one [i.e. assigning state S_{j0} to an unknown $z(\mathbf{x}_0)$, such that transition probability p_{sj0} is maximal]. Here one can see that the spatial pattern of the original section has been well reproduced.

CONCLUDING REMARKS

Markov chain model provides an appropriate approach to modeling geologic spatial phenomena described by a set of discrete states. Under the Markov model, the spatial variability of different ore types are interpreted intuitively and straightforward by spatial state alternations or transitions, which then are described by corresponding transition probabilities. A reliable estimation of transition matrices, therefore, is the first and important step towards the Markov modeling.

The structure analysis based on the k-step transition matrices $\mathbf{P}^{(k\mathbf{h})}$ provides not only such structure parameters such as memory duration, $1-p_{ii}(\mathbf{h})$ if $|\mathbf{h}|$ is small and limited transition probability $\pi_j=\lim_{|k\mathbf{h}|\to\infty} p_{ij}(k\mathbf{h})$. In fact, the Markov diagrams of k-step transition probabilities $p_{ij}(k\mathbf{h})$ against step k reveal the complete spatial auto-correlation and cross-correlation structure of the problem under consideration.

Conditional simulation or *estimation* of the spatial variability or distribution of different ore types is practicable as soon as the transition matrices ($p_{ij}(k\mathbf{h})$) for different lags and directions $\mathbf{h}$ have been estimated. The application of the spatial Markov chain model consists of the following five steps:

(1) state definition and encoding of the sample data;

(2) Markov model analysis and structure analysis based on k-step transition probabilities;

(3) conditional Markov simulation or estimation based on transition probabilities;

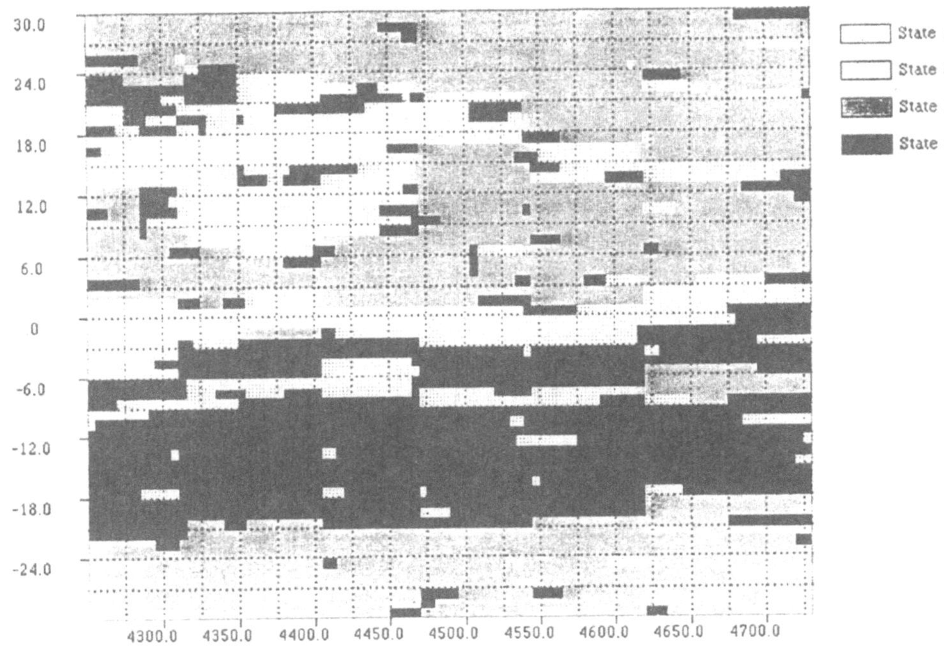

Figure 5A. Image produced by conditional Markov simulation.

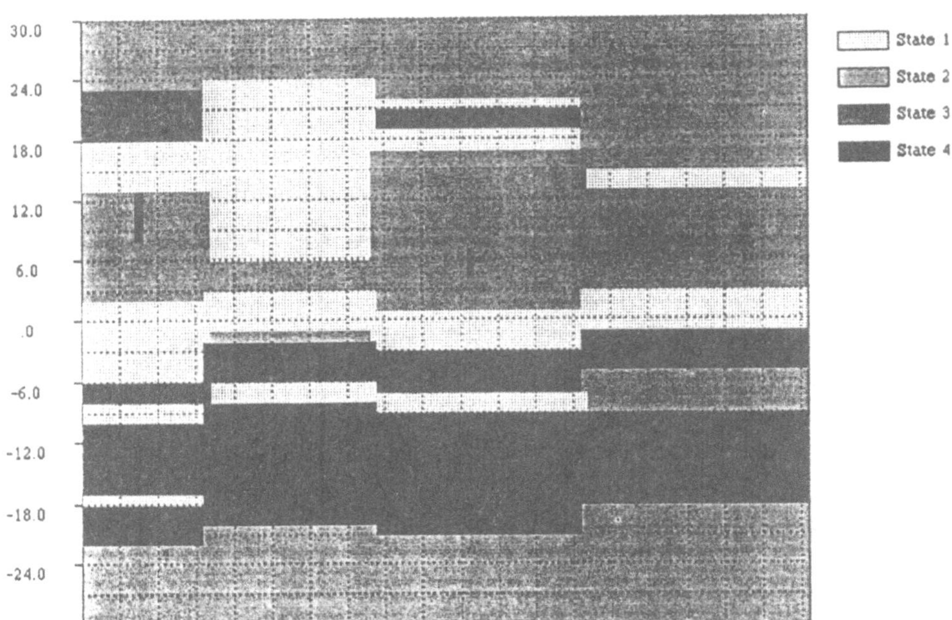

Figure 5B. Image produced by "Markov estimation".

(4) assessment of simulation results; and

(5) graphical representation of the results.

A detailed discussion of algorithms for conditional Markov simulation as well as its applications are given in Luo (1993a). Although geostatistical methods are used mostly for the modeling of spatial phenomena, they are only employable under certain provisions if a spatial phenomenon is described by a set of discrete states, or especially, of qualitative states like rock types or lithofacies. In this situation, the Markov model shows not only the advantages, but the transition probability formalism can be used directly to estimate the bivariate probability distribution defined for a set of class indicator variables (Luo and Thomsen, 1994), because there are, under certain prerequisite, certain relations between transition probabilities and the cross- and variograms of the class indicators (Luo, 1993a, 1993b).

ACKNOWLEDGMENTS

This work was completed in the institute of geology, geophysics, and geoinformatics of the Free University of Berlin. I thank Prof. Dr. W. Skala who encouraged and supervised the author during the research work which, for the most part, was supported by a scholarship from the Free University of Berlin. The author also thanks those who have helped his work, especially Dr. H. Burger and Dipl. Math. A. Thomsen for discussions and useful suggestions.

REFERENCES

Bartlett, M.S., 1971, Physical nearest neighbor models and non-linear time series: Jour. Appl. Prob., v. 8, no. 2, p. 222-232.

Bartlett, M.S., 1975, The statistical analysis of spatial pattern: Chapman and Hall Ltd, London, 90 p.

Bennett, R.J., 1979, Spatial time series - analysis - forecasting - control-: Pion Ltd., London, 674 p.

Besag, J.E., 1972a, On the correlation structure of two-dimensional stationary processes: Biometrika, v. 59, no. 1, p. 43-48.

Besag, J.E., 1972b, Nearest-neighbour systems and the auto-logistic models for binary data: Jour. Roy. Statist. Soc., B34, no. 1, p. 75-83.

Besag, J.E., 1974, Spatial interaction and the statistical analysis of lattice systems: Jour. Roy. Statist. Soc., B36, no. 2, p. 192-236.

Besag, J.E., 1986, On the statistical analysis of dirty pictures: Jour. Roy. Statist. Soc., B48, no. 3, p. 259-302.

Chung, K.L., 1967, Markov chains with stationary transition probabilities (2nd ed.): Springer-Verlag, Heidelberg, 301 p.

de Fouquet, C., Beucher, H., Galli, A., and Ravenne, C., 1989, Conditional simulation of random sets - application to an argilaceous sandstone reservoir, *in* Armstrong, M. ed., Geostatistics v. 2: Kluwer Academic Publ., Dordrecht, p. 517-530.

Feller, W., 1968, An introduction to probability theory and its applications, Vol. 1 (3rd ed.): Wiley International Edition, New York, 509 p.

Feller, W., 1971, An introduction to probability theory and its applications, Vol. 2 (2nd ed.): John Wiley & Sons, New York, 669 p.

Ferschl, F., 1970, Markovketten, Lecture notes in operations research and mathematical systems: Springer-Verlag, Berlin, 168 p.

Journel, A.G., 1983, Nonparametric estimation of spatial distributions: Math. Geology, v. 15, no. 3, p. 445-468.

Journel, A.G., 1988, New distance measures: the route towards truely non-Gaussian geostatistics: Math. Geology, v. 20, no. 4, p. 459-475.

Journel, A.G., and Huijbregts, Ch., 1978, Mining geostatistics: Academic Press, London, 600 p.

Krumbein, W.C., and Graybill, F.A., 1965, An introduction to statistical models in geology: McGraw-Hill Book Co., Inc., New York, 475 p.

Krumbein, W.C., and Dacey, M.F., 1969, Markov chain and embedded Markov chains in geology: Math. Geology, v. 1, no. 1, p. 79-96.

Luo, J., 1984, On some problems of the estimation of experimental variograms: Jour. Geoscientific Information, no. 4, Wuhan, China, p. 130-137 (in Chinese).

Luo, J., 1993a, Konditionale Markov-Simulation 2-dimensionaler geologischer Probleme: Berliner Geowiss. Abh., D, 4, Berlin, 103 p.

Luo J., 1993b. Beziehungen zwischen räumlichem Markov-Modell und Ko-Indikator-Systemen sowie deren praktischen Bedeutungen, *in* Peschel, G., ed., Beiträge zur Math. Geologie und Geoinformatik, Band 5. Neue Modellierungsmethoden in Geologie und Umweltinformatik: Verlag Sven von Loga, Köln, p. 37-43.

Luo, J., and Thomsen, A., 1994, Direct estimation of the bivariate probability distribution of a regionalized variable from its spatial samples: Science de la Terre Ser. Inform. Geol., v. 32, p. 115-123.

Olea, R.A., ed., 1991, Geostatistical glossary and multilingual dictionary: Oxford Univ. Press, New York, 177 p.

Ripley, B.D. 1988, Statistical inference for spatial processes: Cambridge Univ. Press, Cambridge, 148 p.

Switzer, P., 1965 A random set process in the plane with Markovian property: Ann. Math. Stat. v. 36, no. 6, p. 1659-1863.

APPENDIX: THEORETICAL NOTE ON MARKOV CHAIN MODELS

About Markov chain theory there is a lot of literature available, for example Chung (1967), Feller (1968, 1971), or Ferschl (1970). The following are basic definitions of Markov process/chain models.

Denote $\{Z(t) \in \Omega \mid t \in \mathbf{T}\}$ a random process: Ω is the state space which can be (a, b) or [a, b], a continuous interval of real values with a < b, or be a set of discrete states $\{S_1 \ldots S_m\}$. $\mathbf{T}$ is the one-dimension parameter space (e.g. for time $\mathbf{T}=[0,+\infty)$). Suppose $Z(t_i) = z_i \in \Omega$, $i=1\ldots n\ldots$ be values (states) of $Z(t)$ corresponding to time points $t_1<t_2\ldots<t_n$ respectively. $Z(t)$ is a Markov process if the conditional probabilities, for $2 \leq q \leq n$, satisfy the following identity (Ferschl, 1970)

$$p(Z(t_q)\leq z_q \mid Z(t_{q-1})=z_{q-1}\ldots Z(t_1)=z_1) = p(Z(t_q)\leq z_q \mid Z(t_{q-1})=z_{q-1}) \qquad (1A)$$

when Ω=(a, b). If $\Omega=\{S_1\ldots S_m\}$or $\{S_1\ldots S_m\ldots\}$ there is (Feller, 1968)

$$p(Z(t_q)=z_q \mid Z(t_{q-1})=z_{q-1}\ldots Z(t_1)=z_1) = p(Z(t_q)=z_q \mid Z(t_{q-1})=z_{q-1}) \qquad (1B)$$

A Markov chain is a special Markov process defined by Equation (1B). When $\Omega = \{S_1\ldots S_m\}$, $Z(t)$ is a finite Markov chain. The conditional probability relation specified by Equation (1A) or (1B) also is termed Markov property. In Equation (1B), suppose $Z(t_q)=S_j$, $Z(t_{q-1})=S_i$, $S_i,S_j \in \{S_1\ldots S_m\}$, and $\tau=t_q-t_{q-1}$, then

$$p_{ij}(t_{q-1},\tau) = p(Z(t_{q-1}+ \tau)=S_j \mid Z(t_{q-1})=S_i) = p(S_j \mid S_i) \qquad (1B')$$

Equation (1B') is termed the transition probability of process $Z(t)$ from S_i at t_{q-1} to S_j at t_q. τ is the transition step length. If $p_{ij}(t_{q-1}, \tau)=p_{ij}(\tau)$ for any $S_i,S_j \in\{S_1\ldots S_m\}$ holds, then $Z(t)$ is a Markov chain with homogeneous transition probabilities. In this situation the state transition behavior of the Markov chain is described completely by the m by m transition probability matrix $\mathbf{P}(\tau)=(p_{ij}(\tau))$, or $\mathbf{P}=(p_{ij})$ for given step-length τ.

Now consider $\{Z(\mathbf{x}) \in \Omega \mid \mathbf{x} \in \mathbf{R}\}$, a random function or a random field defined in a two- or three-dimensional parameter space $\mathbf{R}$. There are two ways to define a spatial Markov chain model. By one of them, the 2D-Markov property is formulated in the frame of a "nearest-neighbor system" (cf. Bartlett, 1971,1975; Besag, 1972a; Bennett, 1979). There

$Z(\mathbf{x})$ is considered on a regular 2D-grid. Let $Z_{k,l}$ be a realization of $Z(\mathbf{x})$ at the grid point (k,l). The Markov property is defined by the following conditional probabilities (Besag, 1972b, 1974):

$$p(Z_{k,l} \mid Z_{u,v},\ u \neq k,\ v \neq l) = p(Z_{k,l} \mid Z_{k-1,l}, Z_{k+1,l}, Z_{k,l-1}, Z_{k,l+1}) \qquad (2A)$$

or by a more general formula (Ripley, 1988)

$$p(Z_{k,l} \mid Z_{u,v},\ u \neq k,\ v \neq l) = p(Z_{k,l} \mid Z_{u,v},\ (u,v) \in \{Z_{l,k}\text{'s neighboring points}\}) \qquad (2B)$$

Here $\{Z_{k,l}$'s neighboring points$\}$ can enclose also the four diagonal grid points $(k\pm1, l\pm1)$. Markov chain models satisfying Equation (2A) define a group of Markov random functions or Markov fields which are used mainly in image processing for such purposes as classification and estimation of pixels (Besag, 1986; Ripley, 1988).

Another definition of a spatial Markov chain model leads back to an early work of Switzer (1965), who had constructed a plane random process with $m \geq 2$ states, such that the alternation of states along any straightline on the plane considered has the Markov property. From his results, another type of spatial Markov chain model can be derived. Suppose $\{Z(\mathbf{x}_q)=z_q \in \{S_1 \ldots S_m\},\ q=1 \ldots n\}$ be a set of samples ordered along a straightline in such a way that $\mathbf{x}_q - \mathbf{x}_{q-1} = a_q \mathbf{h}$, where $\mathbf{h}$ is the direction vector and $a_1 < a_2 < \ldots < a_n$ are real constants. $Z(\mathbf{x})$ is a spatial Markov chain if

$$p(Z(\mathbf{x}_q)=z_q \mid Z(\mathbf{x}_{q-1})=z_{q-1} \ldots Z(\mathbf{x}_1)=z_1) = p(Z(\mathbf{x}_q)=z_q \mid Z(\mathbf{x}_{q-1})=z_{q-1}) \qquad (3A)$$

One of the obvious advantages of the Markov model defined by Equation (3A) is that regarding a certain direction $\mathbf{h}$, $Z(\mathbf{x})$ can be considered as a one-dimensional Markov chain. As a result, the complete theory of Markov chains/processes can be applied to investigate the spatial structure of $Z(\mathbf{x})$ regarding to different directions under consideration. In fact, the known geologic knowledge or the spatial configuration of available data allows to analysis of the spatial structure of $Z(\mathbf{x})$ merely along certain directions.

AN INTELLIGENT FRAMEWORK FOR GEOLOGIC MODELING APPLICATIONS

Lee Plansky, Keith Prisbrey
University of Idaho, Moscow, Idaho, USA

Carl Glass
University of Arizona, Tucson, Arizona, USA

Lee Barron
University of Idaho, Moscow, Idaho, USA

ABSTRACT

This paper provides a framework that can be used for constructing system models, simulations, or actual devices - real time or otherwise. The framework constructs, in the limit, can be intelligent, adaptive, systems or devices with a common architecture that may be used in a variety of applications. Application examples are given from characterization and exploration. A representation is introduced that generalizes the concept of an artificial intelligent agent, or being. The organization for intelligent systems given in this paper may be applied to systems modeling or control in: waste characterization (buried, stored, or processed), site characterization and restoration, materials processing and handling, regulatory compliance monitoring, environmental monitoring, waste storage or repository modeling, site selection, and facility monitoring and control.

Geologic Modeling and Mapping
Edited by A. Förster and D. F. Merriam, Plenum Press, New York, 1996

INTRODUCTION

The earth sciences, mining and materials processing industries are on the threshold of a technological revolution which could lead to the full automation of exploration, characterization, mining, and materials processing. Integrated, or hybrid, knowledge systems, working together with intelligent sensors and adaptive devices are one of the tools that can help realize this revolution. An integrated knowledge system is an assemblage of computer programs which may work together with electronic hardware and uses the technologies of artificial intelligence including: expert systems, neural networks, classical procedural computing, databases, and information reporting systems. Such systems are new. A recent integrated application in metallurgy uses neural networks and expert systems to control electric arc furnaces in steel manufacturing (Kehoe, 1992). This integrated application is estimated to save two million dollars per year per furnace, where the furnace was a state-of-the-art system before the improvement! The benefits come from using neural networks together with expert systems in the furnace controller. Such developments imply that profit or national defense restrictions may keep many of them from ever being published, subsequently hindering economic growth. Figure 1 shows one way in which neural, expert, and procedural systems may be integrated. A neural network (N) at any node in the diagram may be replaced by a procedural (P) or expert system (E) system, model, or module. There are a finite number (9) of interfaces that must be considered in the construction of such an integrated system, they are: N:N, N:E, N:P, E:P, E:E, E:P, P:N, P:E, and P:P.

DEFINITIONS

Neural Networks

Neural networks are a tool of artificial intelligence. Their uses include: function mapping or fitting, pattern matching, feature extraction, and completion of missing data. They are being used in modeling systems, including rule generation, when information is limited. Neural networks, also known as a parallel distributed processing (PDP) network, neural computing network, or neural net, is a parallel information

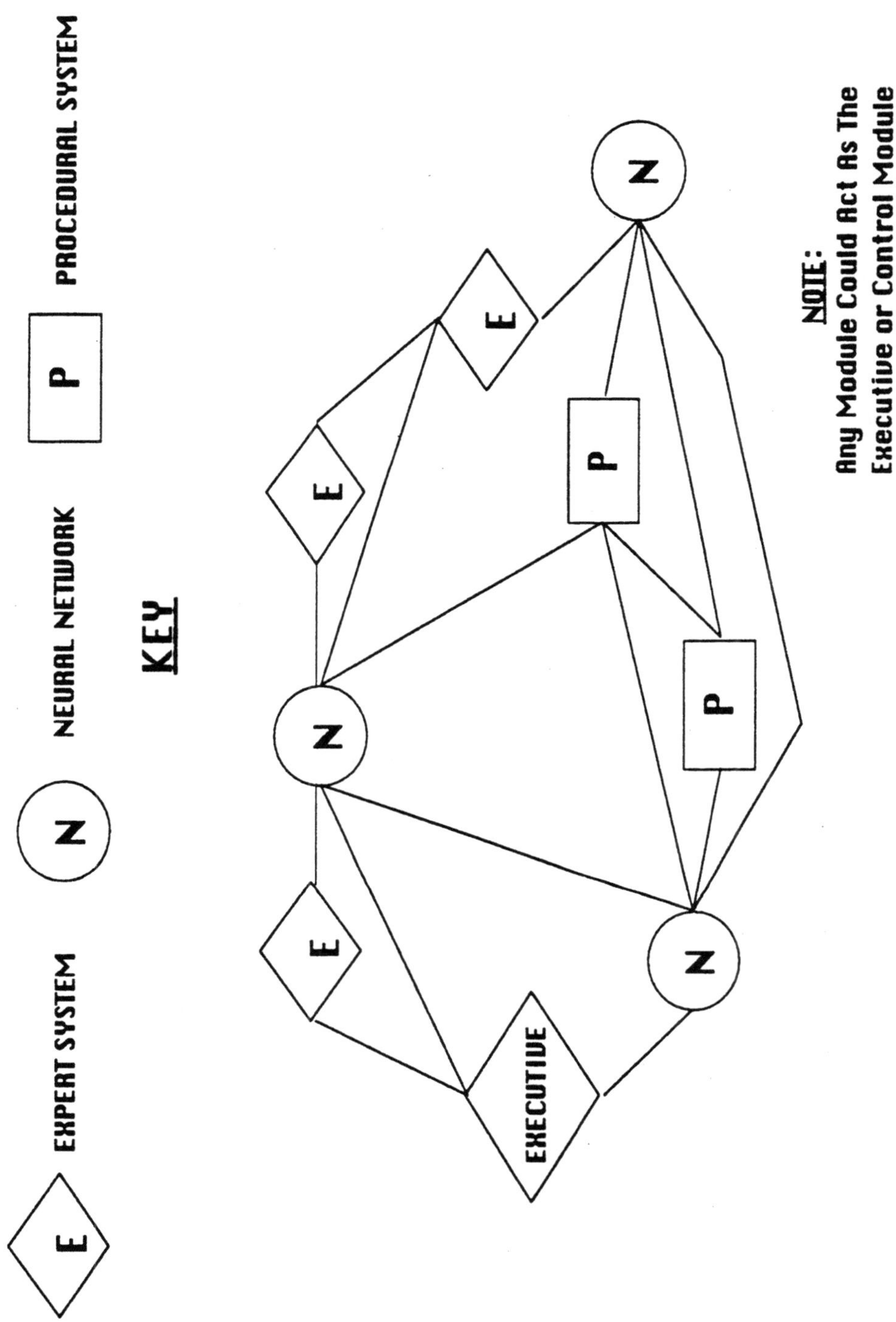

Figure 1. Arbitrary integrated knowledge system.

processing network composed of many simple processing elements. Neural networks can be a complement to expert systems. They can play an important role in the analysis and modeling of the underlying complex processes represented by the data. Neural networks offer a way to generate rules from large amounts of data resulting from the mutual interaction of complex processes. They are effective when the number of rules governing a system or set of processes is large or where data are missing or noisy. Normally, enough data must be available for model construction and model validation when applying neural networks. A neural network can generate rules from the relationships its learns from the data. This ability can be used in place of an expert, especially when dealing with complex processes where a domain expert would be expensive or not available.

Expert Systems

Expert systems are software systems which use automated reasoning procedures or rules to draw conclusions. The most widely used expert system is the so-called rule-based expert system. The heart of the system is the inference engine that uses the rules to draw conclusions from a knowledge base of information and facts. The rules, information, and facts are obtained from human experts in the area discipline. Expert systems cannot be used outside of the domain of their knowledge base. The system usually has an explanation module which explains why or how it arrived at a particular result. In a rule-based expert system, the rules the expert system shell uses to search the knowledge base are in the form of the IFTHEN statements: IF hardness IS 7.0 AND (the) color is white THEN (the mineral is) quartz. The words IF, THEN, and AND are fixed and part of the inference engine. The words hardness, color, 7.0, and white are user defined variables and their values, or data. The word quartz is a rule conclusion.

Procedural Systems

Procedural systems are software systems that consist of a main calling routine and a set of subroutines, or modules, the sum of which perform one or more predefined functions. Procedural systems usually are written in higher level computer languages such as Pascal.

TECHNICAL CONSIDERATIONS

True intelligence requires the ability to adapt to changes. Intelligent systems may be intelligent in the sense that they are programmed to respond in a given way to a known event(s) or event sequence. They, however, are not necessarily adaptive. This applies to artificially intelligent hardware-software-based systems -- without adaptability they respond in a fixed, predetermined manner. Thus, many of today's artificial intelligence methods are simply another type of algorithm, procedure, or model. The key to real intelligent systems lies in making the system(s) adaptive. This requires the implementation of concepts from biology, psychology and the cognitive, learning, and adaptive sciences. These areas are important to real long-term systems applications.

In developing integrated knowledge systems, expert systems can play an important role as the primary way of control, with a sensor system providing the input data to a neural network or its procedural representation, such as a computer program written in the C, Pascal, or FORTRAN computer language and integrated in new or existing systems. An expert system could be replaced by procedural programs, especially if the control functions are simple and the knowledge base or data needed are small. Neural networks excel over classical approaches to sensor data analysis, function fitting, and feature and pattern recognition applications. Neural networks in integrated systems permit the designer to analyze nonlinear processes for which no simple analysis techniques exist. They enable rule, procedure, and function generation of complex processes which "classical" expert systems can not do. This is of interest in process control applications and mining automation. Neural networks can analyze and fit noisy measurement data and determine features, patterns, and functions in the data with ease. They can complement or replace expert systems. New techniques or other software or procedural algorithms also may be included in the framework of this paper. Additional technical considerations for implementing integrated knowledge systems are given in Glass and others (1993). Workers at the U. S. Bureau of Mines, Spokane Research Center are actively developing a hybrid AI roof bolting system for coal mining safety application (Bevan and Hill, 1994; Hoffman, 1994; Howie and Nichols, 1994; Signer and others, 1993).

APPLICATIONS

Integrated knowledge systems are applied here to the conceptual design of a system for the exploration, characterization, identification, and evaluation of earth materials, minerals, soils, or ores in the format of an intelligent, adaptive exploration systems for earth, near-earth objects, or planetary resources. This scenario is selected while the requirements for extraterrestrial systems are severe and demand more ingenuity. They inspire innovation, are challenging, and the need for success makes the impossible, possible. They imply an interdisciplinary approach to problem solving that compels the designer to forge reliable interfaces among different technologies. Humans are assumed not to interact. Such a system also may malfunction, be damaged, lost, or simply disposed. A second example is taken from mining and the characterization of blast fragment sizes using neural networks. This application and its solution is an example of one that may be integrated readily and developed into an intelligent product for market. Many of the concepts and technologies, covered, or implied by this work can be used directly in geologic as well as in industrial applications. A proposed organization for an AI Agent concludes this section.

Applications for integrated knowledge systems include exploration, characterization and mobile (or stationary) monitoring, control, or automatic maintenance systems for use in mining. Exploration and characterization generally require the use of different detection techniques for the target(s) (detection objectives) in a given geologic environment. Measurement becomes more difficult as the geologic environment changes. The nature of different geologic, ore, or waste materials may complicate matters - each material is generally different in properties and behavior. This is in contrast to, but supported by, the present ability to develop such units or devices for specific industrial applications, as witnessed by the arc furnace example noted. Of particular importance to the present work are modular, generic intelligent sensors or sensor systems; such systems have been developed. Modular, intelligent, integrated measurement systems would allow multiuse and interchangability of costly sensor and data-acquisition systems. They have their own computer processors with self-test and repair functions and the ability to include multiple sensors for arbitrary application in different environments. This work also implies the need for intelligent, look-ahead sensor systems.

Characterization and Exploration

Identification of a material implies characterization which requires property measurements and sensor systems. If the material is unknown it is named upon characterization. It then is classified as a known material. Characterization involves determining the nature or attributes of a material and determining its properties. Properties can be taken in the broad sense to include the behavior of a material under external forces. Characterization leads to identifying a material, assuming it is a known one. Exploration is the systematic examination of a region with the goal of increasing existing knowledge of the region. It implies that a target material is sought together with geologic information about the search region that can be used to help locate or identify the material. Geologic information is used in evaluating the potential for the material and its worth.

Given a collection of known materials X = {X1, X2, X3,..., Xm}, two problems relevant to characterization and exploration are: (1) identify an unknown material Uk or, (2) determine a known material Xk. In the latter situation, Xk is a target material(s), or the valuable objective(s) of the search. Characterization requires the determination of the properties of the materials to be identified. The properties of a target material usually are known or defined ahead of time. The characterization, identification and evaluation of minerals, rocks, or ores consists of the steps:

1. Characterize the target material
    1a. Measure the target material properties
2. Identify the target material
    2a. Determine if known material
3. Inventory the target material
4. Evaluate the target material resource value
5. Record and report results.

This can be formulated into a characterization, identification, and evaluation procedure as shown in Figure 2. The procedure assumes that the target material is available, that is, exists or has been located. Its steps can be included in a more general exploration, identification, and evaluation procedure (for known materials) as shown in Figure 3, namely:

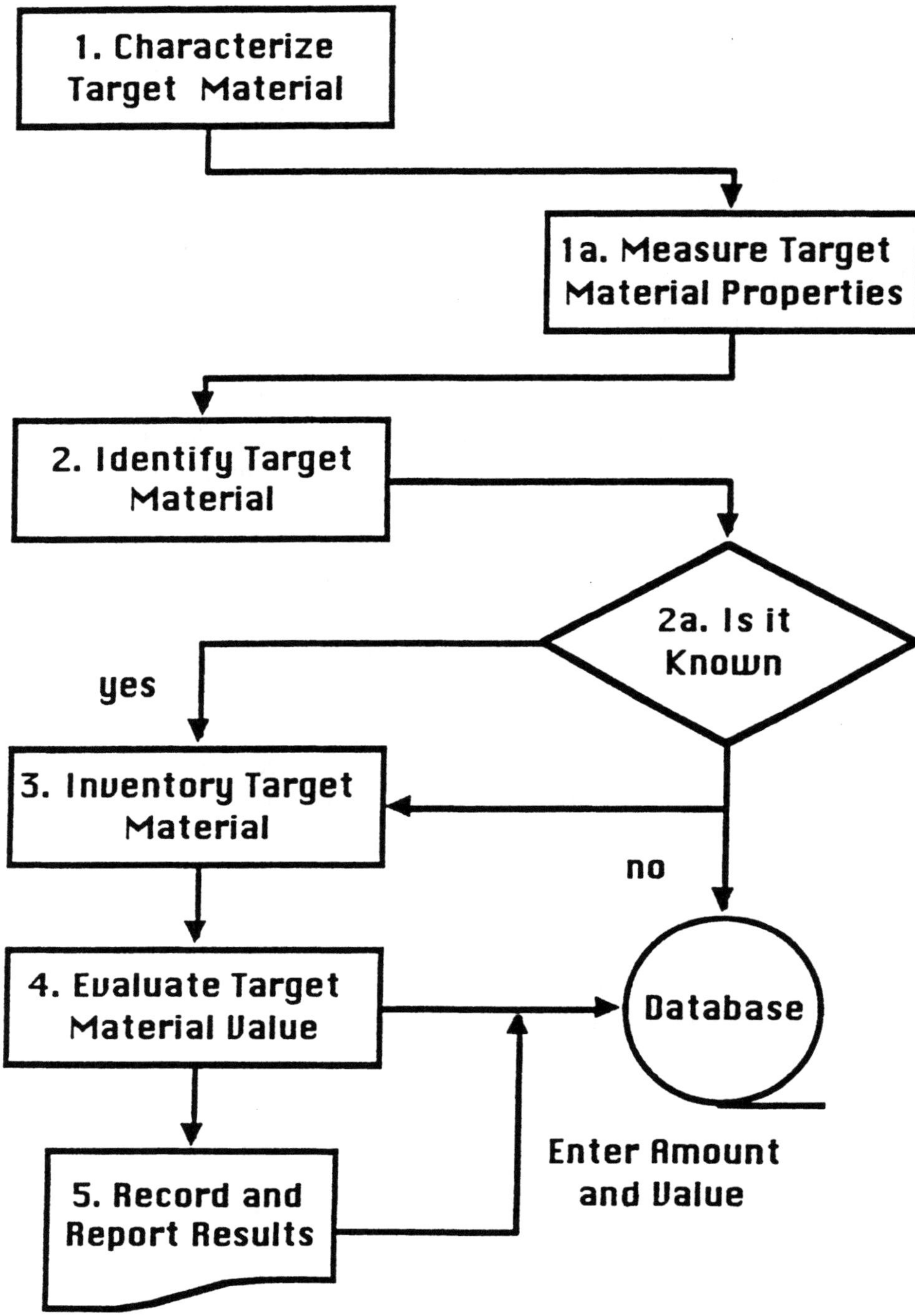

Figure 2. Characterization and identification.

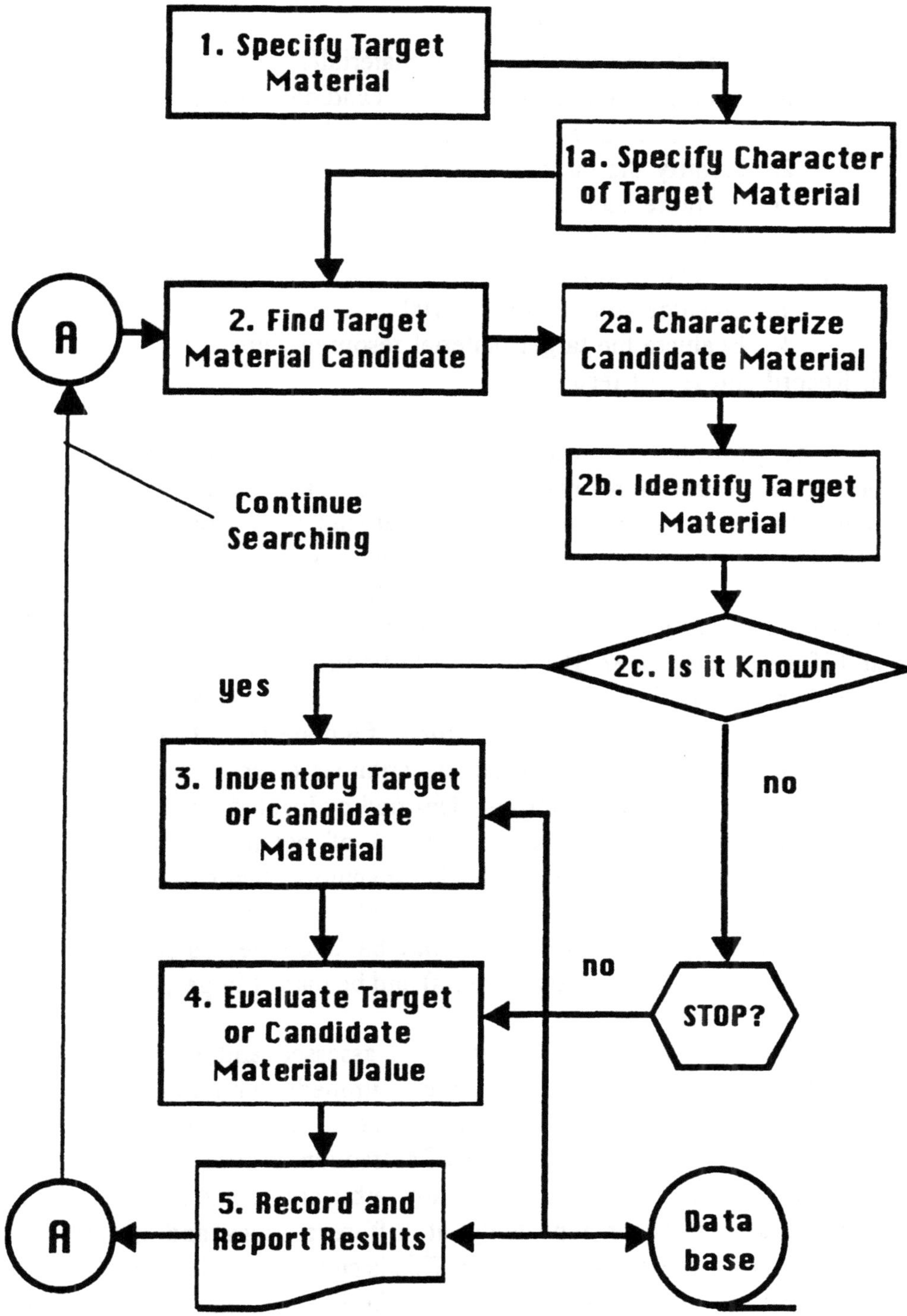

Figure 3. Examination or exploration and characterization and assessment.

1. Determine target material(s)
    1a. Specify character of target material(s)
2. Locate target material(s), (i. e.: locate candidates for a match)
    2a. Characterize the candidate material(s)
    2b. Identify the target material
    2c. Examine candidate material(s) for a match
3. Action if no match (not located, continue or stop search)
4. Action if match (located)
    4a. Inventory the target material
    4b. Evaluate the target material resource value
5. Record and report results.

The procedure shown in Figure 3 requires the definition and properties of the target material. The step "Find Target Material" can represent any search operation or search procedure. These two procedures summarize the two problems noted previously: identifying a known material and locating a known material. The basic questions addressed by both procedures are: Is Uk one of the known materials Xk in the set X and if it is, which one? If a material to be identified is not in the set of known materials, it is characterized and added to the set X of known materials and its properties are recorded. If a material to be located is not located or detected, the search either continues or ends. At the end of a search, an inventory of materials identified or located during the search, including any located target material(s) of potential use, would be produced and a report developed. This inventory then could be used for resource evaluation and future mission planning. Such proceduralized operations for specific applications may be automated as part of special purpose adaptive learning algorithm(s) and be part of larger, more general adaptive systems.

A potential orebody, if located, is assessed and its reserves and potential value are recorded for future or immediate mining. Mining and concentration would normally take place at the location where the orebody is located. The operations of extraction, processing, fabrication, and transport could take place at any arbitrary location. These latter operations may occur repeatedly at different times and in different sequence for different end products. This is summarized in Table 1. Table 2 is a summary of areas to be considered for exploration through end-use. Figure 4 shows the typical sequence followed in exploration or examination. Figure 5 shows the typical sequence followed in the postexploration or post-examination period to mine, extract, transport, and fabricate products for use. Note that characterization implies monitoring but not the converse, monitoring requires characterization; similarly, measurement enables testing.

Table 1. Steps in mining and post-mining activities.

	Mine	Extract	Process	Fabricate	Transport
Mine	X	X	X	X	X
Extract			X	X	X
Process				X	X
Fabricate					X
Transport		X	X	X	
Possible Actions Leading to End Material Use--->					

Blast Fragment Size Analysis

Barron and others (1994) applied neural networks to differentiating and understanding the relationships between mining blast fragment distributions and orebody textures, layering, and crack patterns. Blasting, crushing, grinding, and milling efficiencies could be improved by understanding such relationships (Barron and others, 1994). The problem selected was to recognize fragment size in "muck" piles from digitized images despite daily changes in perspective, lighting, ore size, and ore characteristics. Ideally, blasted ore fragments in truck loads must be sorted by hand to determine size distributions, a prohibitive task. The problem is to determine if neural networks can do the job easier. The procedure used and results are described next (modified after Barron and others, 1994).

Table 2. Exploration considerations for Earth or space-based mining.

Goal: Explore or examine, define objectives or target materials before proceeding.

- 1.1. Locate Target Body or Source of Target Material
  - 1.1.1. Possible Regions to Observe in Transit - List
    - 1.1.1.1. Space Plasmas
    - 1.1.1.2. Above Body Surface
    - 1.1.1.3. Surface Topography
    - 1.1.1.4. Structural Features
    - 1.1.1.5. Surface Materials
    - 1.1.1.6. Deep Regions
    - 1.1.1.7. Minerals
    - 1.1.1.8. Rocks
    - 1.1.1.9. Ore Materials
- 1.2. Possible Actions
  - 1.2.1. Characterize (implies monitoring is possible)
    - 1.2.1.1. Measure Properties
      - 1.2.1.1.1. Overall
        - 1.2.1.1.1.1. volume or extent
        - 1.2.1.1.1.2. orientation
        - 1.2.1.1.1.3. shape
        - 1.2.1.1.1.4. surface form
        - 1.2.1.1.1.5. structure
      - 1.2.1.1.2. Temporal
      - 1.2.1.1.3. Spatial
      - 1.2.1.1.4. Physical
      - 1.2.1.1.5. Chemical
  - 1.2.2. Identify
  - 1.2.3. Inventory
    - 1.2.3.1. Amount
  - 1.2.4. Evaluate
    - 1.2.4.1. Uses
    - 1.2.4.2. Value
  - 1.2.5. Record and Report
  - 1.2.6. Change Location
  - 1.2.7. Locate Next Target Body
- 1.3. Post Exploration
  - 1.3.1. mining and concentration
  - 1.3.2. extraction
  - 1.3.3. fabrication
  - 1.3.4. transport
  - 1.3.5. use

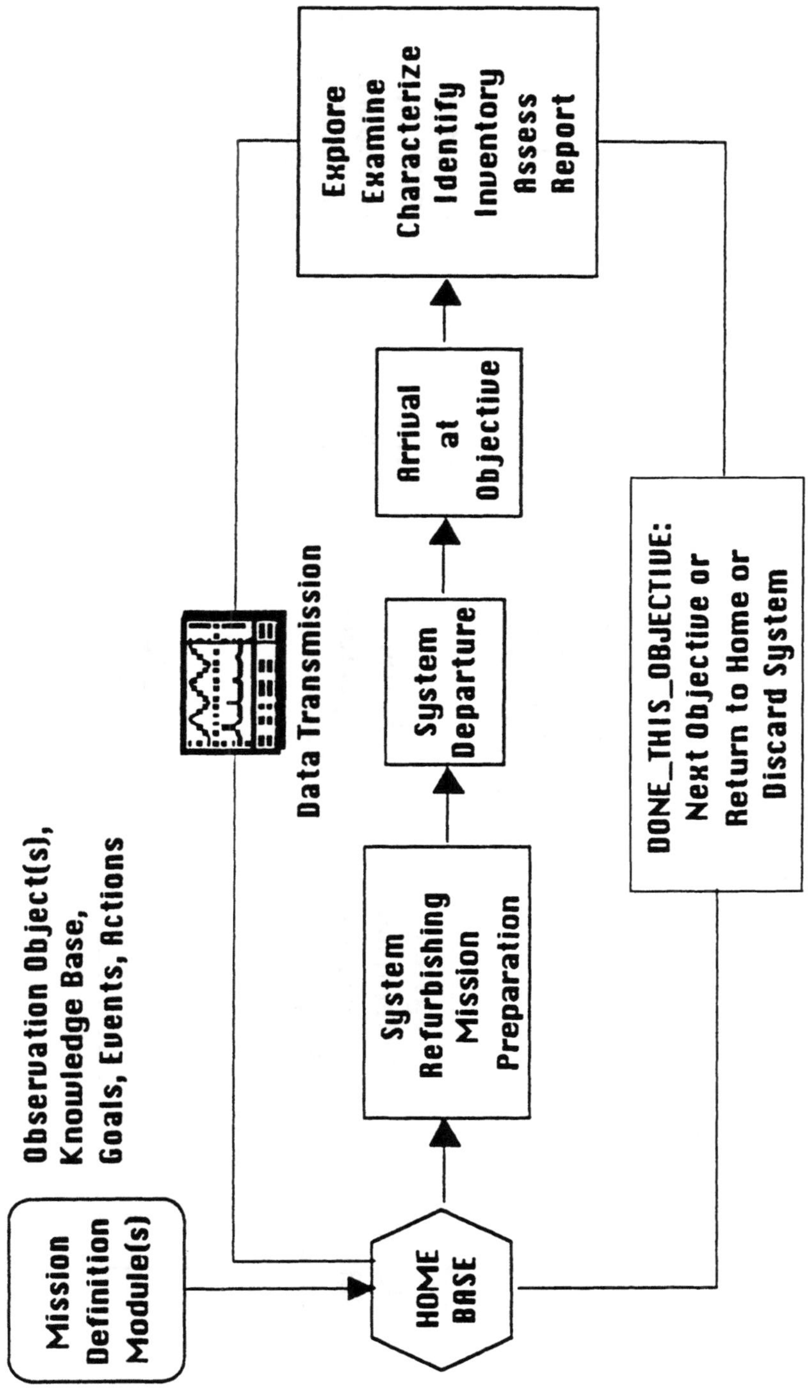

Figure 4. Typical examination, characterization, and assessment sequence.

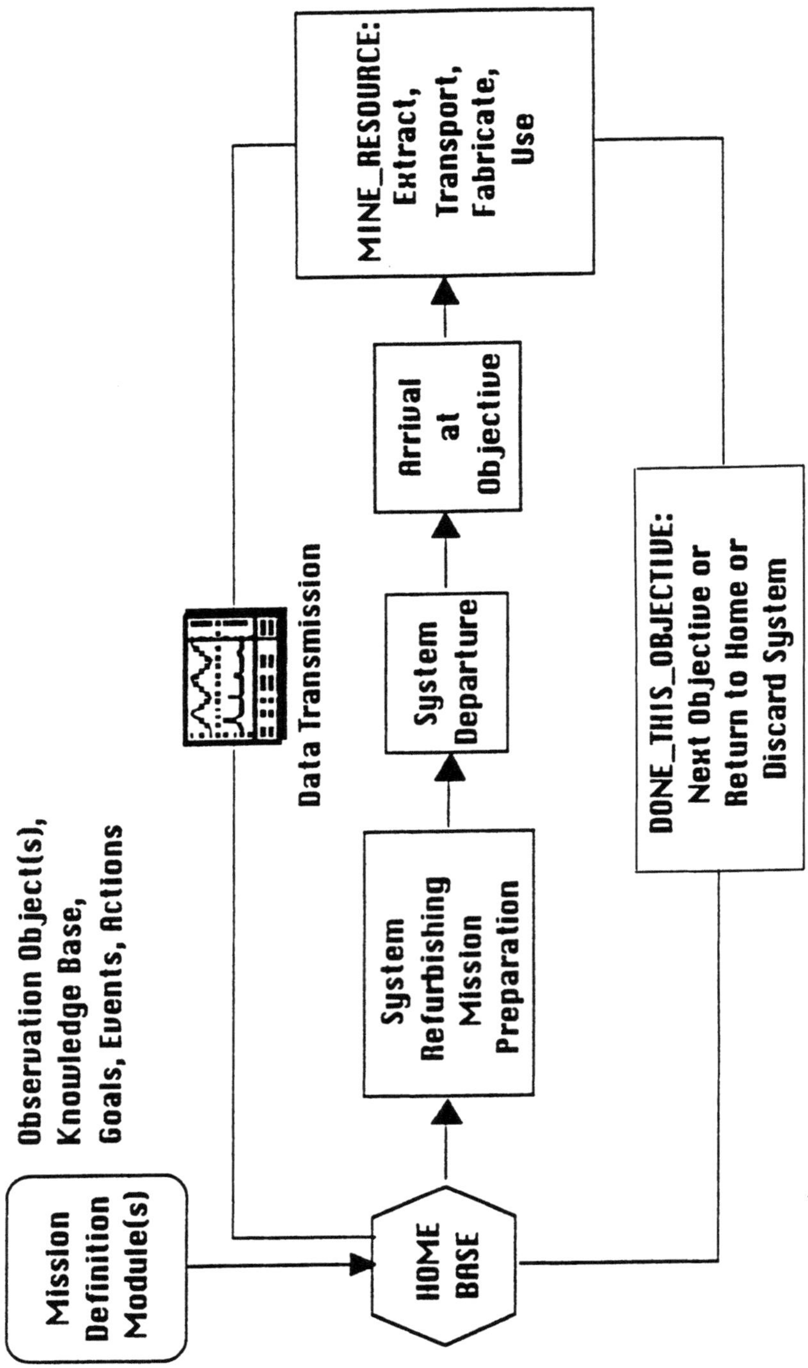

Figure 5. Typical post-examination, characterization, and assessment sequence for mining to transport.

Procedure

Ore piles were photographed and the photographs were digitized. The ore piles were sorted by hand to give the neural network training sets. The established back propagation (BP) neural network algorithm was used. The BP network was trained to recognize correctly ore size by class: 60%, 70%, 80%, 90% finer than 8 inches. In addition to the ability of the neural network to classify correctly the training set, the predictive ability was tested. The procedure included taking characteristic "fingerprints" from the digitized images using the method described by Bottlinger and Kohlus (1993). A two-dimensional Fast Fourier Transform was taken of the digitized image. The Fourier coefficients were shifted to give a large "D.C. Gain" peak in the center of the image. From that center point, the coefficients were averaged over concentric (square) rings. Plotting the average versus ring number gives a unique fingerprint that relates directly to a size distribution in the original image. Figure 6 shows such a typical set of fingerprints taken from the second concentric ring to the thirtieth.

Results

The neural network could be trained to associate a multitude of images to four size distribution classes. Eighty pictures from two different lighting regimes were correctly classified. Furthermore, forty pictures not in the training set were predicted correctly with 98% accuracy. The one misclassified picture was later determined to have digitizing errors.

Different network configurations and fingerprints were examined. Most network configurations seemed to work well. The best network configuration had six input nodes, sixteen hidden layers, and one output node. We used a nonlinear log sigmoidal saturation function, and a modified back propagation training algorithm from the MATLAB® software system. The best results were obtained using a single network to split the four categories of images into two subcategories, and then two more networks to split each of the two subcategories. This meant that the first net divided the images into a fine subcategory containing both 80% and 90% finer than 8 inches, and a coarse subcategory containing both 60% and 70% finer than 8 inches. The fine subcategory then was divided by a second net into final categories of 80% and 90% finer than 8 inches. Likewise, a third net was trained to recognize final categories in the

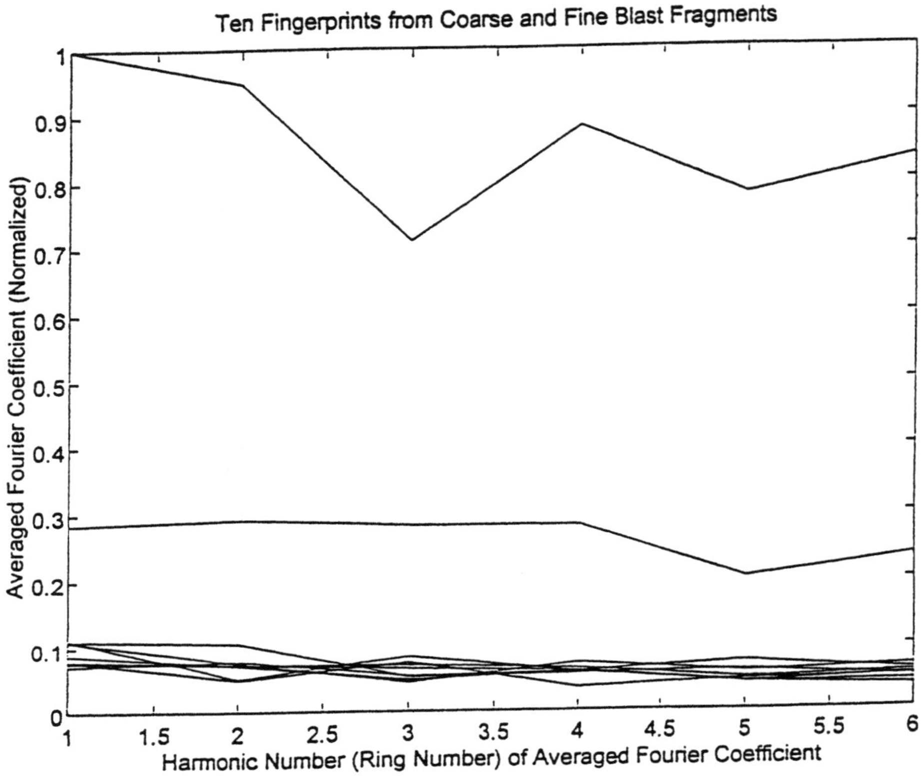

Figure 6. Typical fingerprint results.

coarse subcategory of 60% and 70% finer than 8 inches. Various sections of the fingerprint were used, all with relatively good success. The first six positions of the fingerprint were sufficient for training.

The pattern recognition ability of neural networks is sufficient to determine ore size in a variety of lighting conditions. Lighting conditions can change because neural networks can be repeatedly trained on a calibration pile. Errors introduced by shadows, overlapping of rocks and particles, and other variable lighting conditions are automatically corrected.

An AI Agent Based System Foundation

This subsection describes an intelligent exploration, examination or characterization system based on the "Andral Model" concept (Plansky, 1992), this structure is similar to the "Black Box-Thin Wire" concept in (Rich, 1990).

Intelligent Systems

Adaptation implies the ability to respond to an unknown event or event sequence. In addition, the response must be an intelligent or good one -- good being defined ahead of time. Adaptation also requires the ability to do task planning and task execution. This is a fundamental and important difference between an intelligent being or an "AI Agent" (or Agent), and a so-called "intelligent" system which responds to a series of predetermined events - the latter cannot learn by itself, it must be trained in some way or be given the ability to train itself (thus tending toward intelligence). A subsystem known as an adaptive learning unit (ALU) is required for real intelligent, adaptive systems. An organization for a typical ALU is given in Figure 7.

Vere and Bickmore (1990) have constructed an operating prototype of an Agent for observation and assessment of underwater craft. They conceive of an Agent to be an integrated artificial intelligence device that acts in an environment, its adaptive nature being implicit. The Agent (HOMER) can communicate, reason, plan, perceive, and act in or on its environment. The prototype operates in a simulated, marine environment. Vere and Bickmore (1990) maintain, as we do, that such systems are possible by a resolute effort, with the needed funds and patience on the part of management. Their contribution is of extraordinary significance. The prototype device can: communicate with humans in a vocabulary of 800 English words; use a moderate level of grammar; access an episodic memory; reflect on events; talk, receive commands, answer questions from a human; and, synthesize and execute an action plan. It is a LISPbased expert system with procedural systems as support routines. The structural essence of their Agent is embodied in the work of the present authors.

A General Model

An intelligent construct, the "Andral Model", was first proposed in (Plansky, 1992). Figure 1 is a variant of this construct. Variants of its organization implemented as an adaptive, distributed, artificial intelligent (DAI) system is shown in Figures 8 and 9. Figure 8 shows the upper level organization of a system for exploration/examination or characterization using this concept as an integrated knowledge system as discussed in the first example.

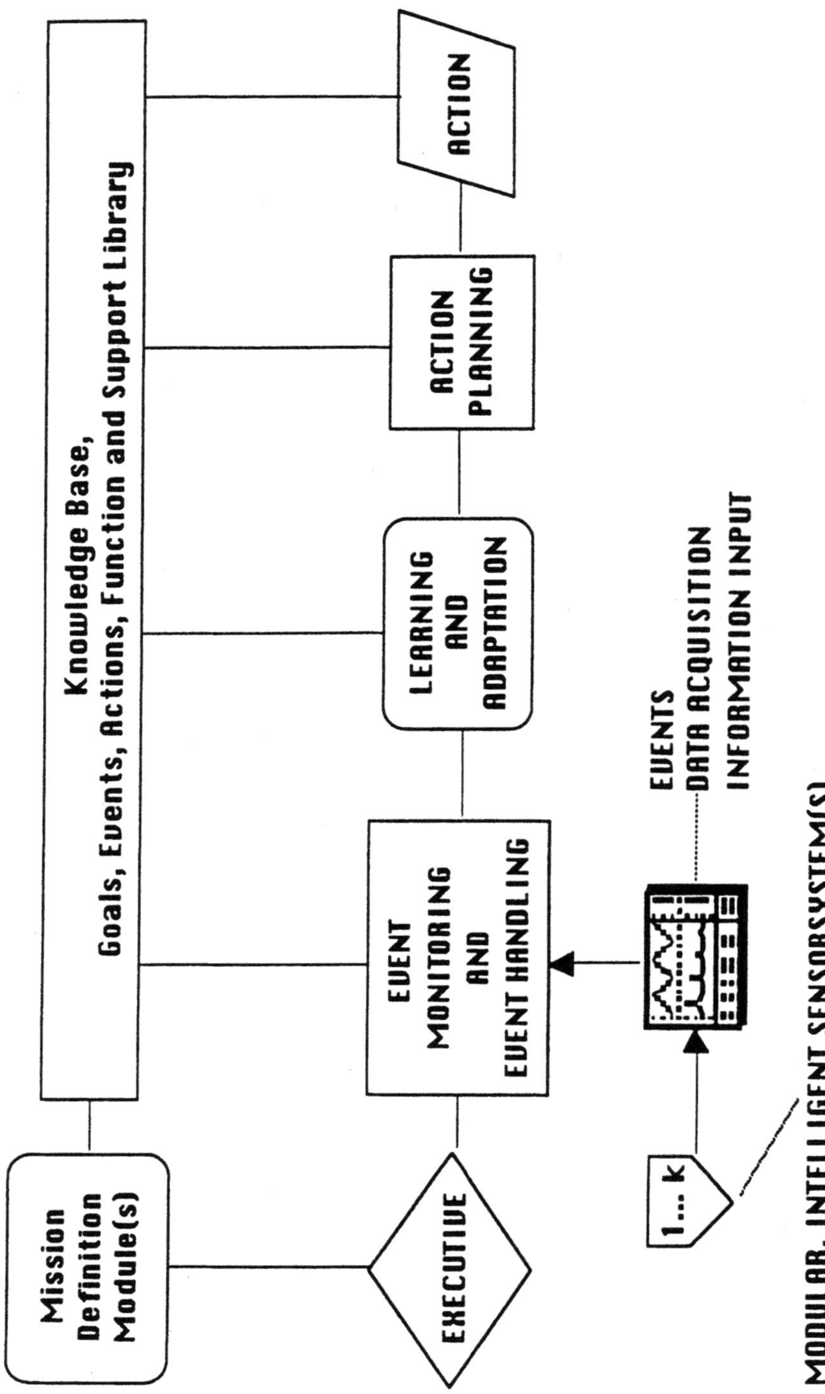

Figure 7. Typical adaptive learning unit.

The Redefinition of an AI Agent

A simple organization and definition for the AI Agent concept is given in (Plansky, 1992). The proposed definition allows Agents of different types and component parts or functions, equivalent to what is shown in Figure 1 complemented by an adaptive learning unit (ALU). It is readily extended to multiple Agents and families of Agents and puts the concept of the AI Agent in a common framework for many applications - small or large, real-time, near real-time, or normal batch-type processing systems. It also includes such systems as HOMER of Vere and Bickmore (1990). An improvement of the proposed definition is: An AI Agent as an ALU and a combination of one or more of:

Parallel and/or sequential andral models
Intelligent sensor systems
Devices
Control systems
Knowledge and information databases
Function and/or software support libraries
Communication systems.

CONCLUDING REMARKS

Human-like intelligent, adaptive systems can be used in many facets of geologic work. They are the next step in removing man from the hazardous environments encountered. The challenge before us is in the immediate development from off-the-shelf hardware and software of integrated knowledge-based systems. If we do not act in this direction, other technologists will - some of the most profitable applications are obvious improvements or extensions of existing systems as shown by the second example in this paper.

It is not easy to change the technological baseline in some industries. The Old Testament (Job Chapter 28) tells of a time when slaves and poor men hung from ropes in darkness using oil lamps for light to search for riches in the Earth. Similar situations exist in our world today - circa 8,000 years later. The greatest advances have been in our century. The next century, hopefully, will outshine the last 8,000 years and approach the dream of geologists, mining engineers, and their fellow scientists of fully automated investigation, resource exploration, evaluation, mining, transport, and processing tools. The door is beginning to open now.

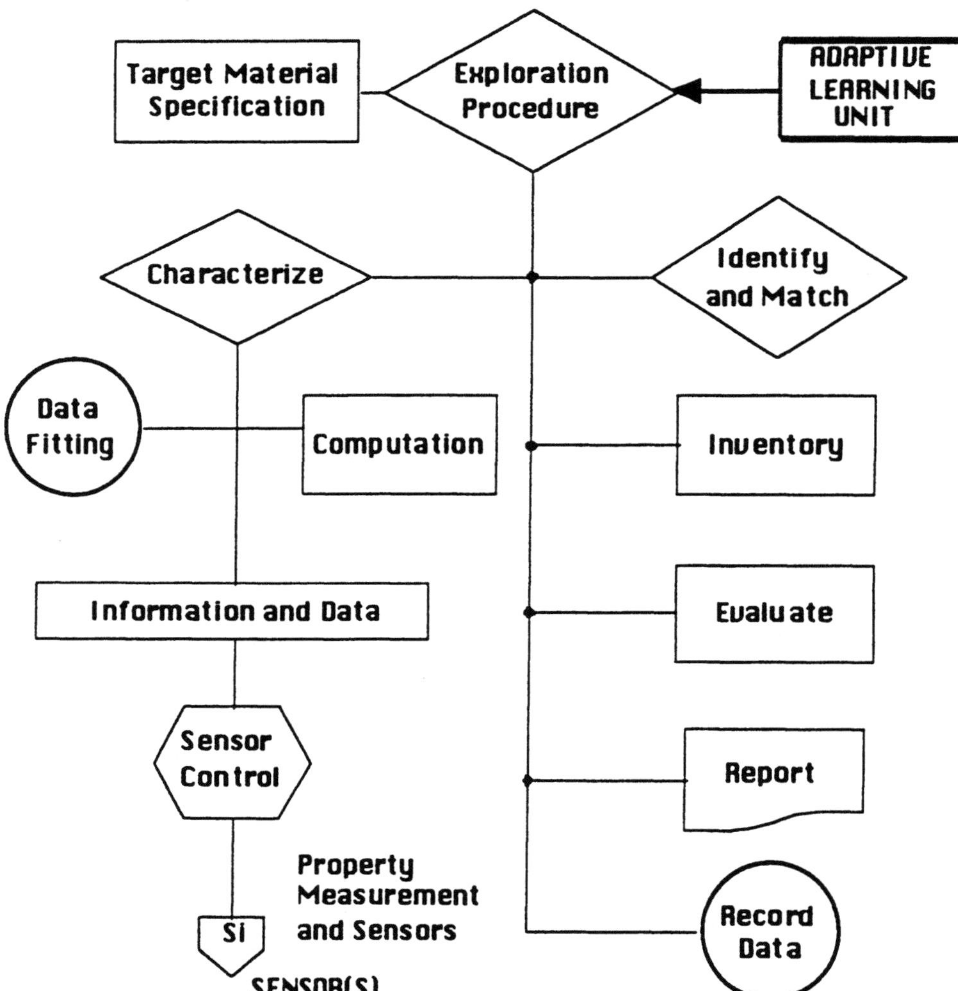

Figure 8. Organization of exploration and characterization system.

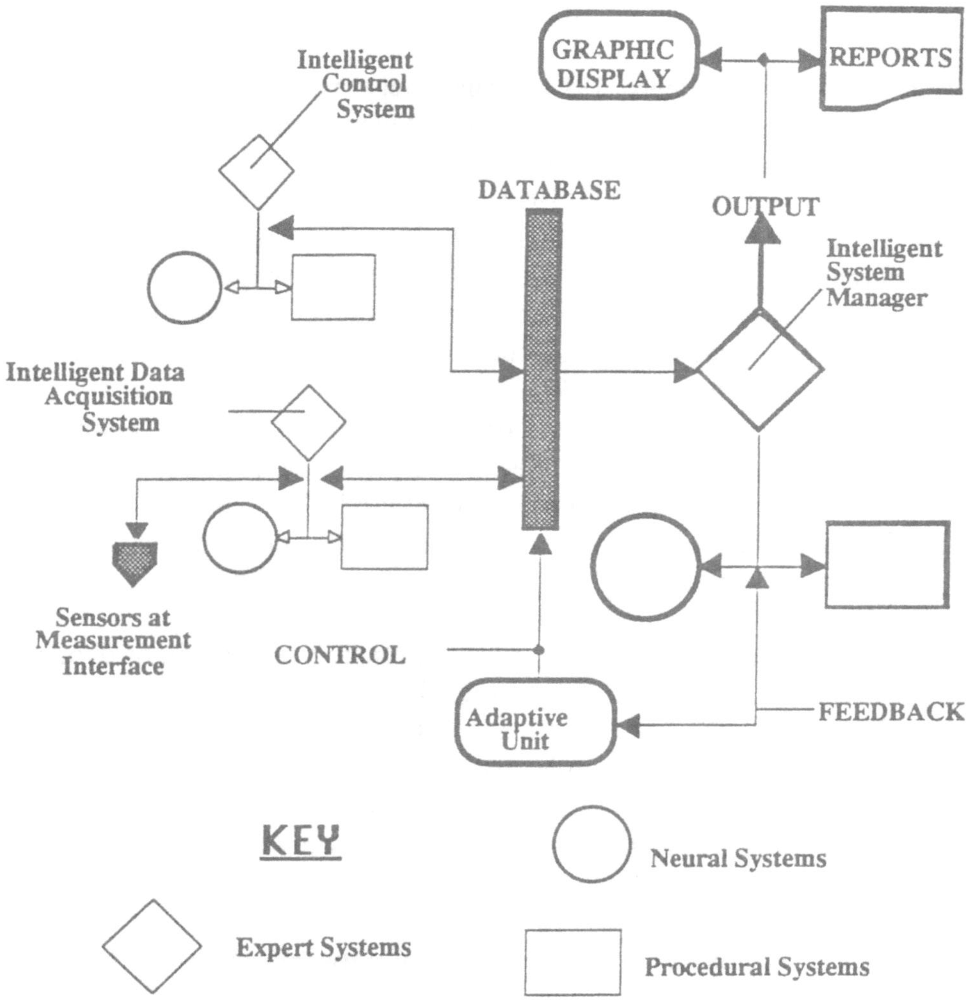

Figure 9. First order, or organization level, of adaptive knowledge system.

ACKNOWLEDGMENTS

Authors Plansky and Barron would like to thank authors Prisbrey and Glass for their support during their advanced degree programs. The first author would like to thank Dr. Dan Merriam for his continued support and encouragement over time and continents, on the lonesome path of a doctoral student - may we all practice and reflect this dedication towards our field and others.

REFERENCES

Barron, C.L., and others, 1994, Neural network pattern recognition of blast fragment size distributions: Jour. Particulate Science and Technology, v. 12, no. 3, p. 235-242.

Bevan J.E., and Hill, J.R.M., 1994. Automating coal mine roof bolting: U. S. BuMines, Spokane Research Center, in press.

Bottlinger, M., and Kohlus, R., 1993, Particle shape analysis as an example of knowledge extraction by neural nets: Particle and Particle Systems Characterization, Deutsche Institut fuer Lebensmitteltechnik, e. V., v. 10, no. 5, p. 275-278.

Glass, C.E., and others, 1993, Complementing the trend in automated mining systems: Proc. Second Intern. Symposium on Mine Mechanization and Automation, p. 609-624.

Hoffman, M.P., 1994, Computer-based monitoring and control of a masttype roof drill: U. S. BuMines, Spokane Research Center, in press.

Howie, W.L., and Nichols, T.L., 1994, Telemetric acquisition of dynamic drill-rock interaction data: Proc. 40th Intern. Symposium of the Instrumentation Soc. America, p. 341-352.

Kehoe, B., 1992, EAF controller passes intelligence test: Iron Age, v. 8, no. 3, p. 28-29.

Plansky, L.E., 1992, Application of integrated knowledge systems to the metallurgy of zircaloy4: unpubl. doctoral dissertation, Univ. Idaho, 172 p.

Rich, E., 1990, Expert systems and neural networks can work together: IEEE EXPERT, v. 5, no. 5, p. 5-7.

Signer, S.P., and others, 1993, Using unsupervised learning for feature detection in coal mine roof: Engineering Applications of Artificial Intelligence, v. 6, no. 6, p. 565-573.

Vere S., and Bickmore, T., 1990. A basic agent: Computational Intelligence, v. 6, no. 1, p. 41-60.

CONTRIBUTORS

Frits P. Agterberg, Geological Survey of Canada, 601 Booth Street, Ottawa K1A 0E8, Canada

Rainer Alms, Geologisches Institut der Universität Bonn, Nußallee 8, D-53115 Bonn, Germany

Neil L. Anderson, Department of Geology & Geophysics, University of Missouri-Rolla, Rolla, Missouri 65401, USA

Stephan Auerbach, Fachbereich Geowissenschaften, Quantitative Methoden, Universität Trier, D-54286 Trier, Germany

Lee Barron, College of Mines and Earth Resources, Department of Metallurgy and Mining Engineering, University of Idaho, Moscow, Idaho 83843, USA

Ulf Bayer, GeoForschungsZentrum Potsdam, Telegrafenberg A17, D-14473 Potsdam, Germany

Geoffrey C. Bohling, Kansas Geological Survey, University of Kansas, Lawrence, Kansas 66047, USA

R. James Brown, Department of Geology & Geophysics, University of Calgary, Alberta T2N 1N4, Canada

Q. Cheng, Department of Geology, University of Ottawa, Ontario K1N 6N5, Canada

John C. Davis, Kansas Geological Survey, University of Kansas, Lawrence, Kansas 66047, USA

M.J. Duggin, Department of Forest Engineering, SUNY College of Environmental Science and Forestry, Syracuse, New York 13210, USA

Andrea Förster, GeoForschungsZentrum Potsdam, Telegrafenberg A17, D-14473 Potsdam, Germany

John A. French, Kansas Geological Survey, University of Kansas, Lawrence, Kansas 66047, USA

Carl Glass, College of Engineering and Mines, Department of Mining and Geological Engineering, University of Arizona, Tucson, Arizona 85721, USA

Eric C. Grunsky, British Columbia Geological Survey, 1810 Blanshard Street, Victoria, British Columbia V8R 3H4, Canada

W.J. (Bill) Guy, Kansas Geological Survey, University of Kansas, Lawrence, Kansas 66047, USA

John W. Harbaugh, Department of Geological and Environmental Sciences, Stanford University, Stanford, California 94305, USA

Jan Harff, Institut für Ostseeforschung Warnemünde, Seestraße 15, D-18119 Rostock-Warnemünde, Germany

Ute C. Herzfeld, Fachbereich Geowissenschaften, Quantitative Methoden, Universität Trier, D-54286 Trier, Germany

Olaf Kahle, GeoForschungsZentrum Potsdam, Telegrafenberg A17, D-14473 Potsdam, Germany

Christian Klesper, Geologisches Institut der Universität Bonn, Nußallee 8, D-53115 Bonn, Germany

Junfeng Luo, Gesellschaft für wasserwirtschaftliche Planung und Systemforschung mbH, D-12526 Berlin, Germany

B. Lünenschloß, GeoForschungsZentrum Potsdam, Telegrafenberg A17, D-14473 Potsdam, Germany

Daniel F. Merriam, Kansas Geological Survey, University of Kansas, Lawrence, Kansas 66047, USA

Ricardo A. Olea, Kansas Geological Survey, University of Kansas, Lawrence, Kansas 66047, USA

Robert Ondrak, GeoForschungsZentrum Potsdam, Telegrafenberg A17, D-14473 Potsdam, Germany

Lee Plansky, College of Mines and Earth Resources, Department of Metallurgy and Mining Engineering, University of Idaho, Moscow 83843, USA

Keith Prisbrey, College of Mines and Earth Resources, Department of Metallurgy and Mining Engineering, University of Idaho, Moscow 83843, USA

Joseph E. Robinson, Department of Geology, Syracuse University, Syracuse, New York 13244, USA

Helmut Schaeben, Fachbereich Geowissenschaften, Quantitative Methoden, Universität Trier, D-54286 Trier, Germany

Esther U. Schütz, Fachbereich Geowissenschaften, Quantitative Methoden, Universität Trier, D-54286 Trier, Germany

Alla Shogenova, Institute of Geology, Estonian Academy of Sciences, Tallinn, Estonia

Agemar Siehl, Geologisches Institut der Universität Bonn, Nußallee 8, D-53115 Bonn, Germany

Jörn Springer, GeoForschungsZentrum Potsdam, Telegrafenberg A17, D-14473 Potsdam, Germany

W. Lynn Watney, Kansas Geological Survey, University of Kansas, Lawrence, Kansas 66047, USA

Claus v. Winterfeld, GeoForschungsZentrum Potsdam, Telegrafenberg A17, D-14473 Potsdam, Germany

Johannes Wendebourg, Institut Français du Pétrole, 1&4 Avenue de Bois-Préau, F-92506 Rueil-Malmaison Cedex, France

D. Yuan, Desert Research Institute, 755 E. Flamingo Road, Las Vegas, Nevada 89119, USA

INDEX

MIX
Papier aus verantwortungsvollen Quellen
Paper from responsible sources
FSC® C105338

If you have any concerns about our products,
you can contact us on
ProductSafety@springernature.com

In case Publisher is established outside the EU,
the EU authorized representative is:
Springer Nature Customer Service Center GmbH
Europaplatz 3, 69115 Heidelberg, Germany

Printed by Libri Plureos GmbH
in Hamburg, Germany